Beyond Manufacturing Resource Planning (MRP II)

Advanced Models and Methods for Production Planning

Springer
Berlin
Heidelberg
New York
Barcelona
Budapest
Hong Kong
London
Milan
Paris
Santa Clara
Singapore
Tokyo

Andreas Drexl · Alf Kimms (Eds.)

Beyond Manufacturing Resource Planning (MRP II)

Advanced Models and Methods for Production Planning

With 93 Figures
and 36 Tables

Springer

Prof. Dr. Andreas Drexl
Dr. Alf Kimms
Christian-Albrechts-Universität zu Kiel
Institut für Betriebswirtschaftslehre
Lehrstuhl für Produktion und Logistik
Olshausenstraße 40
24118 Kiel
Germany

ISBN 3-540-64247-1 Springer-Verlag Berlin Heidelberg New York

Cataloging-in-Publication Data applied for
Die Deutsche Bibliothek – CIP-Einheitsaufnahme
Beyond manufacturing resource planning (MRP II): advanced models and methods for production planning; with 36 tables / Andreas Drexl; Alf Kimms (ed.). – Berlin; Heidelberg; New York; Barcelona; Budapest; Hong Kong; London; Milan; Santa Clara; Singapore; Paris; Tokyo: Springer, 1998
ISBN 3-540-64247-1

Printed in Germany

Hardcover-Design: Erich Kirchner, Heidelberg

SPIN 10674380 42/2202-5 4 3 2 1 0 – Printed on acid-free paper

Preface

The logic of Manufacturing Resource Planning (MRP II) is implemented in most commercial production planning software tools and is commonly accepted by practitioners. However, these people are not satisfied with production planning and complain about long lead times, high work–in–process, and backlogging. As many researchers have pointed out, the reason for these shortcomings is inherent to the methods that are used. The research community is thus eager to find more sophisticated approaches.

This book is an attempt to compile some state–of–the–art work in the field of production planning research. It includes material that somehow dominates the existing MRP II concept. 15 articles written by 36 authors from 10 countries cover many aspects related to MRP II. All papers went through a single–blind refereeing process before they were selected for being published in this book.

When we received papers for this issue, we discovered that MRP II is a topic about which not only management scientists show interest. As the list of authors proves, industrial engineers, computer scientists, and operations researchers from academia as well as practitioners have contributed to this book. This, we hope, makes the book of value for a broad audience.

We thank all authors who submitted papers. And, we are indebted to Dr. Werner Müller from Springer for his support in this book project.

Kiel, December 1997

Andreas Drexl
Alf Kimms

Contents

Chapter 1: Master Production Scheduling

Chapter 2: Material Requirements Planning

Chapter 3: Lot Sizing

Chapter 4: Sequencing and Scheduling

Chapter 5: Production Control

Chapter 1

Master Production Scheduling

A Framework for Integrated Material and Capacity Based Master Scheduling

Jan Olhager and Joakim Wikner
Linköping Institute of Technology

Abstract: Master scheduling is gaining recognition as the major planning process for manufacturing firms, where the appropriate trade-offs can be made among e.g. customer service, manufacturing efficiency and inventory investment, involving both market and manufacturing insights. Master scheduling has traditionally been treated as a material-based planning tool, e.g. in the MRP II (manufacturing resource planning) approach to master scheduling. Capacity issues are then treated in a secondary way, as a capacity check of material plans. However, in, for example, process type industries the main focus is rather on capacity. Taking into account the positions of bottlenecks, order penetration points, and material profiles, the integration of material and capacity plans becomes necessary in many manufacturing environments. This paper presents a framework for the structuring of material and capacity integration taking the factors above into consideration. A matrix is presented where three material based characteristics meet four capacity based characteristics, and the resulting twelve situations depict possible manufacturing settings. In this context, issues such as material and capacity integration, planning and scheduling approaches and multi-level master scheduling are discussed.

1 Introduction

The importance of master scheduling for manufacturing firms has been increasing successively in recent years. Many firms have carried out manufacturing re-engineering programs related to lead times, quality and other manufacturing issues in order to shape the manufacturing environment and increase competitiveness. The traditional competitive goals of a manufacturing firm are usually described as (i) maximum customer service, (ii) minimum inventory investment, and (iii) maximum manufacturing efficiency. These goals differ between manufacturing environments. In a make-to-stock situation, customer service is related to stock availability (zero delivery time and maximum delivery dependability), the inventory investment is focused on finished goods inventory and manufacturing efficiency depends on resource utilisation, primarily machine related. The make-to-order or engineer-to-order firm will relate customer service to a given, agreed upon delivery time (and maximum delivery dependability), inventory investment to shorten manufacturing lead times (minimum work-in-process). Here, manufacturing efficiency in terms of resource utilisation is a subordinate goal, even though time can be interpreted as the most important resource in this case. With shorter lead times, many firms find that the traditional way of planning, e.g. material requirements planning (MRP), is too cumbersome with respect to the time available for managing customer orders. Then, the planning level where the main focus is placed is moving to a higher planning level in a hierarchical manufacturing planning and control system. Typically, this means that the master scheduling process is becoming more and more important. Lower levels will deal with execution rather than planning.

The master scheduling process is usually cited as the primary level where the market and production functions can make appropriate trade-offs among customer service, inventory investments and manufacturing efficiency. The reasons are manifold. Master scheduling is situated between the sales and operations planning level, which is usually based on aggregate data such as product groups, and material (requirements) planning, which is very detailed, thereby making master scheduling the link between strategic and operational issues. Furthermore, dealing with specific products and/or modules, it should be easy for both marketing and production people to be concrete about quantities and timing (see e.g. [8]). Also, it is becoming a widespread notion that volume in aggregate terms is managed at the sales and operations planning level and product mix within product groups at the master scheduling level. Consequently, master scheduling is the tool for manufacturing planning and control (MPC) that can really make a difference.

Master scheduling has traditionally been treated as a material-based planning tool, e.g. in the MRP II approach to master scheduling. The link between hierarchical planning levels such as sales and operations planning, master scheduling, material requirements planning and production activity control is established through material plans. The planning related to capacity has been dealt with as a capacity check, at each respective level. In early MRP applications, this may have been appropriate. However, in for example process type industries the main focus is rather on capacity. The integration of material and capacity plans becomes necessary in many manufacturing environments, even in labour intensive environments such as job shops, if excellence is to be achieved in manufacturing management.

This paper presents a framework for the structuring of material and capacity integration taking the positions of bottlenecks, order penetration points, and material profiles into consideration. A matrix is presented where three material based characteristics meet four capacity based characteristics, and the resulting twelve situations depict possible manufacturing settings. In this context, issues such as material and capacity integration, scheduling approaches and master scheduling issues are discussed.

2 Material and Capacity Interaction

In general, the planning and control of manufacturing activities involves both material and capacity, both being different types of resources. Capacities are used in the transformation process to convert material inputs, such as raw materials and components into material outputs, such as finished goods. Therefore, the interaction of material and capacity resources is the essence of manufacturing. The basic view and focus of the transformation process varies among manufacturing environments. An important issue in this context is the position of the order penetration point (OPP), also known as the customer-order decoupling point. This is where the product or order becomes linked to a specific customer order, typically by the presence of a customer specific configuration. Issues and aspects related to manufacturing strategy are treated in [5]. There are four basic different positions of the order penetration point:

- Make-to-stock (MTS)
- Assemble-to-order (ATO) (or finish-to-order, FTO)
- Make-to-order (MTO)
- Engineer-to-order (ETO)

The first three are also discussed by Berry and Hill [1] as the primary choices at the master planning level, taking manufacturing strategy aspects into account when designing a manufacturing planning and control system. This framework is also described in [3] and [8]. The fourth position (ETO) involves design engineering work related to each new customer order and is added here, whereas the third (MTO) typically involves activities related to parts manufacturing, assembly and possibly also some purchasing after the customer order is received. ATO assumes that semi-finished goods, e.g. in a modular product design, are fabricated based on forecasts and that only assembly remains after the customer order arrives. Similar situations, where the order penetration point is somewhere inside manufacturing can also be found for other finishing activities, such as testing and packaging and therefore does not necessarily have to relate to assembly operations, and thus referred to as FTO. MTS means that finished products are manufactured based on forecasts and placed in a finished goods inventory, from where deliveries are carried out.

Relating these order penetration points to manufacturing strategy and process choice (see e.g. [3]), the view and focus on material and capacity will vary considerably. In an MTS environment, the capacity investment is typically quite substantial and given a prime focus when designing the process. Material is then focussed during operations, where sufficient material must be available to feed the process. Typically, the order winner is price and the sales volume is high, leading to a cost focus related to resource utilisation (see [3]), indicating that capacity levels should be tight. At the other end of the spectrum, ETO is related to very limited investments in capacity, whereas the material value may be substantial and must be given priority. Manufacturing resources are then subordinate to the finishing of the product or order. Order winners here can typically be quality, design, and flexibility (see [3]), criteria that would benefit from some over-capacity. This means that master scheduling typically is dominated by capacity issues (levels, availability and production rates) in MTS environments and dominated by material issues (availability, timing and quantities) in ETO and MTO situations. The ATO case can be viewed as a combination of these two alternatives, and the order penetration point acts as a divider. Upstream or prior to the OPP, the situation is MTS-like, whereas the situation downstream resembles that of MTO.

Material dominated master scheduling is then concerned with the timing of material (orders) at various process stages, whereas capacity checks become secondary. Capacity dominated master scheduling, on the other hand, is concerned with the amount (or rate) of material in the process flow at a given time within the capacity limits. Simultaneously, the goal of customer service is provided through delivery speed and dependability or stock availability.

3 Material Dominated Master Scheduling

Historically, the emphasis in manufacturing planning and control has been on the priority, i.e. the co-ordination of material and is here referred to as the material dominated perspective. Initially the focus was on controlling the end-item using a master production schedule (MPS) and for the material in the engineering bill of materials, MRP has been employed. This approach worked satisfactory for some environments but as MRP was used in more product oriented environments, the explosion of customer specific variants made it necessary to introduce more suitable tools as for example two-level master scheduling. To distinguish among environments from a master scheduling perspective the concept of “V”-, “A”- and “X”-material profiles is introduced. Each letter has a shape corresponding to typical material profiles, i.e. the number of items at each level, with end items at the top, semi-finished goods at the middle and raw materials at the bottom. In some sense, this corresponds to the shape of the average bill of material. A similar classification scheme has been suggested by Umble and Srikanth [7] under their "V", "A" and "T" framework. The underlying assumption is that we want to minimise the number of part numbers involved in master scheduling and thus should try to identify at what level of the material profile we have the least amount of parts, here referred to as the waist of the material profile. This framework has been further elaborated on by e.g. [4] but here we use the original approach.

3.1 Material Dominated: Type "A" (End Item Waist)

Type "A" is commonly associated with a MTS environment where manufacturing is based on forecasts to comply with customers demanding a short (potentially zero) delivery lead time. Thus the OPP is located at the end-item level. Since the end-items are kept in inventory it is important to limit the number of variants to avoid over-investing in the end-items inventory and the distribution network. The range of end-items are manufactured using a convergent structure typical for assembly and thus resembling an "A"-shaped material profile as in Figure 1.

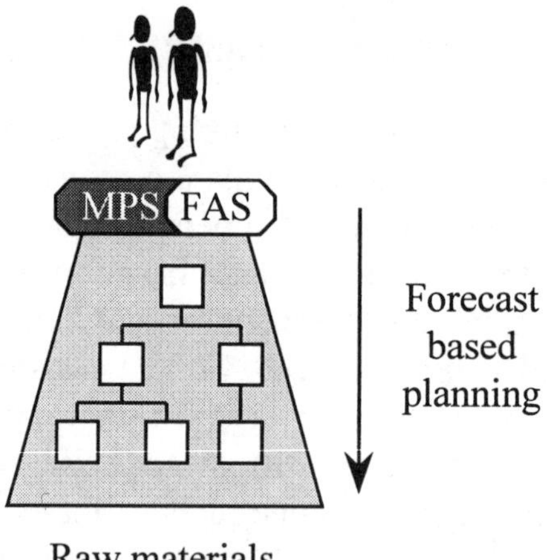

Figure 1. Type "A" material profile.

The two persons at the top of the figure represent the customers. Raw materials are assumed to be outside the control and ownership of the vertical value chain of the manufacturing firm. Since the waist of the material profile is at the same level as the end-item there is no need to use an MPS that is separated from a final assembly schedule (FAS). The focus in this environment is on service levels and machine utilisation, which in turn is dependent on successful forecasting.

3.2 Material Dominated: Type "V" (Raw Materials Waist)

In ETO environments no stock is carried, with the possible exception of raw materials with long replenishment lead times. This is also the case for MTO environments where no sub-assemblies are carried in stock. Manufacturing is typically characterised by a divergent material profile where all orders, except for some of the long lead time raw materials which are purchased based on forecasts, are related to a specific customer order. This results in a "V"-shaped material profile as shown in Figure 2. The OPP is hence located at the waist at the raw material level where the MPS is used. Since almost all action is triggered by a specific customer order the emphasis is on the capability and lead time of the process.

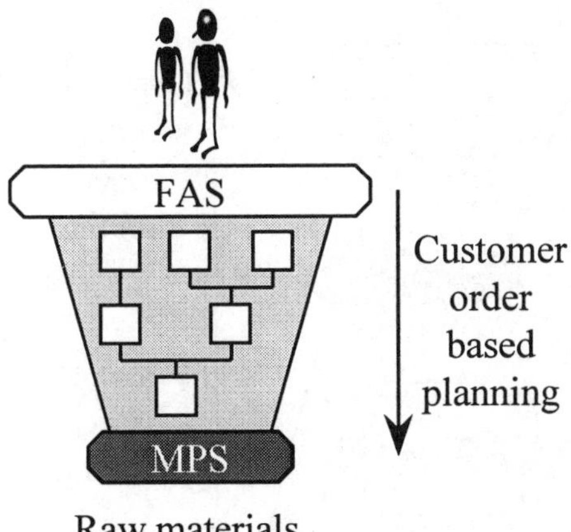

Figure 2. Type "V" material profile.

3.3 Material Dominated: Type "X" (Semi-finished Goods Waist)

When the manufacturing lead time exceeds the delivery lead time that the customer can accept, it is tempting to pursue an MTS strategy. The drawbacks of this decision are mainly twofold relating to the problem of customisation and capital tied-up in inventories. A compromise solution is to offer the customer a limited choice of options that can be assembled according to the customer order (ATO) or related to some other finishing operation, e.g. painting. (FTO). In this case we obtain a situation that can be described as a stacked "A" and "V" situation resembling an "X"-shaped material profile. The modules are produced to demand forecasts based on an MPS as in the "A" case described above. The assembly of the end item configured by the customer is conducted in accordance with the FAS based on the customer order and constrained by the MPS. In this case, a two-level master schedule can be appropriate and useful. Thus the OPP is situated at the module level at the waist of the material profile in Figure 3. This is the typical situation.

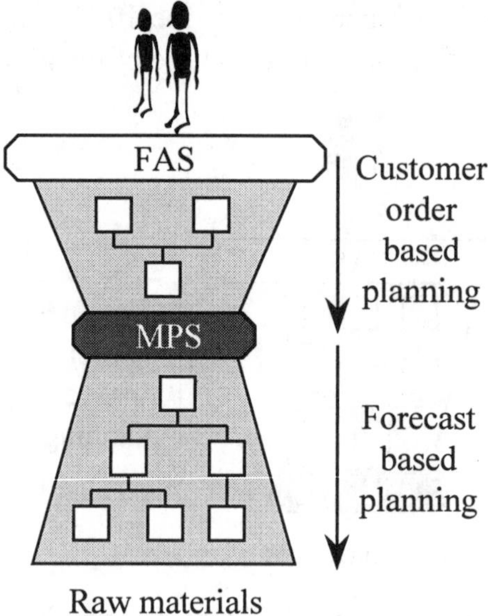

Figure 3. Type "X" material profile.

When a modular product design is used, late variety explosion can be created and the OPP is placed at the material waist. It also works the other way. If the customer delivery lead time is longer than the manufacturing lead time, a suitable semi-finished goods level must be chosen for the shift from forecast-driven to customer order-driven manufacturing. Then, there is a striving for creating a material profile waist at the OPP. Still, there may be exceptions. The market may require very short delivery lead times. Then, even if there is a modular product design implying an ATO situation, the firm may be forced to produce products to a finished goods inventory, i.e. adapting to a MTS situation. Still, a modular product design can offer advantages in terms of a short forecast horizon for final product configuration.

4 Capacity Dominated Master Scheduling

In some industries the availability of capacity is the central problem of master scheduling. In these cases the detailed priority problem is considered afterwards. The capacity dominated environment can be classified according to where the resource constraint is located. Here, we assume that the significant constraint in the system is located at one identifiable and stable position. Dynamic aspects such as moving bottlenecks that can appear at any position are not considered. However, in such cases, the primary concern should be to eliminate the wandering behaviour and instead create a stable and preferably single bottleneck environment.

If the constraint is at the beginning of the material flow (flowing from left to right), it is here referred to as a supply constrained environment and of type "<". When the market is constraining the flow we label this as of type ">". Finally the constraint can be internal, corresponding to e.g. a machine bottleneck, and is then classified as of type "><". The shape of the character only indicates the position of the constraint and does not reflect the relative capacities of other resources. Beside these environments where a specific capacity constraint dominates we also have the situation where the whole system is balanced and we denote this as of type

"=". In these cases the focus is typically on resource utilisation. Classical line balancing problems fall into this category of capacity oriented planning. More recent approaches include the Drum-Buffer-Rope scheduling (DBR) suggested by Goldratt and Fox [2] and Process Flow Scheduling (PFS), see e.g. [6].

4.1 Capacity Dominated: Type ">"(Market Constrained)

When the market is constraining the flow we have a situation where the focus should be on increasing sales using e.g. price cuts or lead time reductions. In this situation all orders with positive contribution would be accepted. Due to the over-capacity of the system, a given point of departure for scheduling would be the delivery dates promised to customers, i.e. the due dates of end items. All other materials would be scheduled using backward scheduling and since the capacity is of no significant concern an infinite loading approach can be employed. Since over-capacity prevails, cautions must be taken not to release too much material into the system as such action would result in excessive work-in-process (WIP) levels and long lead times.

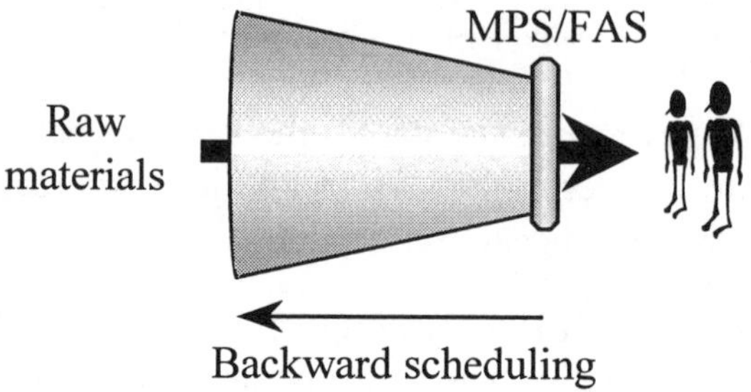

Figure 4. Type ">" capacity profile.

4.2 Capacity Dominated: Type "<" (Supply Constrained)

Here we are confronted with a situation where there is extensive demand for the end products but there is limited access to some kind of resource at the supplier side (which could be perceived as limited access to some raw material). In the long run this constraint must be elevated in some way by for example working more closely with the supplier. As long as this constraint is active the decision rule would be to maximise the contribution per unit of limited raw material. Given the delivery dates of constrained raw material, scheduled by the MPS, forward scheduling of production would be used to comply with the due dates (here represented by the FAS) in the best possible way as in Figure 5. In this case low utilisation of production resources is typically encountered together with large backlogs and thus extensive delivery lead times.

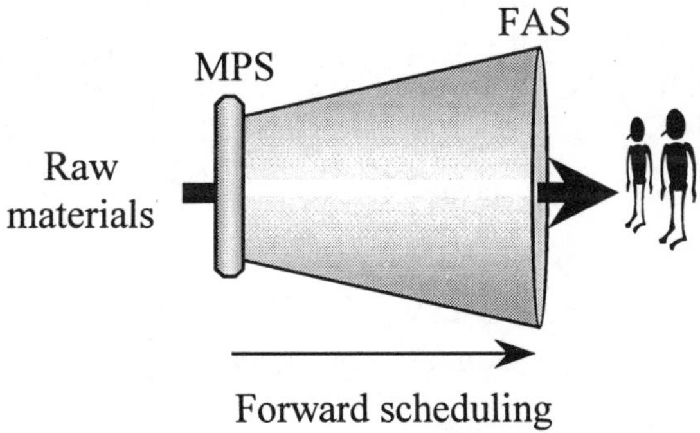

Figure 5. Type "<" capacity profile.

4.3 Capacity Dominated: Type "><" (Production Constrained)

When we are considering an internal constraint, we encounter a scenario that can be considered as a tandem ">" and "<" system. The MPS would exploit the constraint in the best possible way under consideration of the delivery due dates for the end items. Since there is surplus capacity before and after the constrained resource, the lead times through these sections are not too difficult to estimate due to the low level of utilisation. Thus we can use finite scheduling of the bottleneck, represented by the MPS in Figure 6, considering customer order due dates, contained in the FAS, and the estimated lead time downstream. Then we employ forward and backward scheduling downstream and upstream, respectively. This approach would correspond closely to the Drum-Buffer-Rope (DBR) scheduling principle of the OPT concept, suggested by Goldratt and Fox [2]. The decision rule in this case would be to maximise the contribution per unit of bottleneck resource. Obviously, the focus is on exploiting the bottleneck as much as possible. The most common problem in this setting is that it can be difficult to obtain good support from conventional MPC systems for bottleneck identification, scheduling and setting realistic due dates.

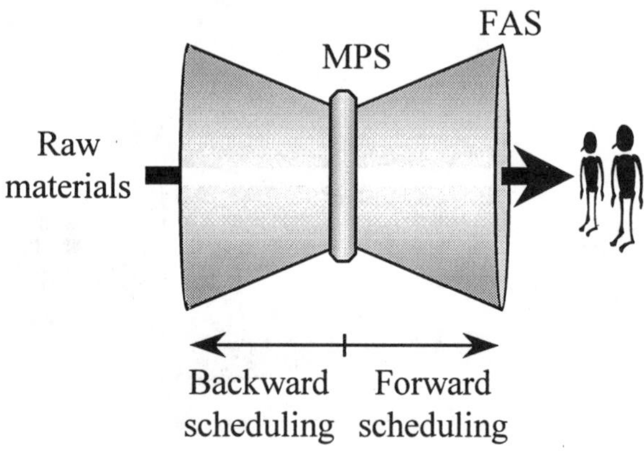

Figure 6. Type "><" capacity profile.

4.4 Capacity Dominated: Type "=" (Balanced)

When the demand is of a more stable character, a level production strategy is suitable. A variable production mix is possible as long as the load on the facilities is levelled. In this case a basic production rate is set based on the average demand rate (and if necessary some safety capacity margin). Thus the foundation is a rate-based master schedule, see Figure 7, guiding a load balancing exercise to maximise the utilisation of the resources. Since the balancing act can be quite challenging the production volume is kept stable, whereas some variations related to the product mix can be handled. Except for the problems created by rate changes, variations in product mix can in some cases generate effects similar to the moving bottleneck syndrome affecting the safety capacity margins. The objective in this setting would be to preserve the high utilisation based on the level load (production volume) by focusing marketing on the product mix flexibility available.

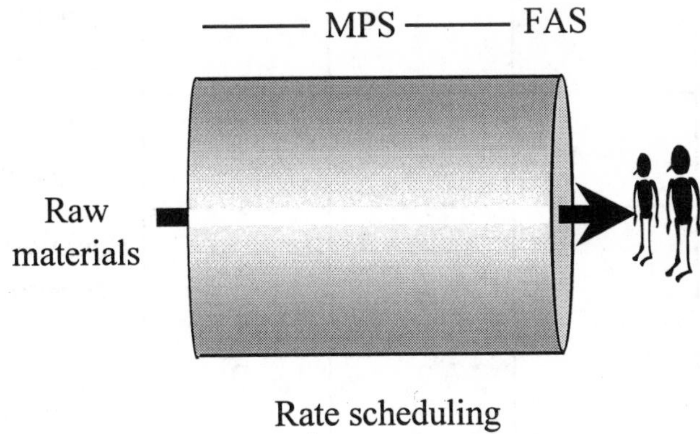

Figure 7. Type "=" capacity profile.

5 Integrated Master Scheduling

Combining the perspectives from both material and capacity dominated master scheduling as discussed above, a three-by-four matrix can be constructed. We call this the material and capacity profile matrix, or the C/OPP-matrix for short, representing the constraint and order penetration point interaction. The constraint issue is related to the capacity profile and the OPP is related to the material profile, as discussed in the previous section. The positions of the constraint and the OPP are fundamental characteristics of the manufacturing planning and control environment. Based on the three material profiles and the four capacity profiles, twelve different situations evolve. These are distinguished by the relative and absolute positions of the constraint and the OPP, as illustrated in the C/OPP-matrix in Figure 8.

In summarising the main properties of the material profile perspective and the capacity profile perspective, respectively, we find some basic keywords and issues related to the two perspectives. The main characteristic of the "A" material profile is forecast-driven

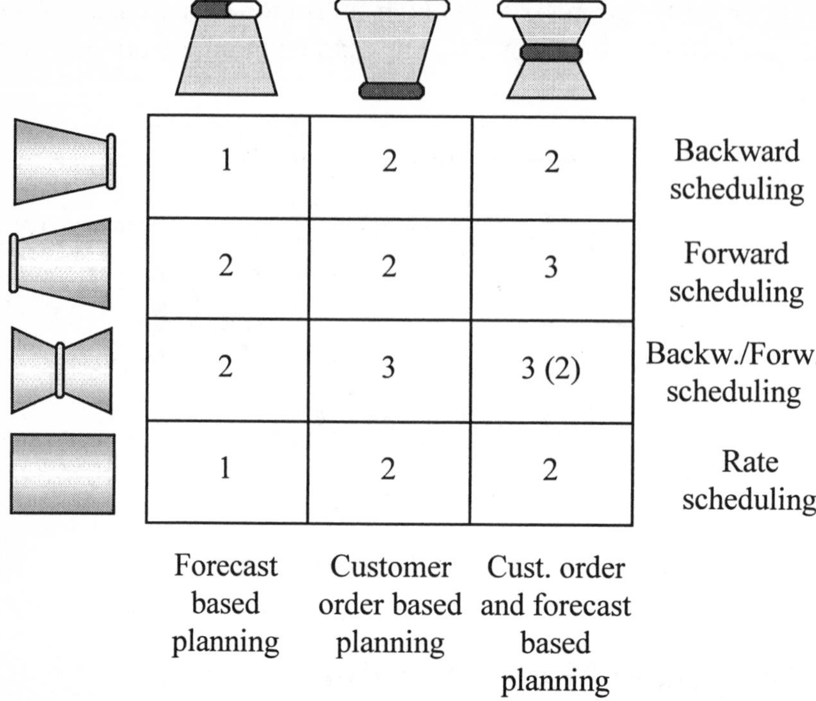

	Forecast based planning	Customer order based planning	Cust. order and forecast based planning
Backward scheduling	1	2	2
Forward scheduling	2	2	3
Backw./Forw. scheduling	2	3	3 (2)
Rate scheduling	1	2	2

Figure 8. The constraint and order penetration point matrix (C/OPP matrix), with key characteristics and the number of levels in a multi-level integrated material and capacity based master schedule.

planning and control of end products. The "V" profile is related to customer order-driven manufacturing. "X", the final material profile is a mix of the previous two, typically related to a modular product design, where the profile waist corresponds to the modules that are produced to forecast and then assembled according to customer order specifications. Thus, the material profiles are strongly related to a manufacturing policy, such as MTS, MTO and ATO, respectively. Among the capacity profiles, the ">" would typically imply MRP-type backward scheduling, whereas "<" would imply forward scheduling in e.g. a "suction" type system with successively increasing relative capacities. The "><" profile is a mix, with OPT-type backward scheduling of operations upstream the bottleneck, and forward scheduling of downstream operations. Finally, the "=" profile corresponds to a balanced line, where rate scheduling would be appropriate. Thus, the capacity profiles are strongly linked to scheduling approaches. Based on these characteristics, the appropriate mix of material and capacity related master scheduling issues can be designed for these twelve manufacturing situations.

A striking property is that the material profile waist corresponds to materials (finished good, module/semi-finished good, or raw material/component), whereas the capacity profile waist corresponds to a constraining resource (e.g. heat treatment or painting). Occasionally, the capacity profile waist can correspond to materials, in such a way that the physical storage capacity is a limiting resource. Thus, this case shows that the waists related to material and capacity can coincide. Also, the total master scheduling problem becomes simplified (i) if the material waist is situated right by the capacity waist, in such a way that the items at the material waist are the ones that are processed by the bottleneck, or (ii) if the waists coincide. This is due to the fact that the number of items and resources is minimised at the point where it really counts for the master schedule.

Another issue for master scheduling is the possibility to use a multi-level master schedule, see e.g. [8]. The number of levels will be influenced by the following issues and their positions relative to one another:

- Customer service, in terms of the planning of external deliveries. This is always related to the end product level, i.e. the top level of material profile and the right hand side of the capacity profile.
- Inventory investment, in terms of the OPP at the material profile waist. The corresponding position can be at the top, middle or bottom level of the material profile, i.e. it can be located at three different positions.
- Manufacturing efficiency, in terms of the constraint at the capacity profile waist. As indicated by the four capacity profiles, the constraint can be at the market or supply side, internal or non-existent. Thus, three positions are possible.

The implications for a master schedule is that the number of levels can be one, two or three depending upon whether these positions, as indicated above, coincide or not. If all waists are located at the “top” level, i.e. the market side constrained capacity profile (“>”), it would be sufficient to master schedule end products only. If one waist is located at a lower profile level, a two-level master schedule will be necessary for the integration of material and capacity considerations at the master level. This is the case for many elements of the matrix. Finally, if both the material and capacity waists are located away from the end products and market side, two possibilities emerge. If these waists coincide, a two-level master schedule would suffice. But if they do not, a three-level master schedule would be necessary for integrated material and capacity based master scheduling. The combination of “><” and “X” profiles offers both possibilities. If, for example, the bottleneck is situated upstream the module level, a three-level master schedule would simultaneously manage (i) customer service and deliveries of end products, (ii) inventory investment and stock availability of modules, and (iii) manufacturing efficiency in terms of bottleneck utilisation. This would truly correspond to an integration of material and capacity planning at the master level.

6 Conclusion

As has been shown, master scheduling is based on different properties of products and processes depending on whether material or capacity is dominating. A general framework for master scheduling should, however, have a more generic base containing both the material dominated and the capacity dominated approaches. Common to both are, however, that they should be demand-driven and thus based on given delivery due dates related to either customer orders or inventory replenishment orders. Significant in material dominated master scheduling is the concern for minimising the number of master scheduled items, which is related to the waist of the material profile. We also noticed the implicit assumption of positioning the OPP at the waist. As has been discussed above this is not necessarily always the case. In the capacity dominated approach to master scheduling the focus was instead on the constraining resource of the system. The properties of the material are more of a secondary concern. Based on this we can observe four significant properties of material and capacity integrated master scheduling. We summarise this below under *the four keys to master scheduling*:

1 Delivery due dates
2 Material profile waists
3 Order penetration points (OPP)
4 Capacity profile waists, i.e. the flow constraints

If the OPP is not positioned at the material profile waist, there may potentially be situations where all four issues are located at different positions, implying that a four-level master schedule could be appropriate. Even if such a multi-level schedule will not be used, the manufacturing planning and control environment will still be very complex. More commonly, the interrelationships among the four keys to master scheduling will lead to a one- or two-level master schedule and only occasionally to a three-level master schedule. The planning and control environment will no doubt be simplified if these issues can be co-ordinated, in such a way that the material and capacity profile waists, including the OPP, coincide or are positioned at the customer order delivery level.

Based on the three material and four capacity profiles identified in this paper, we designed a matrix, indicating twelve different manufacturing scenarios. This constraint/OPP matrix forms the framework for discussing the integration of material and capacity issues for master scheduling. Among the main issues are the choices of planning and scheduling principles and the number of levels in a multi-level master schedule. We have shown that the appropriate integration of material and capacity should correspond to a one-, two- or three- level master schedule depending upon the positions of capacity constraints and order penetration points relative the customer order delivery level. Then, customer service, inventory investment and manufacturing efficiency can be managed in an integrated material and capacity based master scheduling process.

7 Acknowledgement

This research is supported by grants from the Volvo Research Foundation, the Volvo Educational Foundation and The Swedish Foundation for Strategic Research.

References

[1] BERRY W.L. and HILL T. (1992), Linking systems to strategy. *International Journal of Operations and Production Management*, Vol. 12, No. 10, pp. 3-15.

[2] GOLDRATT E. M. and FOX R. (1986), *The Race: For a Competitive Edge*, North River Press.

[3] HILL T. (1994), *Manufacturing strategy: text and cases (2nd Ed.)*, Irwin, IL.

[4] ODEN H. W., LANGENWALTER G. A. and LUCIER R. A. (1993), *Handbook of Material & Capacity Requirements Planning*, McGraw-Hill Inc.

[5] OLHAGER J. (1994), On the positioning of the customer order decoupling point. *Proceedings of the 1994 Pacific Conference on Manufacturing*, Jakarta, pp 1093-1100.

[6] TAYLOR S.G. and BOLANDER S.F. (1994), *Process Flow Scheduling: A Scheduling Systems Framework for Flow Manufacturing*, APICS.

[7] UMBLE M. and SRIKANTH M. L. (1990), *Synchronous Manufacturing*, South-Western Publishing Co.

[8] VOLLMANN T.E., BERRY W.L. and WHYBARK D.C. (1997), *Manufacturing Planning and Control Systems (4th Ed.)*, Irwin, IL.

KNOWLEDGE-BASED SYSTEM FOR MASTER PRODUCTION SCHEDULING

Gürsel A. Süer, Miguel Saiz and Omar Rosado-Varela
Industrial Engineering Department
University of Puerto Rico Mayuagüez

Abstract: This paper discusses a knowledge-based system developed to select the best Master Production Schedule (MPS) that meets the user requirements. The best MPS is selected out of 192 alternative MPS determined by using a Cell Loading software in a Cellular Manufacturing environment. The software is used to help implement a Knowledge-Based Finite Schedule Material Requirements Planning (MRP II) in a jewelry manufacturing plant. The knowledge base system provides the user the flexibility to consider from one up to five different performance measures. The user can assign the lowest or the highest acceptable values for the performance measures considered. Furthermore, the system allows the user to select the order of performance measures he wishes to consider. It was necessary to interview an expert during the development process of this system. The expert offered his real life experience related to establishing MPS and suggested acceptance and rejection values for the performance measures. An experimentation was conducted in order to support the expert recommendations and to observe the behavior of the performance measures in terms of the number of shifts allowed and the number of manufacturing cells used. The results were included in the knowledge base to recommend the user how to make necessary capacity modifications when existing capacity was not sufficient to have acceptable levels of performance measures. A comparison between the MPS selected by the expert and the expert system was also made. Finally, expert recommendations were integrated into the program to help and address the user in the evaluation and selection process of the MPS.

1 INTRODUCTION

Material Requirements Planning (MRP) is probably the most common planning tool used in industry today. It has shown great progress over the years to accommodate the needs of the industry since it was first introduced in the early 1960's. Initially, it replaced Statistical Inventory Control Techniques. Later, rescheduling capability was added thus increasing its acceptance by the industry. The next phase in the development process was at the operational level where a capacity check was added thus obtaining closed-loop MRP capable of showing the impact of material plans on the available capacity. Finally, the last stage in the evolution was to tie planning activities with finance, marketing and accounting activities. This led to Manufacturing Resource Planning (MRP II). A typical flowchart of the MRP II process is given in Figure 1.

2 STANDARD MRP II APPLICATION

MRP II is a hierarchical planning tool. The decisions made at one level impose constraints within which more detailed decisions are made at the lower level. Since feedback from lower level to higher level is allowed, the decisions made at higher levels might be revised. Therefore, it is classified as an implicit hierarchical planning tool. In practice, there are problems with this hierarchical structure. The Master Production Schedule (MPS) is developed using rough-cut capacity techniques such as capacity bills and load profiles. Although, the accuracy of these procedures varies greatly, they fall way short in determining the exact detailed capacity requirements. Therefore, developing realistic MPS becomes almost an impossible task. Once MPS is fed into the MRP II process, planned orders for manufactured parts are generated; they are tested against available capacity by using Capacity Requirements Planning (CRP). Quite often, the unrealistic MPS is detected after CRP. Actions are taken by the MRP planner at the MRP level (modifying MPS schedule) and CRP level (revising available capacity by considering overtime, shift work, temporary workforce, etc.). These

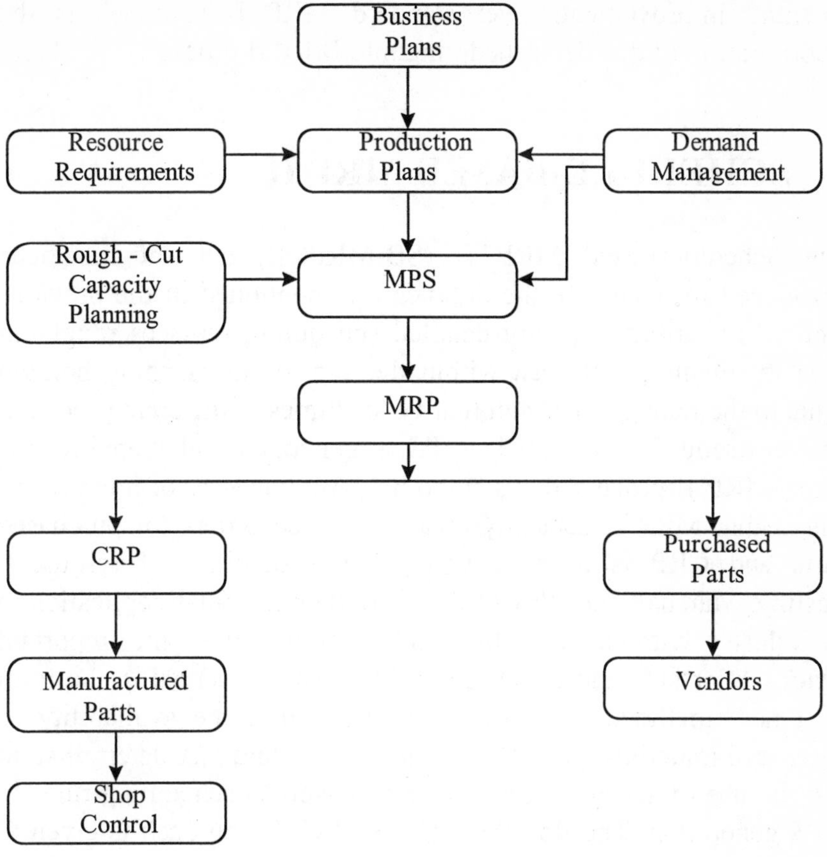

Figure 1. A Typical MRP II Process Flowchart

changes may require MRP to be run again due to chain effect resulting from product structures. These problems result from the inadequacy of the rough-cut capacity planning techniques and the unrealistic nature of the infinite capacity assumption.

Another difficulty with standard MRP II application is that most of the available MRP II software uses only Earliest Due Date (EDD) dispatching rule in establishing priorities for jobs despite the abundance of scheduling research results. As a generic software, it does not consider the shop configuration and its characteristics well enough to improve the quality of scheduling. Therefore, one of the potential improvements over standard MRP II approach is the incorporation of detailed scheduling into MRP II process.

3 SCHEDULE-BASED MRP II

The Schedule-Based MRP II (SB-MRP II) concept has been introduced to overcome the deficiencies mentioned in the previous section, i.e., finite capacity detailed scheduling replaces rough-cut capacity planning (at least within the short-term planning horizon equal to the maximum of cumulative leadtimes of different products) thus ensuring that a realistic MPS is generated and therefore, the entire MRP II process flows smoothly. Another way of interpreting this is that MRP II basically creates planned orders for purchased parts and CRP is used to verify the availability of resources. Hasting, Marshall and Willis [6] reported on a similar application in an industry without presenting and detailing all of the important factors. Süer [14] addressed the problem and presented in the form of a new methodology considering leadtimes, the availability of purchased materials, priority assignment, product mix determination and the use of scheduling algorithms as well as scheduling rules in MPS generation. The flowchart of SB-MRP II approach is given in Figure 2.

Most manufacturing companies are converting their manufacturing facilities into production lines and/or manufacturing cells by using GT principles. This helps to simplify the product structures by creating short or flat Bill of Materials. This in return helps to reduce the work orders for individual work centers. Production rates are established for lines or cells which makes SB-MRP II implementation even more feasible.

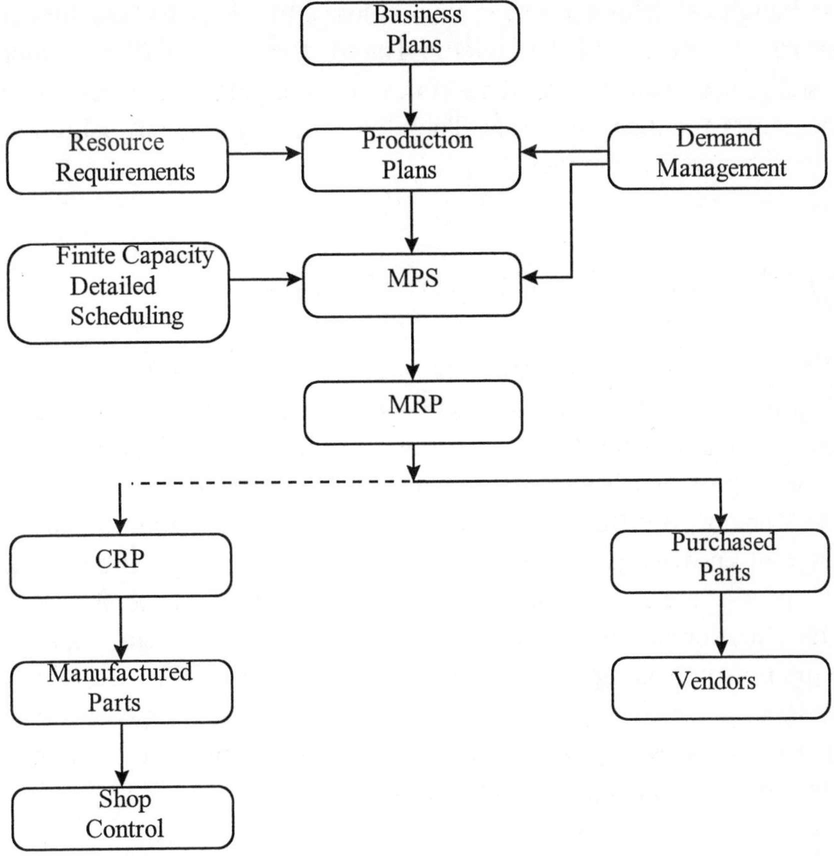

Figure 2. The Flowchart of SB-MRP II Process

4 KNOWLEDGE-BASED SYSTEM IN SB-MRP II

There is an increasing emphasis to incorporate the expertise developed by humans into computer programs. Artificial intelligence (AI) tools developed during the last two decades are increasingly being used to achieve that purpose. Minsky [9] defined artificial intelligence as "the science of making machines do things that would require intelligence if done by men."

The most significant practical product to emerge from thirty years of AI research is the so-called expert system, as mentioned by

Firebaugh [4]. This kind of system belongs to a broader category of programs known as Knowledge-Based Systems (KBS). Expert systems can address problems normally thought to require human specialists for their solutions. Therefore, the key concept behind all expert systems is knowledge.

4.1 Characteristics of Expert Systems

There are some important characteristics shared by almost all existing expert systems, as stated by Firebaugh [4]. These kind of systems perform at a level generally recognized as equivalent to that of an expert or some specialist in the field. The system does not know about everything, instead it knows a great deal about a narrow range of knowledge; therefore, it is a highly domain specific system. The user of the system can halt the processing at any time and ask it why a particular conclusion was reached. Then the system is able to explain its reasoning. A final characteristic is that, if the information or data on which the system is working are probabilistic or fuzzy, then it can correctly propagate uncertainties and provide a range of alternate solutions with associated likelihoods.

4.2 Knowledge Acquisition Methods

One common characteristic of all KBS is that they must be built from knowledge extracted from experts. Among the knowledge acquisition methods, Michaelsen, Michie and Boulanger [8] include the following (1) Being told, (2) Analogy, (3) Example, (4) Observation, discovery, and experimentation, and (5) Reasoning from deep structure. The first method is the most used one for rule-based systems. It is also possible to combine two or more methods to acquire the necessary knowledge.

4.3 Phases of Knowledge Acquisition

Hayes, Waterman and Lenant [7] pointed out some generally accepted phases that characterize the process of acquiring knowledge as: (1) Identification phase, (2) Conceptualization phase, (3) Formalization phase, (4) Testing phase, and (5) Prototype revision phase.

4.4 Literature Review

Fox [5] mentioned ISIS-II, an expert system developed to schedule job shops. It has the same knowledge and understanding of qualitative measures as the human scheduler possesses. It is a constraint directed reasoning system and takes a heuristic approach to generate schedules. Smith and Ow [13] described OPIS, another job-shop scheduling system developed in 1986. The OPIS system, a direct descendant of ISIS-II.

Shen and Chang [12] developed two real time scheduling algorithms for a flexible manufacturing environment. They suggested the integration of planning and scheduling functions, and the use of knowledge to improve the efficiency of an algorithm. Chiodini [2] presented a real time replanning system interfaced with an MRP II system. Parunak [10] developed YAMS, a planning and control system, to illustrate the distributed AI methodology.

Adler, Fraiman, and Pinedo [1] described the development and implementation of an expert system for scheduling in a liquid packaging plant. Bhaskar and Viswanadham [3] developed an expert system for real-time scheduling in flexible manufacturing systems.

4.5 The Proposed Approach

The KBS can be used to test a large set of Master Production Schedules (MPS) against a group of selected performance measures in an SB-MRP II environment. Since the MPS are generated based

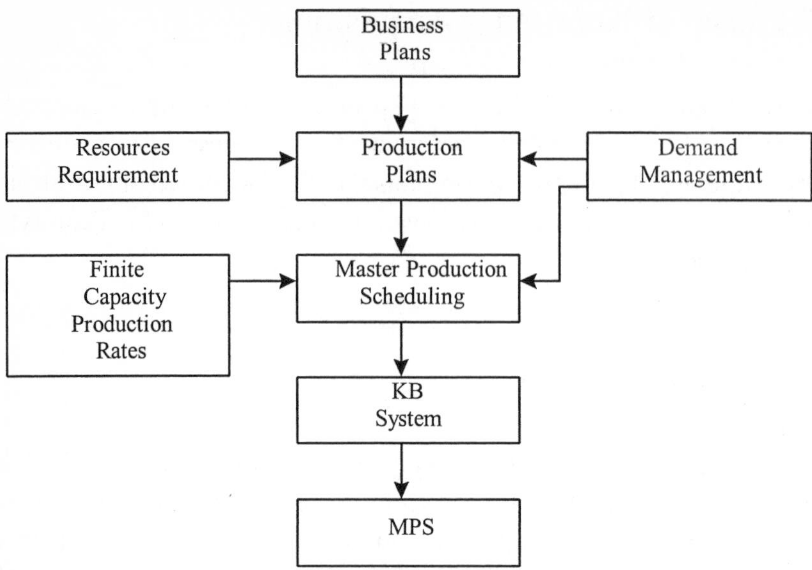

Figure 3. Knowledge and Schedule-Based MRP II Approach

on finite capacity detailed scheduling, the selected MPS can be fed into MRP II process without any capacity concern.

5 DESCRIPTION OF THE SYSTEM CONSIDERED

The proposed knowledge and scheduled-based MRP II has been developed for Avon Lomalinda Inc., a jewelry manufacturing plant located in San Sebastian, Puerto Rico. There are seventeen manufacturing cells grouped into four business units, and a plating area. The Business Unit 2 produces plastic, porcelain, casting, and stamping earrings using the cells allocated to it. This expert system has been specifically designed for the Business Unit 2 of the company.

The cells are arranged and equipped such that inter-cell material transfer follows a unidirectional flow. There are two basic types of inter-cell material transfer: (1) All of the operations are completed within a cell, and (2) Some of the operations are performed in the cell and then, the parts leave the cell to be plated in the plating department. Later, the parts return to the same cell for the remaining operations to be completed. Each product of this category requires one of three types of plating: manual, barrel, or chain, depending upon the processing requirements.

6 SB-MRP II APPLICATION BY USING CELL LOADING APPROACH

In a cellular manufacturing environment, Cell Loading is equivalent to Master Production Scheduling. In this section, the cell loading process developed in the company is explained in detail.

6.1 Cell Loading Rules

Cell loading involves determining to which cell, among the feasible cells, the product should be assigned. The cell loading rules used in the development of the Master Production Schedules are:

1. Search Priority: Cell Priority (CP), Product Priority (PP).
2. Primary Product Rule: Earliest Due Date (EDD).
3. Secondary Product Rules: Number of Feasible Cells (NFC), Number of Cells Required (NCR).
4. Selection Criterion: Maximum (MAX), Minimum (MIN).
5. Primary Cell Rules: Cell Load (CL), Number of Feasible Products (NFP), Product Mix (PM).
6. Selection Criterion: Maximum (MAX), Minimum (MIN).

The rules suggested in this section are combined in all possible ways thus generating 48 possible combinations of which 24 are classified as the product priority and the remaining 24 as the cell priority. Süer et al. [15] described these rules extensively.

For example, the combination PP/EDD /NFC /MAX /NFP /MIN means that this application is a product priority case where first a product is selected based on the EDD rule. If there are several products with the same due date, then the product with maximum number of feasible cells is given the highest priority. The next step is to select the feasible cell to which this product will be assigned. In this case, the product is assigned to the feasible cell that can process the minimum number of products at the time of selection. The possible combination of these rules are summarized in Table 1.

6.2 Different Scenarios

Four different scenarios are generated by adding lot-splitting to the cell loading process. Lot-splitting implies that a product can be assigned to more than one cell if the completion of the product by its due date is essential. When the 48 combinations of rules are applied to the four scenarios one hundred ninety two different ways of generating Master Production Schedules are obtained. The scenarios considered are: (1) Plating Sensitive-Lot-Splitting Allowed, (2) Plating Sensitive-Lot-Splitting not Allowed, (3) Plating Insensitive-Lot-Splitting Allowed, (4) Plating Insensitive-Lot-Splitting not Allowed.

6.3 Cell Loading Program

The purpose of the cell loading program is to create master production schedules. There are two general ways to execute the Cell Loading Program. The first one is to ask the program to generate 192 summarized MPS and the second one is to ask the program to run a detailed MPS. The output of "192 summarized MPS" option is written to a database file that can be accessed from an expert system shell. The performance measures are: number of critical tardy jobs, number of non-critical tardy jobs, maximum tardiness, total tardiness and average cell utilization. A critical job is the one that is due within the next three weeks, whereas a non-

TABLE 1. Combinations of Cell Loading Rules

Product Priority Rules	Cell Priority Rules
1. PP/EDD/NFC/MIN/CL/MIN	25. CP/CL/MIN/EDD/NFC/MIN
2. PP/EDD/NFC/MIN/CL/MAX	26. CP/CL/MAX/EDD/NFC/MIN
3. PP/EDD/NFC/MIN/NFP/MIN	27. CP/NFP/MIN/EDD/NFC/MIN
4. PP/EDD/NFC/MIN/NFP/MAX	28. CP/NFP/MAX/EDD/NFC/MIN
5. PP/EDD/NFC/MIN/PM/MIN	29. CP/PM/MIN/EDD/NFC/MIN
6. PP/EDD/NFC/MIN/PM/MAX	30. CP/PM/MAX/EDD/NFC/MIN
7. PP/EDD/NFC/MAX/CL/MIN	31. CP/CL/MIN/EDD/NFC/MAX
8. PP/EDD/NFC/MAX/CL/MAX	32. CP/CL/MAX/EDD/NFC/MAX
9. PP/EDD/NFC/MAX/NFP/MIN	33. CP/NFP/MIN/EDD/NFC/MAX
10. PP/EDD/NFC/MAX/NFP/MAX	34. CP/NFP/MAX/EDD/NFC/MAX
11. PP/EDD/NFC/MAX/PM/MIN	35. CP/PM/MIN/EDD/NFC/MAX
12. PP/EDD/NFC/MAX/PM/MAX	36. CP/PM/MAX/EDD/NFC/MAX
13. PP/EDD/NCR/MIN/CL/MIN	37. CP/CL/MIN/EDD/NCR/MIN
14. PP/EDD/NCR/MIN/CL/MAX	38. CP/CL/MAX/EDD/NCR/MIN
15. PP/EDD/NCR/MIN/NFP/MIN	39. CP/NFP/MIN/EDD/NCR/MIN
16. PP/EDD/NCR/MIN/NFP/MAX	40. CP/NFP/MAX/EDD/NCR/MIN
17. PP/EDD/NCR/MIN/PM/MIN	41. CP/PM/MIN/EDD/NCR/MIN
18. PP/EDD/NCR/MIN/PM/MAX	42. CP/PM/MAX/EDD/NCR/MIN
19. PP/EDD/NCR/MAX/CL/MIN	43. CP/CL/MIN/EDD/NCR/MAX
20. PP/EDD/NCR/MAX/CL/MAX	44. CP/CL/MAX/EDD/NCR/MAX
21. PP/EDD/NCR/MAX/NFP/MIN	45. CP/NFP/MIN/EDD/NCR/MAX
22. PP/EDD/NCR/MAX/NFP/MAX	46.CP/NFP/MAX/EDD/NCR/MAX
23. PP/EDD/NCR/MAX/PM/MIN	47. CP/PM/MIN/EDD/NCR/MAX
24. PP/EDD/NCR/MAX/PM/MAX	48. CP/PM/MAX/EDD/NCR/MAX

critical job is the one that can be late without causing any significant consequence. The Cell Loading Program can handle 5-7 cells and the number of shifts can be defined as 1, 2 or 3. The length of the planning horizon can be set to eight, ten or thirteen periods.

6.4 Inputs to the Cell Loading Program

The inputs required for the cell loading program are cell feasibility matrix, cell loading rules, production plan and demand forecast, production rates determined based on the bottleneck analysis of the available resources (number of shifts, number of machines of each type, people, equipment, overtime restrictions, etc.)

7 THE ROLE OF KNOWLEDGE-BASED SYSTEM

This role of the knowledge-based system is to help master production scheduler in the MPS selection process. After all of the alternative MPS have been generated and evaluated by the cell loading system, the KBS selects a handful MPS which meet the values set by the master production scheduler with respect to all performance measures, if any. The system uses the reasoning of a typical master scheduler working in such an environment to evaluate and select the best MPS based on company requirements. The KBS does not recommend an MPS that does not meet company requirements. If the KBS cannot find an appropriate MPS, it tells the situation to the user and also advises him what to do next. The KBS developed for the selection of the MPS is incorporated into SB-MRP II Approach, as shown in the flowchart in Figure 4.

7.1 Methodology Used

As mentioned before, the key concept behind all expert systems is knowledge. There are 192 alternative MPS and 64 possible permutations of performance measures thus making this complex problem a good application for an expert system. 64 possible permutations are determined by summing the number of permutations for 1 measure (4), 2 measures (12), 3 measures (24) and 4 measures (24).

The knowledge acquisition methods used in this study are being told and experimentation. The two sources of information used are, (1) the expert knowledge and (2) an extra analysis of the behavior of the system performed by modifying the system size. The knowledge-

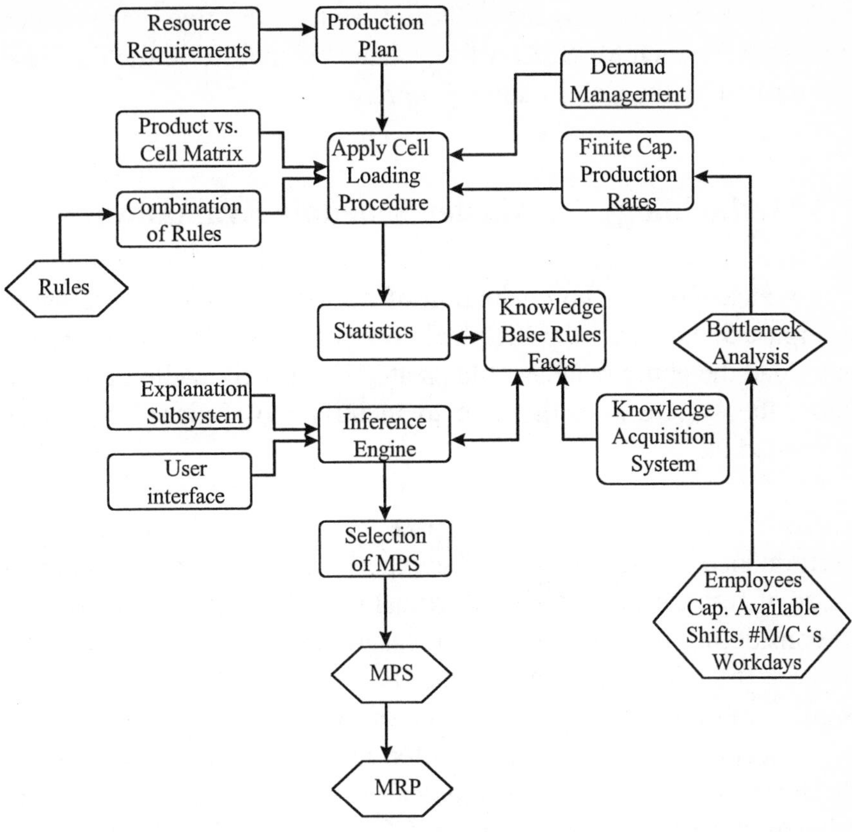

Figure 4. The Use of a KBS in an SB-MRP II Application

based is developed in M.1 Version 3.0, one of the most powerful PC-based knowledge-system software tools available today. Probabilistic information is introduced and propagated by means of an assigned certainty factor. M.1 provides an interface with dBASEIII to access data that has been compiled by other conventional programs and stored as a database file.

7.2 Input Data for Knowledge-Based System

The data required for the knowledge-based system are: (1) The MPS set to be evaluated, (2) the rule combination associated with each

MPS in the set, and (3) the values of each of the five performance measures for each MPS. All of this data is contained in the database file created by the Cell Loading Program.

7.3 Methodology for Master Schedule Evaluation

The typical flow in the evaluation of master schedules is presented in Figure 5. The first step that the knowledge-based system performs is to load the entire database file created by the cell loading program. Next, the system asks the user to identify the scenario that he is interested in.

Later, the system asks the user whether to consider each performance measure or not. The expert system always considers a value of zero for the number of critical tardy jobs. If the user desires to consider a given performance measure, then the lowest (a) or the highest (b) acceptable value must be entered. The lowest or the highest value has to be in the range of ($a_1 \leq a \leq 100$) and ($0 \leq b \leq b_1$), respectively. This range is called the acceptable range and has the form (a_1, 100) for the average cell utilization and (0, b_1) for the other performance measures except for the critical tardy jobs. The values of a_1 and b_1 depend on the performance measures used and they had to be determined. Any value within the acceptable range results in an acceptable MPS according to Avon Lomalinda criteria.

Next, the system needs to know which performance measure, among the ones considered, is to be satisfied the first, the second, and so on. The MPS are evaluated in terms of the first objective function (performance measure) to achieve. If there is a tie between two or more MPS, it is broken by selecting the MPS or MPS that perform better with respect to the next objective function and so on. Finally, in the last step, the system presents the user all the MPS which meet all user requirements, if any.

7.4 When no MPS Satisfies User Requirements

To accept or reject an MPS depends on the values of some or all of the five performance measure values as entered by the user. It informs situation to the user and continues search process by decreasing or increasing the value of the performance measure it is considering, to a minimum (a_2) or maximum (b_2) value (see Figure 6). These values are lower and higher, respectively, than the values presented to the user in the acceptable range or $a_2 < a_1$ and $b_1 < b_2$. The range (a_2, a_1) for the average cell utilization and (b_1, b_2) for the other performance measures is called the permissible range. The minimum value (a_2) is considered only for the average cell utilization, and it is set to 75%. Then, instead of increasing, the system decreases the value of the performance measure to this value. If the system finds an MPS after modifying the performance measure value input by the user, it informs the user about the results. If the user accepts the MPS with this new performance measure value, then the system moves to the next performance measure considered, if any, and the process continues in the normal manner. Otherwise, the search process comes to an end. The expert system recommends to execute the Cell Loading Program again by increasing the system size (the number of cells, shifts or both) whenever it cannot find an MPS that satisfies requirements of the user. Finally, the system returns to the main menu, allowing the user to restart the whole process or quit the system.

7.5 Cell Loading Program and Expert System Integration

After the expert system is built, it is integrated with the cell loading program to create a complete system for MPS development, evaluation, and selection (cf. Figure 7). The first step is to go to the cell loading program and generate the MPS set (192 MPS). Then,

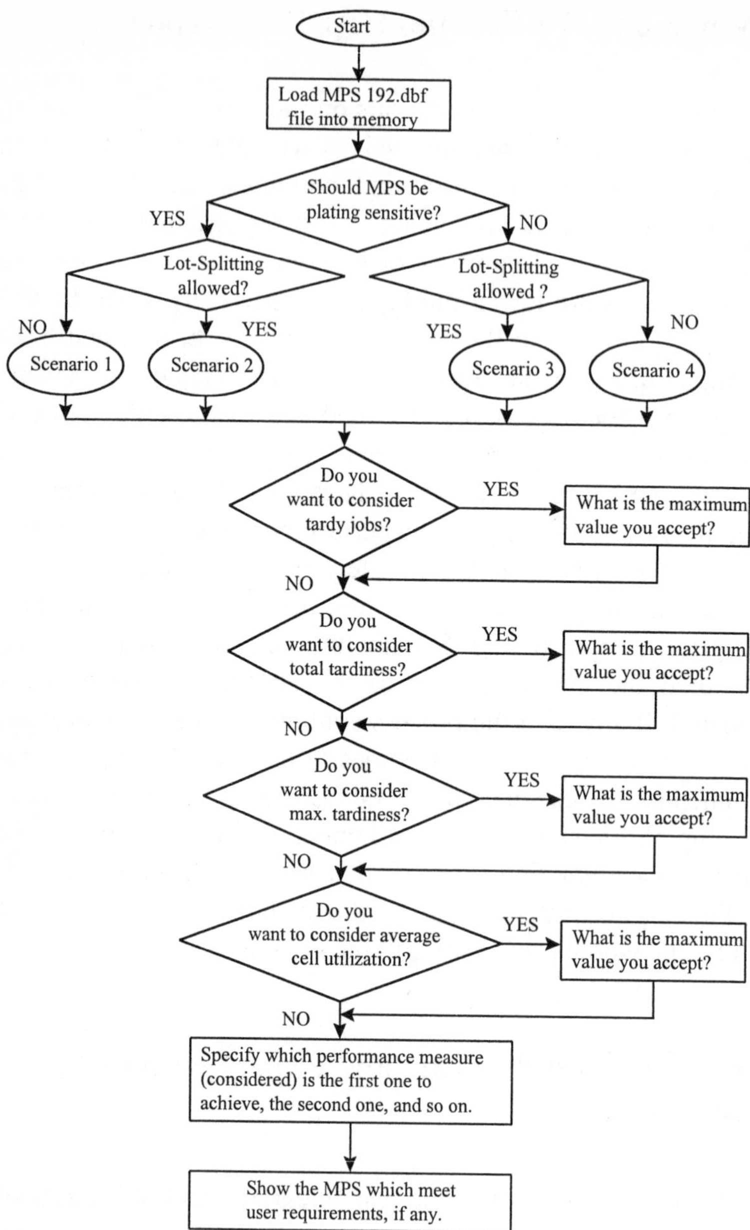

Figure 5. Steps Taken in the Evaluation of MPS Set

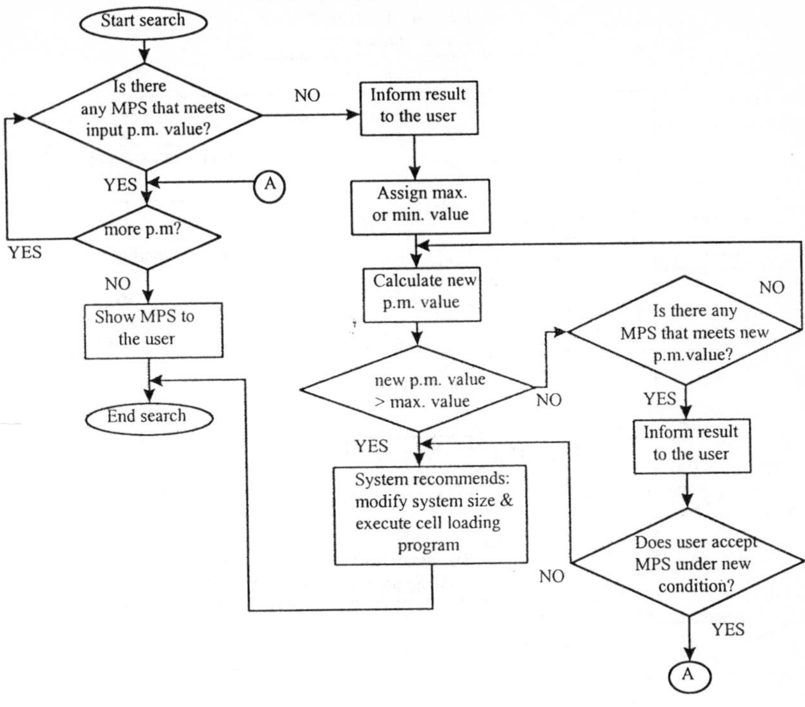

Figure 6. Finding an MPS When no Schedule Meets Given Performance Measures

the user must go to the expert system to begin the consultation process specifying which performance measures to consider, their acceptable values, and in which order to consider them.

If the expert system finds at least one MPS which meets user requirements, the next step is to quit the expert system, and go to the cell loading program again and generate the detailed version of the selected MPS.

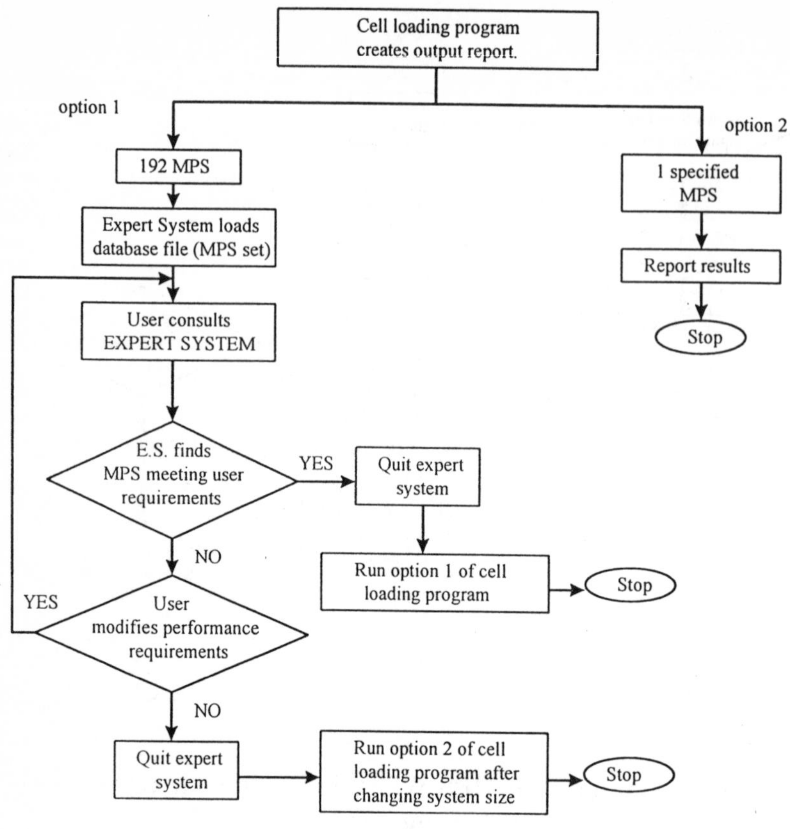

Figure 7. Cell Loading Program & Expert System Integration

7.6 Certainty Factors

An important feature of the KBS is its ability to represent and use uncertain knowledge. The degree to which a fact is believed is indicated by its certainty factor, an integer between -100 and 100.

Each performance measure, except critical tardy jobs, has an acceptable range for the user to select the lowest (a) or the highest (b) value. If the system cannot find any MPS with such requirements, but it finds an MPS with a performance measure value in the acceptable range, then the certainty factor for that results will be 95 instead of 100. The other possibility is that the system finds an MPS with a performance measure value outside the acceptable range (but inside the permissible range). Consequently, the certainty factor for that result will be 80 instead of 100 (cf. Table 2). If a proposition consists of several (simple) propositions connected by and's, then the overall certainty factor of the proposition is the minimum of the certainty factors of the parts. This is done as part of the knowledge acquisition process.

Table 2. Performance Measure Certainty Factors

Condition	MPS's performance measure value **meets** user requirement, or ($0 \leq$ value $\leq b \leq b_1$) and ($a_1 \leq a \leq$ value ≤ 100)	MPS's performance measure value does not meets user requirement but still **inside** acceptable range, or ($0 \leq b <$ value $\leq b_1$) and ($a_1 \leq$ value $< a \leq 100$)	MPS's performance measure value is outside of acceptable range but inside permissible range, or ($0 \leq b \leq b_1 <$ value $\leq b_2$) and ($a_2 \leq$ value $< a_1 \leq a \leq 100$)
Certainty Factor	100	95	80

7.7 Testing the Knowledge-Based System

From a commercial point of view, there has yet to be proposed an infallible method for evaluating the veracity of what the expert system concludes, as mentioned by Pedersen [11]. It is important to understand that errors can never show that a program is correct. Program testing can only demonstrate the presence of errors in a program, it cannot demonstrate their absence. However, it is essential to every user that the system be tested prior to its use through any method available to acquire at least the highest possible degree of reliability. The goal here is to ensure that the advice given is correct.

For testing the system, the master scheduler supplied real examples. The demand data was used in various forms (1) as is (2) 10% increase (3) 20% increase (4) 10% decrease (5) 20% decrease and (6) random increase and decrease in demand. The values of the performance measures established were zero number of critical tardy jobs, greater than or equal to 80% cell utilization and less than or equal to fifty non-critical number of tardy jobs. The expert was asked to select the top five MPS in each scenario for each possible demand level. There was no dominance observed among the rule combinations in terms of the best results. After the testing phase ended, the master scheduler compared the results of the knowledge-based system with his own results. A 100% match was found between master scheduler's choice and KBS's choice.

8 CONCLUSIONS

The performance measures number of critical tardy jobs, number of non-critical tardy jobs, maximum tardiness, and total tardiness are related to each other. This relation was very obvious when the system size was modified, where the improvement for all of them was almost the same for each extended system size. The experimentation provided support to the expert recommendations, and also demonstrated the feasibility of his recommendations. Every recommendation from the expert about acceptable ranges, and other related information were demonstrated as possible to achieve. The experimentation showed that the behavior of the performance measures depends on the system size.

With the creation of this expert system, it is possible to test a given set of master production schedules (MPS) against five performance measures. Having integrated the Expert System and the Cell Loading Program, the best MPS with respect to the desired performance measure(s) can be selected and fed into the MPS module of the existing Avon MRP II software. This capability supports a Knowledge-Based SB-MRP II application for Avon Lomalinda, Inc. purposes.

Toward the end of the completion of this study, the expert interviewed for this project was no longer Avon Lomalinda, Inc. master scheduler because he was promoted to a supervisory position. However, his knowledge about MPS selection process, acceptable ranges, and other valuable information has been retained and translated inside this expert system. This justifies one more time that expert systems is a valid approach for manufacturing planning problems as well.

REFERENCES

[1] Adler, L. B., Fraiman, N.M., and Pinedo, M.L., 1989. "An expert system for scheduling in a liquid plant", Proceedings of the Third International Conference. Expert systems and the Leading Edge in Production and Operations Management, Univ. South Carolina.

[2] Chiodini, V., 1986. "An Expert System for Dynamic Manufacturing Rescheduling", Symposium on Real Time Optimization in Automated Manufacturing Facilities, National Bureau of Standards, Gaithersburg, MD.

[3] Bhaskar, N.E., and Viswanadham, N. 1992. "Expert system for real-time scheduling in flexible manufacturing systems." INF DECIS TECHNOL, Vol.18, No. 3.

[4] Firebaugh, M. W. 1989. Artificial Intelligence: *A knowledge-Based Approach*. PWS- Kent Publishing Company, Boston, MA.

[5] Fox, M. S. 1983. "Job Shop Scheduling: An Investigation into Constraint Directed Reasoning". Ph.D. Thesis, Carnegie Mellon University, Pittsburgh, PA.

[6] Hasting, N.A., Marshall, P., and Willis, R.. 1983. "Schedule-Based Materials Requirements Planning," Journal of Operational Research Society, Vol. 33, No.11, pp. 1021-1029.

[7] Hayes-Roth, F., Waterman, D.A., and Lenat, D. B. 1983. Building Expert Systems. Reading, Addison Wesley, MA.

[8] Michaelsen, R. H., Michie, D. And Boulanger, A. 1985. "The Technology of Expert Systems", Byte Vol. 10, No. 4.

[9] Minsky, M.1975. "A Framework for Representing Knowledge." in The Psychology of Computer Vision. P. Wiston (ed). McGraw Hill, New York, NY.

[10] Parunak, H.V.D. 1987. Distributed Al systems, in Artificial Intelligence Computer Integrated Manufacture, (ed. A. Kusiak), IFS, Kempston, Bedford, UK. Springer, New York.

[11] Pedersen, K. 1989. "Expert Systems Programming." Practical Techniques for Rule-Based Systems. John Wiley & Sons, Inc., New York, NY.

[12] Shen, S., and Chang, Y. 1986. An Al Approach to Schedule Generation in a Flexible Manufacturing System, in Flexible Manufacturing Systems: Operations Research Methods and Applications, (eds. K. E. Stecke and R. Suri), Elsevier, New York.

[13] Smith, S. F., and Ow, P. S. 1986. "The Use of Multiple Problem Decompositions in Time Constrained Planning Tasks." **AAAI-86**, pp. 1013-1015.

[14] Süer, G. A. , 1989. "SB-MRP II: Integration of Production Scheduling Algorithms with Materials Requirements Planning Systems, Ph.D. Dissertation, Wichita State University.

[15] Süer, G. A., Dagli, C., and González., W. 1995. "Manufacturing Cell Loading Rules and Algorithms for Connected Cells." in Planning, Design, and Analysis of Cellular Manufacturing Systems, (eds. A. Parsaei, A. Kamrani & D. Lines), Elsevier Science, New York.

Chapter 2

Material Requirements Planning

Cover-Time Planning: A Less Complex Alternative to MRP

Anders Segerstedt
Mälardalen University College, Sweden

Abstract: This paper presents Cover-Time Planning, a kind of a reorder "point" system which exploits the information available in the bill of material. Cover-Time Planning contrary to an ordinary reorder point system uses time as a decision variable instead of quantity. Cover-Time Planning has like Kanban a built in restriction of work-in-process. Not until previous production is consumed a new production is initiated.

1 Introduction

For a manufacturing company to satisfy the customer with a delivery time shorter than the total lead time at least a part of the manufacturing process must have been started before the customer order arrives. For achieving this mostly two different methods are used reorder point systems or Material Requirement Planning. These methods can shortly be described as follows (cf. Krajewski and Ritzman [1]):

- *Reorder point systems*

For each item a reorder point or reorder level is defined. When the inventory of the item is less than this level replenishment starts for the item through a purchase process or a manufacturing process. The reorder level is decided such that the available inventory will satisfy demand until replenishment arrives. The order size is settled by a calculated economic order quantity and/or decided by the user adjusted to a suitable packing, price, discount, annual demand etc. Behind every reorder point there is an assumption of a future demand (a forecast).

- *Material Requirements Planning (MRP)*

For each end item a master production schedule is created specified with delivery times and order quantities from a forecasted demand. The master production schedule is exploded to downstream items through the product structures (bills of material). Demand is netted towards available inventory. This determines if a new planned order should be created or an old open order should be rescheduled with a new delivery date. The calculation starts with the items on the highest level structural level and continues, level by level, until the lowest structural level is reached. The lowest structural level an item is used in determines at what level the item participates in the calculation. Order quantities are decided by registered fixed order sizes or through some lot sizing technique, e.g., Part Period, Least Unit Cost etc.

Reorder point systems or MRP, which is the best? *Reorder point systems* require a forecast for every individual item, the information in the bill of material is not used; the system may create "lumpy" work load; the system can not present a forward visibility of future receipts and issues and most important the future work load. Due to this *MRP* is often considered to be the better system (cf. Mattsson [2]).

However, there are several problems coupled to MRP: Small changes in demand and real lead times lead to continuous reschedulings in order to correct previous planned replenishments to now existing planned withdrawals. Burbidge [3] says due to changes in demand, and from that necessary changes of the master production schedule, the MRP calculation leads to result where orders are released with delivery dates in passed time. The MRP calculation always strive to keep existing safety stocks full (cf. Orlicky [4]). It is mostly described as a "push" system instead of a "pull" system such as Kanban. Many, especially people in practical work, consider Just-in-Time and Kanban to be what to strive for. (However it is difficult to use Kanban in other part of the production facility and where there is relatively constant demand. For calculations of material requirements in the complete facility MRP or reorder point systems must be used even if in some parts Kanban is used.) In Sweden today some companies change their MRP systems to reorder point systems without any "forward visibility". This must mean that the companies are disappointed with the outcome of the advantages they thought to accomplish when they first installed MRP.

Cover-Time Planning is an idea to a simplification of MRP, based on attempts to avoid some of the practical drawbacks of both reorder point systems and MRP. Cover-Time Planning is basically a reorder point system based on time instead of quantities.

A master production schedule is established for the end items in such a way that a demand rate in units per time period (production days) is supposed to be in order from now until a new possible future demand rate is assumed to reign. Thus it is possible to plan an increase or a decrease of the production. From the master production schedule for the end items and from the information in the bill of material and considering the lead times, demand rates are calculated for all items. From these demand rates Cover-Time Planning can schematically be described with the following calculations:

Supply = Current inventory on hand - Scheduled issues in past time + Scheduled receipts

For how long this already "activated" supply will cover expected demand can be called "cover-time". If the demand rate is fairly constant and the lead time short, cover-time can be determined by:

Cover-time = Supply/Demand rate

If cover-time is greater than the lead time there is no need to another replenishment, comparable to when the inventory on hand is greater than the reorder level.

The following conditions initiate signals for replenishment:

Cover-time + Buffer time < Lead time (including inspection time)

Projected on hand inventory at the end of the lead time (including inspection time) < 0

If the lead time is long and/or the demand rate varies a more accurate calculation of the cover-time must be performed. Figure 1 illustrates that Case 2 requests a more precise calculation of the cover-time.

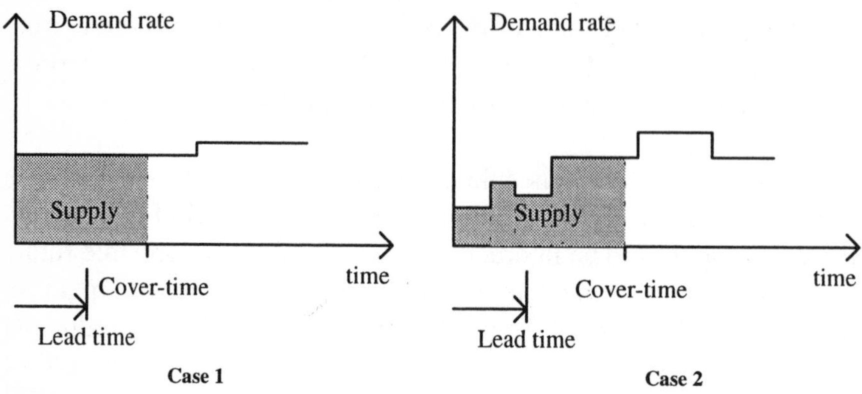

Figure 1: Examples of variations in the demand rate

2 A Small Numerical Example

Cover-Time Planning is probably best, and quickest, illustrated and described by a numerical example. As the demand rate can vary it is not evident how the demand rate should be used in the calculation and therefore how the cover-times should be determined. For illustrating a small numerical example is presented here.

The example consists of four items *A*, *B*, *C* and *D*. Figure 2 shows the assumed product structures in order.

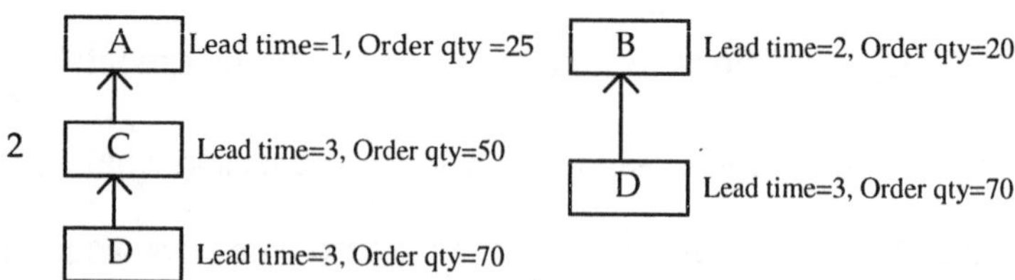

Figure 2: The example's bills of material

A and *B* are end-items. *D* is a component both to *B* and *C*. The manufacturing of one piece of *A* consumes two pieces of *C*, and the production of one piece of *C* consumes one piece of item *D*. For manufacturing one piece of *B* one piece of *D* is used up. Fixed order quantities are used 25, 20, 50 and 70 respectively for *A*, *B*, *C* and *D*.

Expected demand rates for the end-items *A* and *B* are presented in the following tables (Table 1, 2, 3 and 4). The tables also show the current inventory on hand, scheduled issues (with their start dates) and scheduled receipts (with their due dates) due to already existing customer, manufacturing and purchase orders. "Supply" shows the current inventory on hand *minus* scheduled issues in past time (i.e. delays) *plus* scheduled receipts.

	-	1	2	3	4	5	6	7	8	9	10	11	12	13	14	15	16	17	18	19	20
Demand rate		**8**	8	8	8	8	8	8	8	8	**9**	9	9	9	9	9	9	9	9	9	9
Sch. issue	2	1	13	5				6													
Sch. receipt		24																			
Inventory	20	41	28	23	23	23	23	17	17	...											
"Supply"			42																		
Pl. orders							25			25			25			25		25			25

Table 1: **Item A**

Expected demand rate for *A* is eight units per time period from *now, just in the beginning of period 1*, and up to period 9, thereafter the demand rate is nine units per time period. Current inventory on hand is 20 units. Scheduled issues are in this case customer orders. In this example the volume of orders are "short", most orders are expedited the same period they arrive. In period 1 a replenishment of 24 units is scheduled by a manufacturing order. This order is already picked out and in the moment work-in-process as the corresponding scheduled issue for item *C* does not exist. The scheduled receipt of 24 units instead of the ordinary quantity 25 may depend on a rejection for bad quality. "Cover-time" for *A* is determined by:

$$\mathrm{CT}_A = \frac{42}{8} = 5.25 \approx 5.3$$

Cover-time 5.3 is greater than the lead time 2 so a new start of a scheduled receipt is at the moment not necessary. Projected on hand inventory in period 2, now plus lead time of *A*, is positive, so a signal now for preventing a negative inventory is not necessary. A negative projected inventory on hand in period 2 can in period 1 be avoided by a decided replenishment, but not in the next period, due to the lead time of *A*.

Planned orders, the last row in the table, are determined so that "supply" minus "demand rate" plus "planned orders" should not be negative, and the quantity of the planned orders coincides with the standard order quantity. The first planned order is determined such that its due date does not occur before the lead time has passed (for item *A* in period 2). Already scheduled receipts are therefore supposed to be delivered before the first planned order. (The reason for the planned orders will be discussed in section 4.3.)

	-	1	2	3	4	5	6	7	8	9	10	11	12	13	14	15	16	17	18	19	20
Demand rate		**6**	6	6	**7**	7	7	7	7	7	**8**	8	8	8	8	8	8	8	**9**	9	9
Sch. issue																					
Sch. receipt			20																		
Inventory	0	0	20	20	20	20	20	...													
"Supply"				20																	
Pl. orders					20			20			20		20			20		20		20	

Table 2: **Item B**

For item *B* expected demand rate is 6 units per time period in period 1, 2 and 3, from period 4 the demand rate is 7 units per time period, from period 10 it is increased to 8 units and from period 18 it is increased to 9 units per time period. The scheduled receipt in period 2 is delayed and still not started as there is a corresponding scheduled issue for item *D* in past time. Cover-time for item *B* becomes:

$$\text{CT}_B = 3\frac{2}{7} \approx 3.3, \quad \text{as } 20 - 6 - 6 - 6 - 2 = 0$$

Cover-time 3.3 is greater than the lead time 2, which means that no more replenishment is necessary at the moment. Projected on hand inventory in period 3, now plus lead time of *B*, is positive so a signal for replenishment of that reason is neither necessary.

	-	1	2	3	4	5	6	7	8	9	10	11	12	13	14	15	16	17	18	19	20
Demand rate		**16**	16	16	16	16	16	16	16	**18**	18	18	18	18	18	18	18	18	18	18	18
Sch. issue																					
Sch. receipt																					
Inventory	50	50	50	50	50	50	50	...													
"Supply"					50																
Pl. orders					50			50			50			50		50			50		

Table 3: **Item C**

The demand rate for item C is completely determined by the demand rate for item A. The lead time of item A is one therefore the demand rate of C is increased one time period earlier than the demand rate of A is increased. Cover-time for C is:

$$\mathrm{CT}_C = \frac{50}{16} = 3\frac{2}{16} \approx 3.1 \text{as } 50-16-16-16-2=0$$

Cover-time, 3.1, is greater than the lead time, 3, but if one should wait until the next period for initiating a replenishment there will (probably) be a shortage. If a signal for replenishment will not be initiated until the next period, it is necessary to present a signal now. Precisely as with an ordinary reorder point system the reorder level must be sufficient for demand during the lead time including inspection time. Projected inventory on hand in period 4, now plus lead time of C, is positive, so a replenishment for that reason is not necessary.

The first calculated planned order should have a due date in period 4, which is in line with the calculated cover-time. It must be started in period 1, as the lead time of C is 3 periods.

	-	1	2	3	4	5	6	7	8	9	10	11	12	13	14	15	16	17	18	19	20
Demand rate		**22**	**23**	23	23	23	**25**	25	**26**	26	26	26	26	26	26	26	**27**	27	27	27	27
Sch. issue	20																				
Sch. receipt																					
Inventory	92	72	72	72	72	72	72	...													
"Supply"					72																
Pl. orders					70			70		70			70			70		70			70

Table 4: **Item D**

The demand rate for D is governed by the demand rates for B and C. (16 in period 4 for C and 6 in period 3 for B make 22 in period 1 for D.) In this case cover-time is:

$$\mathrm{CT}_D = 3\frac{4}{23} \approx 3.2 \quad \text{as } 72 - 22 - 23 - 23 - 4 = 0$$

Cover-time, 3.2, is greater than the lead time, 3, however initiating a replenishment can not wait until the next period, then a future shortage will (probably) occur. The first calculated planned order has a due date in period 4 for preventing shortages, this planned order should start in period 1 as the lead time of D is 3 time periods.

3 Cover-Time Planning in Formulas

3.1 Notations and assumptions

- d_{it} forecasted demand rate, units per time period required of item i in period t
- D_{it} external demand, required quantity of item i in period t
- H_{ij} number of units of item i required for the production of one unit of item j
- IO_i initial inventory on hand for item i, physically available at the beginning of period 1
- P_{it} manufactured or purchased quantity (for receipt) in period t of item i
- CT_i cover-time
- BT_i buffer time
- τ_i lead time for item i, including time for both production, queuing and also *inspection time* (i. e. time between two succeeding checks if a replenishment is necessary)
- t_0 the earliest time period, often in bygone time, used in the planning calculation
- T the maximum (latest) time period used in the planning, therefore it also determines the *planning horizon*

The items are assumed to be ordered in such a way that the following important conditions are fulfilled for each item *i* and *j*:

$$H_{ij} \neq 0 \quad \Rightarrow \quad i > j \tag{1}$$

If item *i* is a parent item to item *j*, item *i* has a lower "item number" than item *j*.

3.2 The Master Production Schedule

Cover-Time Planning, as MRP, starts with the master production schedule. Contrary to MRP its master production schedule does not contain a special delivery quantity in a special time period, D_{it}, instead it contains and average quantity per production day, a demand rate, d_{it}. The natural time period to work with for Cover-Time Planning is production days. The demand rate is supposed to be in order until in a future time period a possible other demand rate is supposed to reign forwards. The master production schedule in past time disappears automatically, production in past time is impossible and this is also reflected in the master production schedule.

(This is mentioned because it is not evident when using MRP that the master production schedule in past time should be deleted. The master production schedule is often a type of commitment and a special manufacturing order has often been started just to satisfy a special delivery in the master production schedule (MPS) and if this demand is deleted the MRP calculation will suggest current production to be postponed to the future. Observe that the MPS of MRP consists of real and fictive customer orders but the MPS of Cover-Time Planning is a pure forecast for future demand rates.)

The external demand rates are exploded ("broken down") to create a total demand rate, d'_{it}, for every component item:

$$d'_{it} = \sum_{j=1}^{i-1} H_{ij}\, d'_{jt+\tau'_j} + d_{it} \tag{2}$$

3.3 Replenishment or Not, According to Cover-time

From the item's expected demand rate we calculate, for every item, how long the inventory on hand, IO_i, replenished with already decided scheduled receipts will last. The time for which already decided replenishments will last is called "cover time", CT_i.

(Starting production creates costs, and not starting production also creates costs. An important question for production management is how long do we dare to wait before we have to start production? If we wait too long there will be shortage costs, and if we wait too short there will be (at the moment) unnecessary production expenditures and capital costs for inventories and work-in-process. Trying to answer this question is a background to Cover-Time Planning. The term "run-out time" (cf. Bitran and Hax [5]; Karmarkar [6]) is similar to "cover-time". The Swedish consulting company Mysigma used "cover-time" as a measure for inventory turns and priorities. This is also a background to the idea of Cover-Time Planning.)

Let us define S_i as:

$$S_i = IO_i - \sum_{t=t_0}^{0} \sum_{j=1}^{i-1} H_{ij}\, P_{jt+\tau_j} - \sum_{t=t_0}^{0} D_{it} + \sum_{t=t_0}^{T} P_{it} \tag{3}$$

In words, S_i is the current physical stock *minus* scheduled issues in bygone time *plus* already scheduled receipts. P_{it} means a scheduled receipt quantity, an *open order*, of item *i* in period *t*. This open order can also cause a scheduled issue of quantity $H_{ji}\, P_{it}$ for item *j* in period $t - \tau_i$ (if the order is not yet started). D_{it} is a possible customer order of item *i* in period *t*. Scheduled issues in past time refer to demand in the past. Current stock must be adjusted with existent delays for deciding if expected demand from now and toward the future is covered. If S_i is negative then there is at the moment a shortage of item i, in such case cover-time is defined to be negative.

$$S_i \leq 0 \quad \Rightarrow \quad \mathrm{CT}_i = \frac{S_i}{d'_{i1}} \tag{4}$$

$$S_i > 0 \quad \Rightarrow \quad \mathrm{CT}_i = t_m + \frac{S_i - \sum_{t=1}^{t_m} d'_{it}}{d'_{it_m+1}} \tag{5}$$

t_m is the greatest time period (in the future) with an expected available inventory still positive:

$$t_m = \max_{S_i - \sum_{u=1}^{t} d'_{iu} \geq 0} (t), \quad t \in \{1,2,\ldots,T\} \tag{6}$$

For avoiding shortages a replenishment of item i may be necessary if one of the two following conditions is fulfilled:

$$\mathrm{CT}_i - \tau_i - \mathrm{BT}_i < 0 \tag{7}$$

$$\mathrm{IO}_i - \sum_{t=t_0}^{\tau_i} \sum_{j=1}^{i-1} H_{ij} P_{jt+\tau_i} - \sum_{t=t_0}^{\tau_i} D_{it} + \sum_{t=t_0}^{T} P_{it} < 0 \tag{8}$$

Equation (7) is analogous to the reorder point check in an ordinary reorder point system. Instead of a safety stock a buffer time, BT_i, is used for preventing shortages when the demand varies. In case the current cover-time is greater than the lead time there is no need to register a new order for manufacturing or purchase, one can still wait. If "Supply" is less than "scheduled issues" within the lead time, then a new replenishment is required, even if cover-time is greater than the lead time. Equation (8) is especially important for customer-order production, i. e. when assembly/manufacturing not starts depending on a forecasted demand rate instead it starts when the customer order arrives. Often the modules (components) of the end items are easier to forecast and therefore possible manufacture before the customer order arrives. Due to this two rules then when using Cover-Time Planning one can choose to forecast demand rates for end items or for underlying modules.

All items that need replenishments according to equation (7) and (8) are reported on a special signal report (on screen or paper). This report corresponds to the report from MRP which signals the planned orders currently needed for start. The signal report of Cover-Time Planning can be created daily or even more often. The computer calculation for its creation is simple, the demand rates are fetched from the latest "breaking down" of the master production schedule and the calculations are performed together with scheduled issues and receipts. The signal report presents suggestions for order sizes, a fixed order quantity, a calculated economic order quantity or a calculated order quantity based on "cycle time", e. g. every 10th production day.

4 Special Comments on Cover-Time Planning

4.1 Rescheduling

The time-phasing system of MRP means that the system plans that a receipt should occur precisely before a planned issue. The time-phasing ability of MRP is based on rescheduling. If an open order could not be handled as an asset in period 2 instead of as earlier stated in period 5, then in order to avoid a shortage the MRP-system may be forced to suggest a planned order with a completion time in period 2. This order must then start and be finished in a much shorter time than the ordinary lead time and in a much shorter time than the planned order with a completion date in period 5. This is unrealistic. It would result in disorder. If the MRP system is to work properly, a new planned order for the same item must have a due date later than those of already open and planned orders. However the MRP calculation often suggest a rescheduling which involve an unrealistic shortening of the lead times impossible to achieve in the practical application.

When using Cover-Time Planning the approach is as follows. A planned order will immediately become an open order. A planned order (except for customer orders) should be released at once and the order should be finished as soon as possible. Components to the manufacturing order are then picked out from inventory as soon as possible and the production will start at once. If this is not required, then the order should not be released. The demand is not strong enough, there is still time to concentrate available capacity on production of other items. When demand changes in the future, the priority to complete orders can be different than it was when they first were started. "Shortage"- and "Order hunting"-reports should be generated by the computer system for preventing shortages of components, subassemblies and delays of customer orders. When working with Cover-Time Planning there exist no planned orders (except for an approximate calculation of future work load, see section 4.2), all orders are open and "sharp" orders, and a rescheduling is carried out only at the user's discretion and not automatically.

Using Cover-Time Planning the time setting of orders and operations can be performed exactly with the same methods used for MRP. In the (production) calendar one registers the production days, and with help of the calendar and the lead times of the items, scheduled start and completion days can be calculated for the orders and their operations considering the number of actual working days in different weeks.

4.2 Less of Push and More of Pull

After a registration of an order for purchase or manufacturing another start of the item will not happen until a withdrawal of the item has occurred (or the forecasted MPS has been changed). If the demand is covered a new signal for replenishment will not be created, the total amount of inventory and work-in-process for each item is restricted. If there is no withdrawal of the end-item, then there are no withdrawals of subassemblies and in turn no withdrawals of their components, precisely as with Kanban. However, contrary to Kanban, Cover-Time Planning does not control work-in-process (WIP) at every work center.

Spearman et al [9] say: "that controlling WIP levels at every work center in the production process may be more restrictive than necessary. Setting WIP levels for all work centers may also complicate the problem of managing the bottleneck process, since one must set the number of cards (when using Kanban) for the bottleneck process and other processes in a manner that avoids bottleneck starvation. If the bottleneck moves, due to a change in product mix, say, the card counts must be adjusted accordingly." Spearman and Zazanis [10] argue that the effectiveness of pull systems does not result from pure pulling as in Kanban but from limiting WIP and WIP variability. For every item in a production system controlled by Cover-Time Planning a maximum quantity of the inventory and the (upstream, replenishing) WIP is settled by the lead time and the demand rate in use for the item. For MRP all demand on the upstream level does not disappear completely, until available inventory is greater than zero during the entire planning horizon for the item. The planning horizon in an MRP installation should be at least as long as the longest cumulative lead time and even a bit longer for preventing “rolling horizon disturbances”. This means that MRP also has an ability to control WIP, but with a long planning horizon it is slower and leads to a greater WIP variability than for Kanban and Cover-Time Planning.

(*An explanation to the long planning horizon of MRP*: If the planning horizon is shorter than the longest cumulative lead time there will be suggestions of replenishments in past time when there is an increase in the MPS. Most MRP calculations act and treat (correctly) the last demand in the MPS as the last requirement ever, which means that projected inventory on hand after the last demand is planned to be zero (or equal to the possible safety stock). This means that when the planning horizon rolls on in the next MRP calculation a possible new demand in the end of the MPS is a “disturbance“ to the previous “zero“. The longer the planning horizon is, the smaller the part of these end disturbances is of the total demand.)

MRP does not immediately consider stops or delays in production. Orders are released according to planned orders deduced from the master production schedule. Even if produced component items cannot be used at the current moment because one other item is missing, production continues. This construction makes it possible for manufacturing processes to be working at a level equal to their total capacity, but the production is not in balance. Production of end-use items may be too small, while the inventory of subassemblies and their components is increasing.

When using Cover-Time Planning the buffer created by starting manufacturing orders according to planned orders in the Factory plan of MRP, and not considering current disturbances, has gone. One can believe that this increases the variation in work load in different parts of the manufacturing process. Comparative simulation studies (Segerstedt [7], [8]) between MRP and Cover-Time Planning, on the contrary, show that with an increased coefficient of variation for stochastic demands and stochastic lead times, MRP appears to create a larger variation in WIP than Cover-Time Planning.

4.3 Future Work Load

The suggested planned orders which are generated by MRP are used together with already open orders and registered setup and production times for every item to calculate the work load in different work centers (Capacity Requirements Planning). For intermediate and long range planning this is a useful tool, although these calculations in practical installations often show a heavy work load for bygone and first periods. Orlicky [4] complains that in many real installations the master production schedule contains many quantities in bygone time, i. e. delays against the master production schedule, delays in completion of open orders on lower structure levels. When these heavy work loads for MRP in bygone and first periods really shall be produced in different work centers is very difficult for the user to state. However, the work load calculation gives the user a useful hint of the future demand of capacity, which an ordinary reorder point system does not provide.

The current demand rate determines how many working days already released orders will last, and then a new order must replenish inventory. The prevailing demand rate decides how many working days this order quantity will last etc. until the end of the calendar used (as shown in the numerical example section 2). If no shortage should occur the following condition needs to be satisfied.

$$\sum_{t=t_0}^{T} P_{it} \geq \sum_{t=t_0}^{T} t\, d'_{it} - \mathrm{IO}_i \tag{9}$$

For these "planned orders", parent items and components are not time-phased (contrary to MRP). These calculated proposals together with already released orders can then be used together with registered setup and production times for every item to calculate the work load in different work centers. The fact that replenishment is also a withdrawal on the underlying structure level in the bill of material is not considered in these calculations. The error in time phasing is in most cases negligible from a practical point of view, the important thing being that the work load is correct according to the volume of setup and production times. By applying this procedure, the same types of work-load reports as with MRP can also be presented by Cover-Time Planning.

4.4 Just-in-Time and Cover-Time Planning

Just-in-Time production requires accurate information. When using MRP this requires a high frequency of MRP calculations. Unfortunately in most installations the complexity of MRP creates the most time-consuming calculation in the company's computer. Love and Barekat [11] point out, the well known practical experience, that the MRP calculation may take many hours of dedicated computer time and has to be performed when all other activities have ceased. The number and complexity of the transactions involved in calculating material requirements may limit the frequency of recalculations to one per week or month. The process of "closing the loop" (MRP calculation, Work load analyses, Adjustment of MPS, MRP calculation etc.) is also constrained by practical problems.

When using Cover-Time Planning no special preparation is needed before the computer calculation and the calculation is simple compared with MRP. Observe that the master production schedule of Cover-Time Planning, d_{it}, does not intervene directly in the planning calculations as the master production schedule, D_{it}, of MRP does. Therefore, to avoid production both of old forecasts, D_{it}, and new customer orders, also D_{it}, a continuous adjustment of the master production schedule is not necessary for Cover-Time Planning as it is for MRP. New demand rates for the items are only necessary to recalculate every time the forecasts for end-items and spare parts have been changed. The signal report from Cover-Time Planning can easily be produced once or twice a day (from "saved" demand rates, inventory on hand and existing open orders) and thus it gives information for Just-in-Time production.

5 Summary and Conclusions

Cover-Time Planning is a reorder system, which use the information available in the bill of material. The external demands are exploded to upstream items without lot-sizing disarrangement, in the same way as in a base stock system (cf. Silver and Peterson [12]). Contrary to a traditional base-stock system Cover-Time Planning is based on installation stocks.

With Cover-Time Planning, like MRP, a decrease or increase of production for a specific product can be planned. In spite of its generally accepted superiority over reorder point systems, MRP has been criticized for being complex and difficult to handle, needing much computer time and capacity and exhibiting an inferior performance. Cover-Time Planning is a simplified technique, a technique more similar to Kanban but without manual handling of Kanban cards. Cover-Time Planning lacks the automatic time-phasing ability of MRP but uses the information available in the bill of material and can present a vision of the future work load. Nothing indicates that Cover-Time Planning will operate in an inferior manner compared to an ordinary reorder point system. The decision variable is time instead of quantity. Cover-Time Planning contains possibilities that are improvements of a traditional reorder point system. Segerstedt [7], [8] shows in simulation studies that MRP has difficulties to handle variations in demand and lead times. A simpler robust technique like Cover-Time Planning is therefore possible to use without inferior results concerning amount of capital investments in inventory and WIP. Especially, Cover-Time Planning will suit assembly production, where components are used in several end items and when there are variations in demand.

Cover-Time Planning might prove to be a technique in computerized Production and Inventory Control for achieving practical Just-in-Time production. Because of its simplicity, its speed in computer calculations, its stability against variations and its ability to handle customer-order production

References

[1] KRAJEWSKI, L. J. and L. P. RITZMAN (1996), *Operations Management Strategy and Analysis* (4 ed.), Addison Wesley Publishing Company, New York

[2] MATTSSON, S.-A. (1987), *Metodkoncept för MPS* (Methods for Production and Inventory Control: in Swedish), Nr 87206, Mekanförbundets förlag, Stockholm

[3] BURBIDGE, J. (1980), What is wrong with materials requirement planning?, *Production Engineer*, October, 57-59

[4] ORLICKY, J. (1975), *Material Requirements Planning*, McGraw-Hill, New York

[5] BITRAN, G. R. and A. C. HAX (1977), On the design of hierarchical production planning systems, *Decision Sciences*, Vol.8, No1

[6] KARMARKAR, U. S. (1981), Equalization of Runout Times, *Operations Research*, Vol. 29, No. 4

[7] SEGERSTEDT, A. (1991), *Cover-Time Planning - An Alternative to MRP*, PROFIL 10, Linköping

[8] SEGERSTEDT, A. (1995), *Multi-Level Production and Inventory Control Problems - Related to MRP and Cover-Time Planning*, PROFIL 13, Linköping

[9] SPEARMAN, M. L., D. L. WOODRUFF and W. J. HOPP (1989), A Hierarchical Control Architecture for Constant Work-in-Process (CONWIP) Production Systems, *Journal of Manufacturing and Operations Management*, **2**, p. 141-171

[10] SPEARMAN, M. L. and M. A. ZAZANIS (1992), Push and Pull Production Systems: Issues and Comparisons. *Operations Research*, **40**, No. 3

[11] LOVE, D. and M. BAREKAT (1989), Decentralized, distributed MRP: Solving control problems in cellular manufacturing. *Production and Inventory Management Journal*, Third Quarter, 78-84

[12] SILVER, E. A. and R. PETERSON (1985), *Decision Systems for Inventory Management and Production Planning* (2 ed.), John Wiley & Sons, New York

Handling Multiple-Variant Production: Methodologies for Parametrisation and Configuration of Products

Karsten Schierholt and Eric Scherer
Institute of Industrial Engineering and Management (BWI), ETH Zürich

"Any customer can have a car painted any colour that he wants so long as it is black."
(Ford 1922)

Abstract: The buyer's market that developed over the past decades demands for a large number of customized product variants. At the same time, companies face an immense pressure to reduce costs. These contradicting goals dominate many product strategy discussions. Should variants be avoided or should companies rather seek to master their variants efficiently?

This paper presents a possible solution that combines both strategies mentioned. The multiple-variant problem is tackled through a flexible definition of product structures which forms the basis for product configuration. Industrial examples demonstrate how such a strategy can be implemented successfully.

1 Increasing Product Complexity and Customer Influence through Multiple-Variant Products

One of the major changes most manufacturing enterprises faced in the passed decade was the shift from standardized mass production towards supplying customer oriented, individualized products. As a consequence this basic shift of market orientation a second shift could be observed: from homogeneity towards heterogeneity. While for a long time enterprises were able to accommodate market demands with one, single manufacturing strategy nowadays it becomes more and more necessary to provide a set of different, even contradicting strategies to succeed on the market.

With customer satisfaction becoming the main objective of manufacturing, individualized production has risen in consumer as well as in capital goods manufacturing. An enterprise has to provide the customer with a product tailored to his demand. This not only includes a first step from Ford's standardization principle towards a limited set of choices but has to cover all thinkable - or even unthinkable - variations of a product. Besides an increase of variants (Figure 1) manufacturers are facing an increasing complexity of the product itself, consisting out of more and more parts, and a shortening of product life cycles [20]. This results in a tremendous increase of data and information and enforces several new requirements towards manufacturing management, IT support and organization.

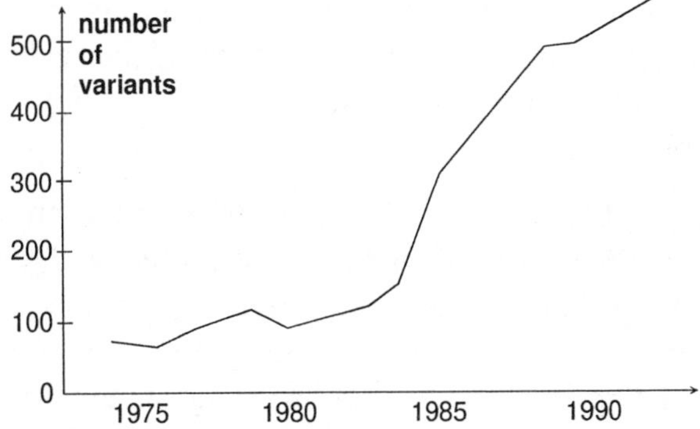

Fig. 1: Increasing number of variants for a specific sub-assembly in automotive manufacturing (Source: BMW, see [19, p. 102]).

To further complicate the problem, the term 'product' - for a long time related to either a single physical item or a defined set thereof - has experienced a tremendous enlargement [13]. Besides the mere physical item a product nowadays may include software, installment, a service program, tailored guarantee, training, etc. This all leads to several new functions to be fulfilled by an enterprise but mainly increases the amount of data and information to be handled significantly.

1.1 Product Variants - an Economic Perspective

The recent years have seen a large increase in customer dominance. Enterprises have to focus on the customer and have to fulfill his demands. As one of the results the number of variants have increased tremendously in many cases driven by marketing or sales departments in their quest to satisfy the customer. To use the economic terms, the broadened offering of variants satisfies the *economy of scope.*

To ensure an enterprises' competitiveness, the newly discovered *economy of scope*, i.e., to cover a large variety of customers and their demands respectively, and the traditional principle of *economy of scale* need to be dealt with jointly. As management principle it is therefore necessary to provide the customer with a well suited service fulfilling his demands while simultaneously trying to produce goods in high numbers. This requires, that no more variants are offered than the customer is able to distinguish. In many cases changes of colors and names are already suited to offer the customer a broader scope of products while changes in the technical design and the manufacturing processes often do not lead to a distinguishable variant.

Another aspect of variant production is to find an appropriate method for product cost calculation. Costs arising through variant production are often not allocated with the respective variants directly but are distributed over all products in the product family. In many cases the real costs of manufacturing an exotic product are much higher than the costs actually allocated with this product. On the other hand, the costs of standard products are calculated too high. Competitors that only produce these standard products and which do not offer such exotic variants thus gain an advantage. Since the fraction of standard products is decreasing constantly while customer oriented product variants increase in number, a cost accounting system that does not allocate the variant costs to the specific variants might put a company in jeopardy (Figure 2).

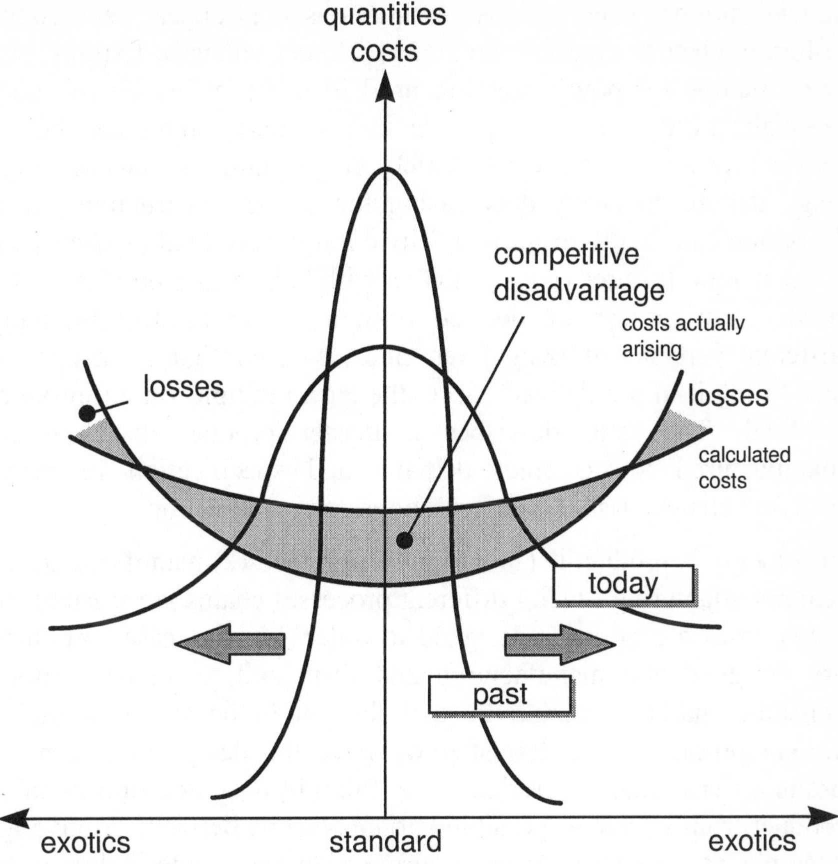

Fig. 2: Cross subvention between standard and exotic products [16].

1.2 Types of Variants

In the following a variant should be defined as a (customer) specific product that is part of a family of products. A family of products is a set of products with a similar form or function and an identical or similar production process, consisting out of a high percentage of identical or similar components. The definition of a specific 'variant' and 'product family' may vary and depends on the situation of an enterprise.

For many manufacturing enterprises this means, that a master version of a product is offered but sold in many different variations. Such variation can be classified into three kinds: adaptations, specializations or extensions. Adaptations may appear as variable coloring, electric supplies satisfying different voltages, fixtures, etc. (e.g., pumps for power steering used in different brands of cars). Specializations or special devices satisfy the customer's requirements such as security, safety or special conditions of usage (e.g., laptops in heavy duty casing for use at construction sites). Variation can finally result in a fully configurable final product with some unique features (e.g., a kitchen with build water purifiers). If a product variant can be derived from a master product in many different versions utilizing fixed rules, e.g., configuring a kitchen and fitting it in pre-defined space, the term multiple-variant product is used. The term describes a master product that can be manufactured to customer demand and specification in many different characteristics, configurations and combinations.

In principal a product is (1) designed, and then (2) manufactured. In real manufacturing several different processes chains are feasible. In many cases a product is designed to order. In other cases products are designed and manufactured and then sold to an anonymous consumer market in a make-to-stock fashion. In the case of multiple-variant products in a first step we have the design of a generic product. This master product is a full physical description of a product from which all possible variants can be derived. To give an example, in automotive manufacturing a car has a generic design. In case of an order, the customer has to describe various features according to his demands, e.g., color, size of the engine, extras like air-conditioning or four-wheel drive. Based on the specification provided by the customer it is possible to generate the car, i.e., to manufacture it. The process of transforming the generic model of the car into a customer specific instance is highly automated and utilizes rules as they have been defined during the generic design. Such a rule could define that a four-wheel drive is only possible in combination with a specific engine size. A full classification of the process chain of multiple-variant products in contrast to others does not seem possible, still *Figure 3* provides some examples.

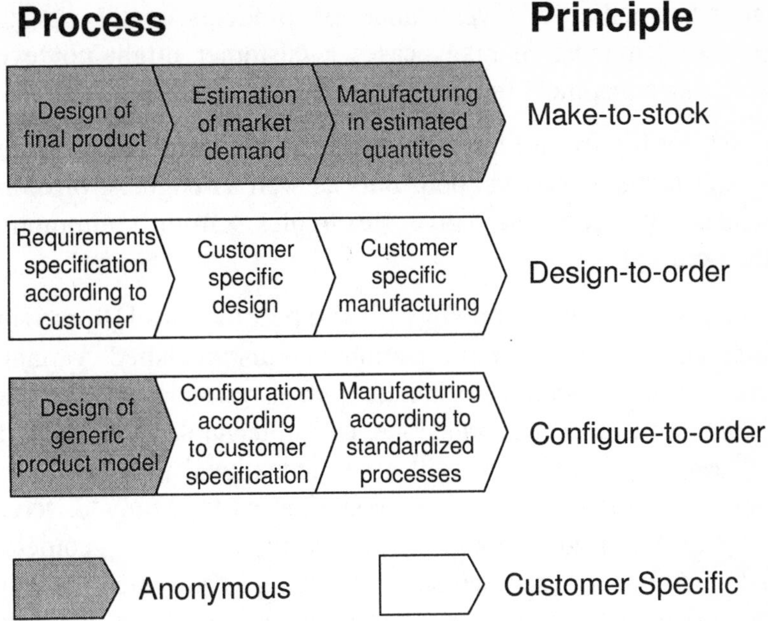

Fig. 3: Examples of process chains and related production principles.

1.3 Tackling the Variant Problem

During the 1980s the increasing range of products and respective variants led to the concept of Flexible Manufacturing Systems (FMS) and therefore was a major part of the philosophy of Computer Integrated Manufacturing (CIM). The approach of handling increasing product complexity by solely relying on manufacturing technology failed. By focusing on manufacturing technology, other possible solutions on the organizational level were left out of sight.

Nowadays it is necessary to concentrate on organizational measures first. To tackle the variant problem, in a first instance variants are to be *avoided*, in the second they must be *managed*. To avoid variants it is usually necessary to *minimize* the number of existing variants. For the minimization of the number of possible variants several measures are necessary:

- Examine whether a given range of products really fits the customer demands. In many cases a customer might not even distinguish a technical feature of an offered variant as such.
- Identical or highly similar parts should be standardized. Such are low cost items as screws, bolt, nuts as well as engines, breaking systems, steering to give examples from automotive manufacturing.

The principles of group technology [2] and part families [10] provide the foundation for the minimization of distinguished variants. Thereby it is the objective to handle former variants as a product family and de facto treat them as one. Group technology is based on a set of design principles as well as organizational measures. In a first stage, it aims to build sets of similar parts. Many different features can be used to describe similarity such are geometry, material, manufacturing processes, machining requirement, quality aspects, or even the necessity of external operations. A set of similar parts is termed a part (or product) family. In the 1960s Opitz established the basis for a standardization of part family features through his well known part feature key. After establishing part families it is possible to handle any variant within a family uniformly throughout the whole production process chain from design to manufacturing and delivery. Ultimately it is the aim to reduce the number of explicitly handled variants significantly. Warnecke therefore suggests to create an explicit variant as late as possible in the manufacturing process [19, p. 106]. This means that a variant even if it exists in logical terms is not handled explicitly as a variant for as long as possible. Other implementations of the avoidance strategy include the attempt to further standardize components and to consider the increased use of pre-assemblies [15].

In contrast to the strategy of *minimization* and *avoidance* the strategy of *management* tries to handle all existing variants in an optimal fashion. Here it is necessary to be aware of all possible variants as early as possible and react according to this knowledge. Here a generic approach seems useful, i.e., to have a generic product model with parameters and rules which are predefined and can be utilized to derive a specific variant in case of an individual customer order. This generic approach seems to be most appropriate for multiple-variant products.

1.4 Aspects of Information Technology

Tackling multiple-variant products on the first hand requires several organizational measures as they can be found in the principles of group technology. Besides organizational measures an integrated solution to handle the respective data and information through information technology (IT) becomes necessary. Computer aided production planning and control systems (CAPPC)[1] play a central role. The design of CAPPC systems during the passed two decades concentrated on the implementation and further development of *functions*. In the beginning it was a major aim to support all functions as suggested by the classic theory of production and inventory control [11; 21]. This led to large scale systems offering a wide quantity of functions and support to the user. Interestingly, the focus on functions for the longest time was closely coupled to a focus on mass production and a lack of customer focus.

The need for change of organizational structures as pushed by changing markets in the very same terms concerns the need of change in CAPPC systems. Up to now, most implemented CAPPC systems are programmed in a procedural language. As a result of the systems design process, the focus of these systems lies on their functionality while their data structure usually is a side product of the design process. Until today, the data structures of many CAPPC systems still remains unknown or undocumented. For the future, it will be necessary to pay more attention to the master data and information structure of CAPPC systems. This necessity can be easily seen, if one examines the principle structure of a modern IT

---

[1] The authors like to use the term CAPPC to describe a set of IT functions related to the management, planning and control of production systems. In most cases such systems are refered to as MRP or MRP II systems even if the underlying principles are not related to the MRP or MRP II philosophy. In industrial reality the CAPPC can be a subsystem of a larger enterprise management and information system.

application based on a layer concept (Figure 4): the logical data structure as the bottom layer of an IT system provides the foundations on which the functions are build upon; in an industrial implementation typically only a few functions will be utilized regularly out of those provided and several more possible.

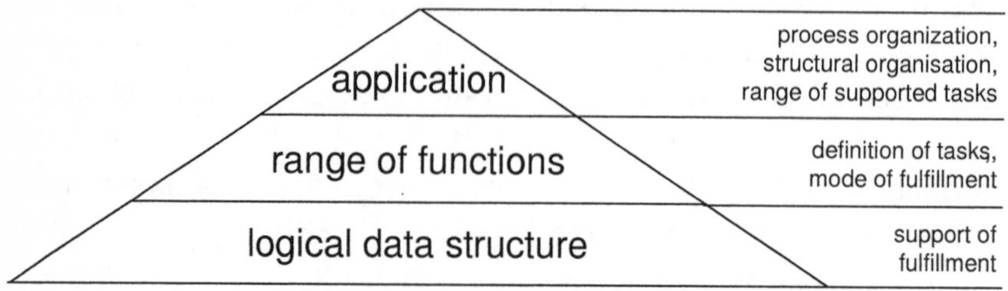

Fig. 4: Principle structure of an IT application [13].

As experience shows in many implementation projects, it is virtually impossible to add any functions if they are not founded in the implemented data structure. On the other hand, changes in the data structure with the aim of accommodating new functions can have a largely negative impact on the original functionality. A proper logical data structure provides the basis to handle product variants independently in the different units of an enterprise, e.g., design, engineering, logistics, purchase, manufacturing, while simultaneously dealing with the same object.

In the following sections a possible solution to the multiple-variant problem will be indicated. Therefore mainly the logical data structure of CAPPC will be examined and some underlying principle derived which will provide the foundations to handle variant throughout the overall process chain of a manufacturing enterprise.

2 Product Structures

2.1 Requirements for Product Structures

2.1.1 Problem Description

Manufacturers often give customers many choices for certain product parameters or optional components in a product. Parameters can be classified into two kinds, (1) those with a finite number of possible values and (2) those with continuous value ranges. Parameters with a continuous value range impose special problems to the efficient description of product structures and will be examined later.

Parameters with a finite set of values are used in cases where one value has to be chosen and only a limited number of possibilities is available. Common parameters of this type are the color or the material of a certain product part. The limitation of choices can be imposed by technical reasons or by the supplier's product variety. In addition, standardization aspects are often responsible for reducing the variety of parameter values. Bolts, for example, have standardized diameters so that the nuts fit on them and tools of the right size are available.

Optional components that may or may not be part of the final product can be handled as parameters with two possible values: "is part of" and "is not part of". Even with only two choices, every option doubles the number of variants in a product family.

Assuming that the parameters and options of a product that can be chosen by the customer are independent of each other, the number of possible product variants grows extremely fast and easily reaches the order of thousands or millions. Every combination of parameters will lead to a new product variant. These variants are very similar to each other, in the product structure represented in the bill-of-materials (BOM) as well as in the associated operations plan[2].

---

[2] The terms *operations plan* and *routing* are used similarly for defining the actions to be taken in a production process whereas the operations plan also describes the

The following example demonstrates how even simple products might need to be produced in a huge number of variants to meet the customer demands.

2.1.2 An Example: The Fire Protection Flap

The fire protection flap (Figure 5) is built into a ventilating shaft. In the case of a fire, the flap closes to stop the airflow and thus reduce the oxygen supply. Since the dimensions of the ventilating shaft are determined by the building geometry, the fire protection flap has to be produced in practically any possible dimension. Besides length, width and height of the flap, the customer also has the choice between several types of closing mechanisms and options for the connecting profile to the neighboring modules of the ventilating shaft [14]. Some other parameters can also be chosen but are left aside in this example to limit the complexity.

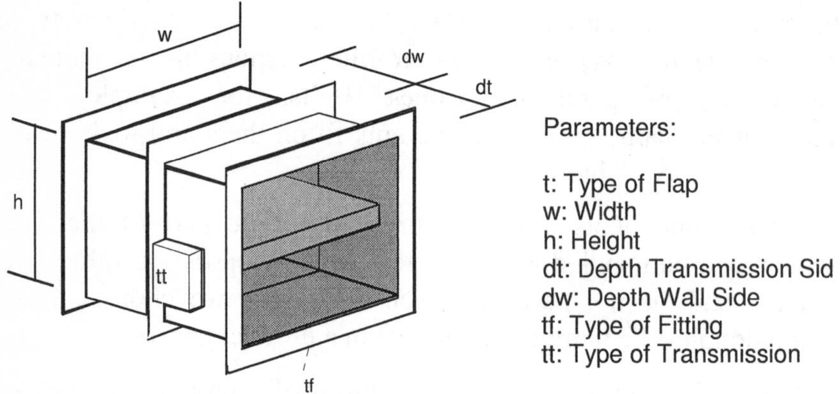

Fig. 5: The fire protection flap.

In the given case, product parameters are independent of each other. Any combination within the range of possible values can be chosen. Through its special use, every fire protection flap is produced as a

---

utilized resources and the administrative information used in the production process. Within this text, we use the term operations plan and describe the process of developing an operations plan as *operations planning*.

single product. The dependency of parts on the specified parameters, requires that most of the parts are produced on demand. Only few parts for common parameter combinations are hold on stock. In total, there are about 30 to 50 BOM-positions for each flap. Some values in the operations plan (machine number, machining time, etc.) also depend on the customer selection of parameters.

The fire protection flaps are produced in widths between 15 cm and 250 cm and in heights between 15 cm and 80 cm. If the customer is allowed to chose values from the given range in intervals of 5 cm, and in addition has the choice between two different drives, more than a thousand different variants of the fire protection flap could be designed. This number is still a low guess, taking into account that 5 cm intervals are probably not practicable and that other parameters like the connecting profile are not considered in the calculation.

In most cases, only a small fraction of all possible variants of a product family is actually produced. This fraction decreases with the growing number of variants. Product data is traditionally kept only for those variants that have already been produced. If there is need for designing a variant that has not been produced before, the necessary BOMs and operations plans are developed by taking the data of an existing variant with very similar parameters and changing the relevant values. In many cases this is not done utilizing the CAPPC system but by adding alterations, amendments or remarks on the operations plan and BOM of a similar product. Thus the handling of the variant product becomes solely a task of the manual organization and is highly error prone. In many cases errors occur in materials handling which relies on the data provided by the CAPPC system while manufacturing uses the hand written notes.

This method keeps the number of stored variants low compared to the number of possible variants but does not solve the evolving data maintenance problem and - in case of manual remarks - leads to redundancies and lack of data at the same time. The BOMs and the operations plans of variants are still extremely similar. If a module that is used in many or even in all variants is replaced, the BOMs of all these variants have to be adjusted to keep the production data up to date. This is a costly effort considering that many variants are produced only once and never again afterwards. The data of variants that are not used any longer would still be maintained.

If, on the other hand, every variant was updated only when being used, the maintenance effort would made only where necessary, but in this case most of the stored data would not be up to date and always tracking changes is again extremely vulnerable to mistakes.

2.1.3 Future Requirements

With the huge number of possible product variants it seems impossible to have BOMs and operations plans stored in the CAPPC system for each single variant. Not only is the space needed to store the associated data enormous. The management and especially the maintenance of this data grows to an immense task.

Neither of the mentioned data handling methods generally utilized in industry today solves the real underlying problem, the problem of high data redundancy within the production data of the different variants. As long as a separate BOM and a separate operations plan is stored for every variant, this problem cannot be solved. Other ways of representing the variant product structure have to be used.

2.2 Bills-of-Materials for Variant Products

Many solutions are suggested to reduce the data redundancy in BOMs. These solutions can be divided into two kinds. Static product descriptions try to find a condensed description of the product variants but still store data for every used variant. Dynamic product descriptions, on the other hand, define the product variants not until they are actually needed. They are derived from the description of a generic master product for each concrete customer order [8, p. 86].

Before describing concepts of static and dynamic product data representations, some requirements on the representation of variant products are defined.

2.2.1 Requirements on the Data Management of Variant Products

Zimmermann lists a number of requirements on the product data management of variant products [22, p. 44]. The most important ones are explained shortly.

Due to the similarity, large portions of the BOMs and the operations plans are identical in all variants. A redundancy-free storage of product data, which stores every fact only once, not only reduces the required storage space but also effects the changes that are necessary in cases of insertions, modifications and deletions within the product data.

An unequivocal assignment between a variant option and the corresponding variants must be possible to enable inquiries to the system about the use of certain parts or modules. These inquiries should help to determine the effects of changes in parts and modules on the products in the product family.

The used product data representation should be capable of handling multiple level variant BOMs. These may have several options for one position in the BOM where each option is a variant module itself.

The product data should be stored in a way that an automatic generation of a variant's product data is possible from the stored generic data. Configuration methods that support this task are described later. Support should be available for the development of a new variant if this variant cannot be generated automatically.

2.2.2 Static Product Representation

All concepts for the static representation of product data build on the fact that in most BOMs some positions exist that are identical in all variants.

The *Supplement Bill-of-Materials* divides positions of the BOM into two categories. Of those positions that are identical in all variants just one version is kept. All other positions of the BOM are maintained for every variant. This method is effective if only few parameters in the BOM are subject to change. If, on the other hand, one position is no longer fixed and becomes a parameter as well, this position would have to be added to all existing variants. Product families that continuously develop are not well suited for this kind of representation [6, p. 24].

Plus / Minus Bills-of-Materials describe product variants in relation to a master product structure. For every variant, only those positions are stored that are needed in addition to the master product structure (plus) or that are not needed in this specific variant but are present in the master product structure (minus). The advantage of plus / minus BOMs is that existing BOMs do not have to be changed if new variants are developed, even if positions change that have been identical in all variants so far [6, p. 27].

Type Bills-of-Materials simply represent tables of BOMs in which several variants are represented. Variants can be compared easily. Still, the redundancy problem is not handled at all. The clarity of the representation for products with many variants is lacking. This structure should not be used if more than 15 to 20 variants exist [8, p. 93].

In a survey conducted among almost 50 German middle to large size manufacturing companies in 1990, Lingnau finds that the concept of type BOMs is the most widely used [9, p. 202]. Nearly 40% of the questioned companies used this concept. Less than 30% used supplement BOMs and less than 10% represent their products with plus / minus BOMs. Besides those companies that have no BOMs for variant products at all, there are about 10% of the companies that use rule-based concepts for product representation. These concepts are described in further detail in the next subsection.

Other ways of reducing the complexity problem of product representation using static descriptions focus on representing the product in a hierarchical structure that minimizes the combinatorial complexity arising through parameters with many possible values. A measure mentioned by Kurbel would be to use one generic BOM. Some positions of this BOM relate to a general assembly part of which special forms exist. In this case, a variant is defined not only by the BOM but also by the further specification of the part used in this specific case [8, p. 93].

Static BOMs help to organize the variant data and support the management of BOMs and operations plans for variant products but cannot fully meet the above mentioned requirements. They do not reduce the number of stored variants. For every variant, some data has to be maintained. Changes in those positions of the BOMs that differ among the variants still cause a large maintenance effort.

Production data for every new variant has to be defined explicitly. This is of special importance in cases where the customer focus and, as a consequence, the number of variants are so high that a variant is produced only once. In this case, all cost associated with the development of the product and production data, which is usually derived from a similar variant and then included in the product family, has to be covered by the profit produced with this variant. Products with many variants of low production volume and low profit margins are therefore not well suited for a static product description.

2.2.3 Dynamic Product Representation

Dynamic representations of products are used when the product and production data is constructed instantaneously at the time of demand. There are no fixed product structures kept for each variant. Rather, one generic product structure in the form of a maximum BOM is used to represent the whole product family. A maximum BOM lists all choices for a certain position of the BOM parallel to each other. With specified customer input, one selection is chosen for each of the options and thus the resulting BOM is constructed [1].

Such a representation of product data avoids the problem of a vast number of combinations. Any change in the design of variants leads to only one change in the generic product structure. New options can be integrated easily and can be used instantaneously. This method is mostly used for handling very large number of variants in extremely customer oriented environments with no stock keeping of products due to unpredictable customer demands [8, p. 93].

When ordering a product of this family the customer has to specify the options of his choice. This can be done using an explicit questionnaire. The customer gives the parameter for each position of the BOM with several choices. For matters of organization, many CAPPC systems use extended article numbers for representing variant products. The first part of this article number represents the product family. In the second part, which may be specific for every product family, the variant parameters are specified [1].

All the methods for handling dynamic product representations described so far are very efficient in reducing the amount of data storage needed for the variants as well as in reducing the maintenance effort to a minimum. This can only be done as long as all the options that are chosen by the customer are absolutely independent of each other which is not always the case. There could be a requirement that, for example, when ordering a furniture, the colors of all modules of the furniture have to be identical. In such a case, the customer should not define the same color several times for each module. Also it must be taken care by the system that no invalid parameter combinations such as different colors for the modules appear. Static product descriptions do not have this problem. Variants are defined only in the cases where a realization is possible.

Extensions to a generic product structure help to handle this problem. One example are rule-based BOMs [9, p. 204] which associate each variant of a position in the BOM with a rule describing the conditions under which this specific position option is to be part of the product variant [14]. Each variant position, i.e. each position of the BOM that has more than one option, is thus connected to a decision table. Customer specified parameters are used to determine the choice from this table. This allows that not all options of a variant position need to be specified explicitly by the customer but that any related parameters can be set as a basis for evaluation by the rules.

In the example of the above mentioned furniture, the color parameter is set once by the customer but might be evaluated several times to determine the color of each module. Parameter values that are not directly used for selecting options of variant positions can also be used. The working of this method is explained with some examples in a later section.

The use of rules in combination with dynamic product structures allows new ways of describing products. Rules specify the condition, i.e. the setting of certain parameters, under which an option of a variant position is used in the BOM of the variant. Parametric descriptions of products are now easily implemented. Even dependencies between the selection of an option of one variant position and possible choices of options in other positions can be expressed by using general parameters that are referred to by several rules. Parameters may also have continuous values which would explode the complexity of any static product representation. Still, a drawback is that the rules and decision tables are part of the generic BOM. A separation of the product structure and the knowledge about the product structure would be preferable.

With a growing number of product parameters, decision tables can reach a level where the maintenance becomes a task as difficult and error prone as the maintenance of BOMs with a growing number of variants [22, p. 188]. A larger number of product parameters leads to more and more complex rules in the decision tables. With an increasing number of special cases for certain combinations of parameters, the decision tables need to be changed more frequently. The cases of rest in the decision table, i.e. those special cases that are not covered automatically, grow rapidly and have to be handled manually. The use of expert systems based on rules or constraints helps to reduce the described problems and to improve the handling of complex product descriptions [3].

2.3 Developing Customer Oriented Product Variants

Product configuration "is the effort to make products in such a way that they are useful to people" [12]. The profit to the customer is the larger the more he can influence the design of the product. Manufacturers who accept this customer demand must also be ready to produce a variant of a product only once or in a very small batch. The design cost of a new variant must be kept to a minimum.

Configurators use customer input for automatically designing product variants. Frayman and Mittal define the configuration task as a special kind of design activity [5]:

"Given a fixed, pre-defined set of components, an architecture that describes certain ways of connecting these components, and requirements imposed by a user for specific cases, either select a set of components that satisfies all relevant requirements or detect inconsistencies in the requirements."

This definition covers a broad range of configuration implementations. The construction of product descriptions using a generic product representation in connection with rules would certainly fit into this definition. Many other ways such as e.g. separating the knowledge about the product structure or the product components and knowledge about the relation between them are plausible and might lead to a system in which the knowledge base and the rules are better maintainable than in a rule-based BOM.

In some cases, products are so vaguely defined that a general product structure does not exist. Here, the interface definitions between components become very important. Knowledge about the product structures must be kept in addition to the knowledge about inconsistencies between the components.

Configurators can be used as early as in the sales process for determining inconsistencies in the customers' product requirements. The product parameters specified by the customer might lead to a product which technically cannot be produced. In such a case the customer can change the product requirements in a way that a consistent description is assured.

3 Dealing with Generic Product Structures

Two examples are shown in this section that demonstrate how a variant management can be implemented. The standard methods described in the last section usually need to be extended to fulfill the specific manufacturers' needs.

3.1 Variants in Furniture Manufacturing

The customer structure of a German furniture manufacturer has changed dramatically over the past years. Being one of the country's largest office furniture manufacturer before 1991 with only few but high volume customers, the company now sells individually customized products to small office furniture dealers with orders of usually one to five positions. These changes required the company to fully restructure and flexibilize the production program. Before the reorganization, only few variants were stored in the CAPPC system. These variants were also advertised in the sales brochure [1].

Today a maximum BOM is used as a generic product structure from which the specific BOMs for the manufactured products are generated. Similar positions are listed in the maximum BOM of which one is selected by comparison of a code with special positions in the article key of the ordered product. The product is specified through the definition of the article key shown below (Figure 6) by the customer (and the sales person) without having an explicit BOM for this specific article key stored in the CAPPC system.

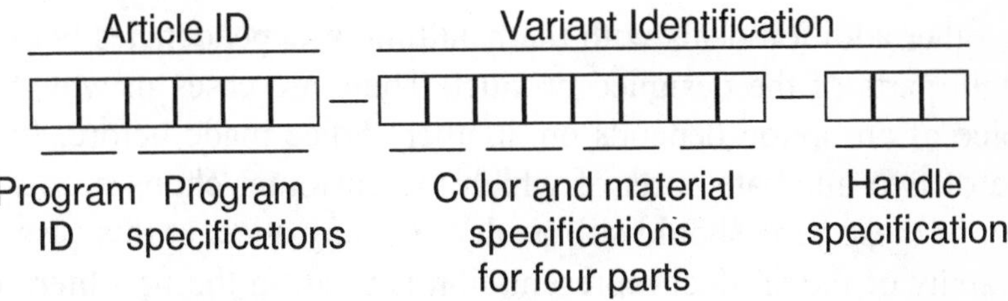

Fig. 6: Article key of a variant product [1].

This way of handling variants has several advantages over the method used before 1991:

- The effort spent in the development and maintenance of the generic product structures was reduced by an order of magnitude. A production program of this flexibility could not be handled with the old approach.
- The number of special orders, i.e. orders that cannot be represented by the BOMs stored, was reduced.
- The data volume stored in the CAPPC system was reduced.

The furniture manufacturer has implemented two additions to the concept of maximum BOMs as it was explained so far. The first one deals with the further reduction of redundant data. In the maximum BOM exist a number of options for every variant position that differ only in a few digits of the part ID. Table plates, for instance, differ only in the color. The idea is now to use one representative position in the maximum BOM for all versions of the plate and have the respective digits of the part ID left blank. In the article key specified by the customer one part is directly related to these digits. The contents is copied directly from the article key to the part ID. This procedure makes it even easier to change the product's variant options. Adding a new color to the production program requires no changes at all to the maximum BOM. Only a part representing the new color has to be added to the CAPPC-system and the sales force has to be informed of the new code for the additional color.

The other addition deals with the handling of dependencies between certain parts of the designed product. There are cases in which the choice of an option depends on another choice made before. In the case of two alternate parts of which the customer chooses one, the color of these parts should still be defined only once. In this case, the hierarchy of the product representation is used. In the first hierarchy level, the customer chooses which of the options he prefers. The color is taken from the article key and assigned to both options in the lower second hierarchy level. This concept is reasonable only for choices of little complexity and interdependency. Many dependencies lead to extremely large maximum BOMs that have similar problems of exponential growth through further dependencies as static BOMs have with additional options.

3.2 VAR

VAR is a product configurator using rule-based BOMs. It is integrated into the CAPPC-system EXPERT/400 [14]. The idea was to allow BOMs to have as many variants as possible. The basis for the "variant generator" has been the extension of the CAPPC-system BOM database with rules . These rules determine whether an option of a variant position is used in the BOM of the specified variant.

Data fields may not only have a constant value but can also be combined with a formula varying in parameter values of the variant. Schönsleben explains how the application in variant CAPPC-systems is very similar to that in a conventional expert system [14]. He also describes how the approach is realized in EXPERT/400, i.e., how variants and relations are modeled.

Parts of a BOM as used in VAR is shown in Figure 7.

Pos	Variant	Quantity/ Unit	Part ID	Part Description	Condition
130	01	2 pcs	295191	...	conditioned: If Type = 1 AND Width ≥ 150
130	05	2 pcs	295205	...	conditioned: If Type = 2
140	01	2 pcs	295477	...	conditioned: If Type = 2
150	01	1 pc	296589	...	conditioned: If Type = 1 AND Height < 150
150	03	1 pc	295108	...	conditioned: If Height ≥ 150
150	05	2 pcs	295108	...	conditioned: If Type = 2 AND Width > 1300
155	01	1pc	494798	...	conditioned: If Type = 1 AND Height < 150

Fig. 7: Extract of a bill-of-materials as used in VAR

When designing a new variant, every option of a position of the BOM is checked whether it fulfills the condition. If it does, this variant is chosen and the next position is checked. It is allowed to have positions that might not be used. Position 140 in the above example, for instance, will not appear in the BOM of any variant with a type that is set to 1 by the user. In that sense the rule-based BOM consists of a decision table for each position in which the decisions are triggered by a rule interpreter that is integrated in VAR. The benefits and shortcomings of such an implementation have been discussed earlier.

3.3 Future Aspects of Configuration Systems

Configuration systems support the tendency to move the sales process further towards the customer. Still, most of the configuration systems require the customer to give a very technical description of the desired product as input to the configurator. The customer often has difficulties to fully understand all technical terms and the sales person may interpret the customer's descriptions incorrectly.

One step towards designing a sales process with improved customer orientation would attempt to use a product description in the customer's language for the configuration task. The customer wants the product to have a certain functionality. Therefore, instead of technical parameters, he is more likely to describe a product in terms of functional requirements.

Schwarze introduces a Specification Mapping as the first stage in the configuration process. The goal of this stage is to transform the functional knowledge provided by the customer to a structurally concrete and precise product specification. External knowledge from the customer is transformed to meet company-internal requirements. Internally, the technical and the production-oriented knowledge is most important [17].

Functional knowledge given by the customer is not easily handled by an automated system. Customers have a very different understanding of terms, such as heavy, tall etc., that seem identical to a computer system. The vague knowledge given as input to the system has to be transformed into crisp values that are used for the technical configuration process. Schwarze suggests the use of fuzzy logic for the integration of vague customer knowledge in the Specification

Mapping [17]. Membership functions are defined for the expressions of a linguistic variable. In the following fuzzy set calculations are used to evaluate given rules and to decide whether certain technical components should be included in the final product or not.

The described steps towards a more customer oriented configuration process are certainly not sufficient as a final solution. Many other attempts are made to use the capabilities of configuration systems to sell customized product variants efficiently. During the last years the combination of electronic media with customer oriented sales systems has been developed and led to a large number of new applications [17; 7]. Configuration is therefore no longer a technical instrument for deriving BOMs of variant products from a generic product description. With its extensions it develops towards a marketing instrument that accepts the customer as the source of the product manufacturing process.

References

[1] Behrendt, O. (1997), Variantenmanagement als Grundlage einer marktorientierten Produktflexibilität. Aachener PPS-Tage, 4th - 5th June 1997, Aachen, Forschungsinstitut für Rationalisierung.

[2] Burbidge, J.L. (1979), Group Technology in Engineering Industry. London, Mechanical Engineering Publications.

[3] Faltings, B., Weigel, R. (1994), Constraint-based Knowledge Representation for Configuration Systems. Technical Report No TR-94/59, Laboratoire d'Intelligence Artificielle - Département d'Informatique, Ecole Polytechnique Fédérale de Lausanne.

[4] Ford, H. (1922), My Life and Work. London, Heinemann.

[5] Frayman, F., Mittal, S. (1987), COSSACK: A Constraints-Based Expert System for Configuration Tasks. In: Sriram, D. and Adey, R.A. (eds.): Knowledge Based Expert Systems in Engineering: Planning and Design. Southampton, Computational Mechanics Publications, pp. 143-166.

[6] Glaser, H., Geiger, W., Rohde, V. (1991), Produktionsplanung und -steuerung; Grundlagen - Konzepte - Anwendungen. Wiesbaden, Gabler.

[7] Klein, S., Schubert, P. (1996), Künftige Entwicklungen des Internet. Thexis **13** (4), pp. 30-34.

[8] Kurbel, K. (1995), Produktionsplanung und -steuerung. 2nd edition, München, Oldenbourg.

[9] Lingnau, V. (1994), Varaintenmanagement. Berlin, Erich Schmidt Verlag.

[10] Opitz, H. (1969), A Classification System to Describe Workpieces. Pergamon Press.

[11] Plossl, G.W. (1985), Production and Inventory Control. Principles and Techniques. 2nd ed. Englewood Cliffs, Prentice-Hall.

[12] Rodgers P. A., Patterson, A. C., Wilson, D. R. (1994), Product Performance Assessment Based on the Definition of Users' Needs. In: Proceedings of the International Conference on Data and Knowledge Systems for Manufacturing and Engineering, Hong Kong, pp. 252-257.

[13] Scherer, E. (1995), Entwicklungspfade in der Informatik-gestützten Produktgestaltung. In: Schönsleben, P. (ed.): Die Prozesskette Engineering. Zürich, vdf Hochschulverlag.

[14] Schönsleben, P. (1988), Flexibilität in der computergestützten Produktionsplanung und -steuerung. Halbergmoos, Angewandte Informationstechnik.

[15] Schuh, G. (1989), Gestaltung und Bewertung von Produktvarianten. Düsseldorf, VDI.

[16] Schuh, G. (1993), Wettbewerbsvorteile durch Prozesskosten-senkung. In: Proceedings of the 28thth Conference "Normenpraxis" - DIN Jahrestagung. Stuttgart, VDI.

[17] Schwarze, S. (1996a), Configuration of Multiple-Variant Products. Zürich, vdf Hochschulverlag.

[18] Schwarze, S. (1996b), Produktvertrieb via Internet. Technische Rundschau, Nr. 44, pp. 32-35.

[19] Warnecke, H.-J. (1992), The Fractal Company – a Revolution in Corporate Culture. Berlin, Springer.

[20] Wiendahl, H.-P., Scholtissek, P. (1994), Management and Control of Complexity in Manufacturing. *Annals of CIRP, Vol. 43/2,* pp. 533-540.

[21] Wight, O. (1974), Production and Inventory Management in the Computer Age. Boston, CBI Publishing.

[22] Zimmermann, G. (1988), Produktionsplanung variantenreicher Erzeugnisse mit EDV. Berlin, Springer.

Chapter 3

Lot Sizing

A PERFECT ALGORITHM FOR THE CAPACITATED LOT SIZE MODEL WITH STOCKOUTS

Michael Bastian, Michael Volkmer[1]
Lehrstuhl für Wirtschaftsinformatik, RWTH Aachen, Germany

Abstract: For the single–product capacitated dynamic lot size problem with stockouts, a procedure is proposed that detects the optimal first lot size at the earliest possible moment. The proposed algorithm uses the data structure of a dynamic tree, in which *long run* potential optimal policies are represented by paths and each production decision by a node. Starting with a horizon $t = 1$ the procedure successively increases t and eliminates in each iteration t all decisions that are not part of an optimal policy using data of the first t periods.

1 Introduction

Consider the problem of lot–sizing one or several items on a single machine when demand is time–varying. The MRP terminology distinguishes *small bucket* from *big bucket* models reflecting different assumptions on the length of the basic time–period (cf. [8]).

The **Discrete Lot–Sizing and Scheduling Problem** [10, 12] is solved by a small bucket model. It presumes short time–periods of an hour or a shift. Regarding this problem, production of a certain item always lasts an integer number of periods without changeover, and no setup costs occur when the same item was produced in the previous period.

In a big bucket model, period length is a week or even a month. Therefore, it seems reasonable to assume that there are several items produced in one period, and even if an item is produced in consecutive periods, one has to take into account the setup

[1]Since April 1997 Business Engineering Center, SAP AG, Walldorf, Germany

cost for both periods. In the presence of capacity constraints this second type of model is called the **Capacitated Dynamic Lot Size Problem** [5, 6].

Most MRP–systems follow a sequential solution approach where lot–sizing is first done independently for single items or a group of similar items using the large bucket model without capacity constraints. Modifications due to capacity constraints are postponed to a later stage. If, on the other hand, upper bounds are introduced on the lot sizes of the single–item model, undesirable solutions may be excluded in an early stage. These bounds may reflect limited storage or a maximum lot size by reason of worn out tools, for example.

What happens when the demand of some period exceeds the capacity bound? Some models propose that demand may be backlogged (at some cost) and satisfied in a later period. In these models we may be faced with solutions that unduly try customer's patience. Many models do not permit backlogging, i. e. if the demand of a period is too high then at least part of it has to be produced in an earlier period (if feasible). Thus, if demand is high in the first periods, there may even not exist a feasible solution.

In todays highly competitive markets it seems more realistic to presume that demand which is not satisfied on time is lost demand and causes stockouts. Facing a particularly high demand in some period t the decision rule would be to check whether this demand could be produced in an earlier period and made available in t at a cost less than the stockout cost. In addition, this approach promotes a forward solution method: high demands in the far future will not influence the policy of the initial periods.

This paper considers the single–item capacitated lot size problem with stockouts. A forward algorithm is presented that determines the optimal first decision using information from the smallest possible number of periods. Period $t \geq 1$ is called a **forecast horizon**, if data for the first t periods are sufficient to ascertain the optimality of the first lot size. If period t is a forecast horizon, the optimal first lot size is independent of the data in the following periods $t+1, t+2, \ldots$. Furthermore, it is important to determine the earliest period t^* where the first lot is fixed and data of the periods $t^*+1, t^*+2, \ldots$ do not influence the optimality of this decision, i. e. $t^* := \min\{t \mid t \geq 1, \quad t \text{ forecast horizon}\}$. An algorithm that finds the optimal first lot utilizing only data from the first t^* periods is called **perfect** according to Lundin and Morton [11].

Perfect procedures are known for the uncapacitated dynamic lot size problem (DLSP) [2, 7, 9, 11], for the discounted DLSP [3] and for a dynamic location/relocation problem [4]. For the capacitated DLSP with stockouts (CDLSPS) Sandbothe and Thompson [13] developed two forward procedures to find the optimal initial decision, distinguishing between constant and variable capacity constraints. This distinction is not necessary for the algorithm proposed in this paper. In addition, the model is generalized, because all data may vary over time. Since the algorithms of Sandbothe and Thompson are not perfect, the main purpose of this paper is to provide a perfect procedure.

In the next section the mathematical formulation of CDLSPS is presented. Subsequently optimality conditions are derived. In Section 4 the tree structure is introduced and in section 5 the algorithm is developed. An example illustrates the algorithm in the next section. After that complexity considerations follow. Finally a special case of the problem will be discussed.

2 Model formulation

The CDLSPS for T periods, CDLSPS(T), can be stated as follows:

$$\min \sum_{t=1}^{T} A_t\delta_t + p_t x_t + h_t I_t + s_t S_t$$

$$
\begin{aligned}
I_{t-1} + x_t - d_t + S_t &= I_t && \forall\ 1 \leq t \leq T && (1)\\
x_t &\leq C_t\delta_t && \forall\ 1 \leq t \leq T && (2)\\
x_t, I_t &\geq 0 && \forall\ 1 \leq t \leq T && (3)\\
\delta_t &= \begin{cases} 1, & \text{if } x_t > 0 \\ 0, & \text{otherwise} \end{cases} && \forall\ 1 \leq t \leq T && (4)
\end{aligned}
$$

where:

A_t	:	setup cost in period t
p_t	:	production cost per unit in period t
h_t	:	holding cost per unit in period t
s_t	:	stockout cost per unit in period t
d_t	:	demand in period t
C_t	:	production capacity in period t,

the decision variables are:

x_t	:	lot size in period t
I_t	:	inventory level at the end of period t ($I_0 = I_T = 0$)
S_t	:	number of stockouts in period t

The goal is to minimize the sum of the setup costs, variable production costs, holding costs and stockout costs. Constraint (1) presents the relationship between inventory, production, demand and stockout. Constraint (2) ensures that the capacity constraints are satisfied and that setup costs are added when production is positive. As $I_t \geq 0$, backorders are prohibited. Concerning the parameters of the model, we assume:

$$h_t > 0, \quad p_t > 0, \quad s_t > 0 \qquad \forall \quad 1 \leq t \leq T \tag{5}$$

$$p_t < s_t \qquad \forall \quad 1 \leq t \leq T \tag{6}$$

$$p_{t+1} < p_t + h_t \qquad \forall \quad 1 \leq t \leq T-1 \tag{7}$$

$$s_{t+1} < s_t + h_t \qquad \forall \quad 1 \leq t \leq T-1 \tag{8}$$

All parameters are positive. Inequality (6) states that stockout costs are higher than variable production costs in any period t. Constraints (7) and (8) exclude *speculative motives* from the optimal inventory policy. If (7) were violated, then it would pay to produce some demand of later production cycles just because of sharply increasing variable production costs (and not because of capacity problems in the future). Without assumption (8) it might pay not to satisfy a demand in period t (although there is some inventory at hand!), because stockout costs in a later period are so high that it is preferable to save and store that stock.

3 Properties of optimal solutions

A well known property of optimal solutions of the classical Wagner–Whitin model [14] is $I_{t-1}x_t = 0 \;\forall\; 1 \leq t \leq T$. In the case of capacity constraints, this condition cannot be maintained, but a weaker property holds ([1, 13]) that is also valid for CDLSPS(T).

Theorem 1 *Let $(x_1, .., x_T)$ be an optimal solution for CDLSPS(T) under the assumptions (5)-(8), then:*

$$I_{t-1}(C_t - x_t)x_t = 0 \qquad \forall \quad 1 \leq t \leq T. \tag{9}$$

Proof: Assume $I_{t-1}(C_t - x_t)x_t > 0$ and let r be the last production period before t. For small values of $\varepsilon > 0$ a change of policy $\hat{x}_t = x_t + \varepsilon,\; \hat{x}_r = x_r - \varepsilon$ is feasible and yields the following change of cost:

$$+\varepsilon p_t - \varepsilon(p_r + h_r + h_{r+1} + \ldots + h_{t-1}) \overset{(7)}{<} 0.$$

Hence, the original policy was not optimal □

Suppose t is a production period. Then Theorem 1 gives some a priori information on the optimal production level x_t:

$$\text{Case 1:}\;\; I_{t-1} > 0 \;\;\Rightarrow\;\; x_t = C_t$$

$$\text{Case 2:}\;\; I_{t-1} = 0 \;\;\Rightarrow\;\; 0 < x_t \leq C_t$$

For reasonable values of the cost parameters, a produced unit will be on stock at most for a certain number of periods. The following Lemma generalizes a result of Sandbothe and Thompson [13]:

Lemma 1 *Let $r, k \in I\!N$, $r + k \leq T$, and $p_r + \sum_{i=r}^{r+k-1} h_i > s_{r+k}$. If $x_r > 0$ and $I_i > 0 \;\; \forall\, i = r, r+1, \ldots, r+k-1$ in a solution of CDLSPS(T), then this solution is not optimal.*

Proof: In this case, producing a unit in period r and storing it for at least k periods leads to higher costs than taking stockout costs into account since $p_r + \sum_{i=r}^{j} h_i \overset{\text{ass.}}{>} s_{r+k} + \sum_{i=r+k}^{j} h_i \overset{(8)}{>} s_{j+1}$ $\forall\, j \geq r+k$. Consequently, such a policy cannot be optimal □

We get some more insight into the structure of optimal policies, when we consider subpolicies between two periods with vanishing stock. Define a **major production cycle (MPC)** to be a sequence $\{t_1 < t_2 < \ldots < t_m \leq \bar{t}\}$, where t_i are production periods $\forall\ 1 \leq i \leq m$ such that $I_{t_1-1} = I_{\bar{t}} = 0$ and $I_l > 0\ \ \forall\ \ t_1 \leq l \leq \bar{t}-1$. Notice first that stockouts may occur only in the last period $\bar{t}$ of a major production cycle. Theorem 1 ensures that in a major production cycle of an optimal policy all production lots equal the capacity bound of the corresponding period except maybe for the first one. Lemma 2 indicates how the first lot of a MPC can be determined.

Lemma 2 *Given a MPC $\{t_1, t_2, \ldots, t_m, \bar{t}\}$ of an optimal policy, optimal production quantities are obtained by the rule $x_{t_i} = C_{t_i}$ $\forall\ \ i \geq 2$ and x_{t_1} equals the largest feasible value satisfying $\sum_{i=t_1}^{\bar{t}} x_i \leq \sum_{i=t_1}^{\bar{t}} d_i$.*

Proof: $I_t > 0\ \forall\ t_1 \leq t < t_m \overset{\text{Theorem 1}}{\Rightarrow} x_{t_i} = C_{t_i}\ \forall\ i \geq 2$.

For x_{t_1} we distinguish two cases:

i) $x_{t_1} > \sum_{i=t_1}^{\bar{t}} d_{t_i} - \sum_{i=t_1+1}^{\bar{t}} x_{t_i} =: z$ would imply $I_{\bar{t}} > 0$, a contradiction.

ii) $x_{t_1} \leq z$ and $\exists y \leq C_{t_1}$, $x_{t_1} < y < z$.

$p_{t_1} + \sum_{i=t_1}^{\bar{t}-1} h_i \leq s_{\bar{t}}$, because of Lemma 1. Increasing x_{t_1} would reduce stockout in period $\bar{t}$ without raising costs □

In a forward procedure, however, neither the end of a major production cycle nor the production periods are known in advance, and the corresponding size of the first lot of this cycle, x_{t_1}, can not be decided upon when considering period t_1. The idea is to allow a range of values for x_{t_1} until the end of a MPC is determined. Further decisions within the cycle will be quite simple: either there will be no production ($x_t = 0$) or production at the capacity bound ($x_t = C_t$). The corresponding subpolicies are generated for increasing values of t.

For all values of x_{t_1} satisfying $\sum_{i=t_1}^{t} x_i \leq \sum_{i=t_1}^{t} d_i$ the end of a MPC is reached, i. e. $I_t = 0$. For all values in the range of x_{t_1} not satisfying the inequality, we continue the MPC with this subrange, unless we can show that the resulting policies are dominated.

For ease of exposition, the following notation is introduced:

$$D_t := \sum_{i=1}^{t} d_i, \quad X_t := I_t + D_t = \sum_{i=1}^{t} x_i + \sum_{i=1}^{t} S_i, \quad H_t := \sum_{i=1}^{t} h_i \quad (10)$$

$$E_t := \sum_{i=1}^{t} D_i h_i \qquad \tilde{p}_t := p_t - H_{t-1}, \qquad \tilde{s}_t := s_t - H_{t-1} \qquad (11)$$

D_t is just the cumulative demand of periods 1 to t and X_t may be viewed as a state at the end of period t. The equation reveals that demand is either satisfied (by production) or it is not satisfied (i. e. stockout), and each unit of ending inventory has been produced in the past.

Notice, that $X_t \geq D_t$ and the maximum value for X_t is: $X_t^{max} := D_t + \sum_{i=1}^{t} C_i$.

Let $K_t^*(X), D_t \leq X \leq X_t^{max}$, denote the minimal cost to end up in state X at the end of period t. $K_t^*(D_t)$ solves CDLSPS(t), and $K_t^*(X)$, $X > D_t$, is useful to compute $K_\tau^*(X)$ for $\tau > t$. There is a choice to reach X via production in period t at cost $K_t^+(X)$ or without production at cost $K_t^o(X)$. If $X > D_t$, also holding costs have to be considered:

$$K_t^*(X) = h_t(X - D_t) + \min\{K_t^+(X), K_t^o(X)\}, \quad X \geq D_t \quad (12)$$

Without production, X may be reached from states $Y = X_{t-1}$, $LB := \max\{D_{t-1}, X - d_t\} \leq Y \leq \min\{X, X_{t-1}^{max}\}$, hence

$$K_t^o(X) = \min_{LB \leq Y \leq X} \{s_t(X - Y) + K_{t-1}^*(Y)\}, \qquad X \geq D_t \qquad (13)$$

($K_{t-1}^*(Y) := \infty$ for $Y > X_{t-1}^{max}$).

$K_t^+(X)$ has two branches. In case $X_{t-1} = D_{t-1}$, $I_{t-1} = 0$ and production may be at level $0 < x_t \leq C_t$. Let $Y := D_{t-1} + x_t \leq X$. The cost to reach state X, $D_t \leq X \leq D_t + C_t$, (without holding costs of period t) is: $K_{t-1}^*(D_{t-1}) + A_t + p_t(Y - D_{t-1}) + s_t(X - Y)$.

As $p_t < s_t$ by (6), Y should attain its largest feasible value $Y = \min\{X, D_{t-1} + C_t\}$ and the first branch of $K_t^+(X)$ becomes:

$$K_{t-1}^*(D_{t-1}) + A_t + p_t \min\{C_t, X - D_{t-1}\} + s_t \max\{0, X - C_t - D_{t-1}\}.$$

In case $X_{t-1} > D_{t-1}$, production has to be at level C_t by (9). The corresponding branch of $K_t^+(X)$ is:

$$A_t + p_t C_t + \min_{\max\{D_{t-1}+C_t, X-d_t\} \leq Y \leq X} \{K_{t-1}^*(Y - C_t) + s_t(X - Y)\}.$$

It is defined for $\max\{D_{t-1} + C_t, D_t\} \leq X \leq X_t^{max}$. These two branches overlap for $\max\{D_{t-1}+C_t, D_t\} \leq X \leq D_t+C_t$. But in this case, $Y = D_{t-1} + C_t$ is a feasible choice in the second branch for which the two branches become identical. Hence, the second branch dominates the first one in the overlapping region and we obtain:

$$K_t^+(X) = \begin{cases} K_{t-1}^*(D_{t-1}) + A_t + p_t(X - D_{t-1}), & X \in R_1 \\ A_t + p_t C_t + \\ \quad \min_R\{K_{t-1}^*(Y - C_t) + s_t(X - Y)\}, & X \in R_2 \\ \infty, & X_t^{max} < X \end{cases} \tag{14}$$

$$\begin{aligned} \textit{where}\quad R_1 &:= \{X | D_t \leq X \leq C_t + D_{t-1}\} \\ R_2 &:= \{X | \max\{D_{t-1} + C_t, D_t\} \leq X \leq X_t^{max}\} \\ R &:= \{Y | \max\{D_{t-1} + C_t, X - d_t\} \leq Y \leq X\} \end{aligned}$$

Theorem 2 *$K_t^*(X)$ is an increasing piecewise linear function that is continuous for all feasible states $X > D_t$.*

Proof:

i) Monotonicity $(K_t^*(X - \varepsilon) \leq K_t^*(X))$

Consider the optimal policy leading to X, and reduce the first lot x_{t_1} of the last (yet uncompleted) MPC by ε. This saves production and holding costs and yields an upper bound for $K_t^*(X-\varepsilon)$. The solution is feasible as long as ε is small enough to guarantee $I_\tau \geq 0 \; \forall \tau, t_1 \leq \tau \leq t$. (For larger values of ε, the last MPC changes, but the procedure may be continued).

ii) Continuity and piecewise linearity (by induction)

As $I_0 = D_0 = 0$, the starting conditions are:

$$K_0^*(X) = \begin{cases} 0, & X = 0 \\ \infty, & \text{otherwise} \end{cases}$$

$$K_1^o(X) = \begin{cases} s_1 D_1, & X = D_1 \\ \infty, & \text{otherwise} \end{cases}$$

$$K_1^+(X) = \begin{cases} A_1 + p_1 X, & D_1 \leq X \leq C_1 \\ A_1 + p_1 C_1 + s_1(X - C_1), & X \in R \\ \infty, & D_1 + C_1 < X \end{cases}$$

where $R := \{X \,|\, \max\{D_1, C_1\} \leq X \leq D_1 + C_1\}$.

Because of (12), this implies for $D_1 < C_1$

$$\begin{aligned} K_1^*(X) &= h_1(X - D_1) \\ &+ \begin{cases} \min\{s_1 D_1, A_1 + p_1 D_1\}, & X = D_1 \\ A_1 + p_1 X, & D_1 < X \leq C_1 \\ A_1 + p_1 C_1 + s_1(X - C_1), & C_1 \leq X \leq D_1 + C_1 \\ \infty, & D_1 + C_1 < X \end{cases} \end{aligned}$$

and for $D_1 \geq C_1$

$$\begin{aligned} K_1^*(D_1) &= \min\{s_1 D_1, A_1 + p_1 C_1 + s_1(D_1 - C_1)\} \\ K_1^*(X) &= h_1(X - D_1) \\ &+ \begin{cases} A_1 + p_1 C_1 + s_1(X - C_1), & D_1 < X \leq D_1 + C_1 \\ \infty, & D_1 + C_1 < X \end{cases} \end{aligned}$$

In both cases, $K_1^*(X)$ is piecewise linear and continuous on $(D_1, D_1 + C_1]$. Assume now that $K_{t-1}^*(X)$ is piecewise linear and continuous on $(D_{t-1}, X_{t-1}^{max}]$, and consider $K_t^o(X)$,

$K_t^+(X)$ and $K_t^*(X)$. For each value $X \geq D_t$, (13) and the piecewise linearity of $K_{t-1}^*(X)$ imply that $K_t^o(X)$ either equals $K_{t-1}^*(X)$ or it may be written in the form

$$K_t^o(X) = K_{t-1}^*(y_i) + s_t(X - y_i),$$

where y_i is an endpoint of one of the linear segments of $K_{t-1}^*(Y)$. Let $\{y_1, \cdots, y_k\}$ denote the set of endpoints of the linear segments of $K_{t-1}^*(Y)$. Then the k functions

$$L_i(X) = \min\{K_{t-1}^*(X), K_{t-1}^*(y_i) + s_t(X - y_i)\}, \ 1 \leq i \leq k$$

are continuous and piecewise linear, which is also true for

$$K_t^o(X) = \min_{1 \leq i \leq k}\{L_i(X)\}, \ X \geq D_t.$$

Consider now $K_t^+(X)$ as given by (14). Its first branch is the sum of piecewise linear continuous functions. Its second branch is also piecewise linear and continuous by the same argument used for $K_t^o(X)$. Furthermore, if the first branch exists ($D_t \leq C_t + D_{t-1}$), then $K_t^+(X)$ is continuous at the point $X = D_{t-1} + C_t$. Therefore, $K_t^+(X)$ is piecewise linear and continuous on $(D_t, X_t^{max}]$.

By (12), $K_t^*(X), D_t < X \leq X_t^{max}$, is the sum of a linear function and the minimum of two piecewise linear and (for feasible values of X) continuous functions.

This proves the theorem □

If $K_t^*(X)$ steeply increases in some area, it is possible to identify ranges that will never be part of an optimal policy, because stockout costs are cheaper:

Lemma 3 *Let $K_t^*(X + \varepsilon) \geq K_t^*(X) + s_{t+1}\varepsilon$. Then, for any horizon $T \geq t + 1$ there is an optimal solution to CDLSPS(T) that does not use state $X_t = X + \varepsilon$.*

Proof: Suppose, $X_t = X + \varepsilon$ is a state in an optimal solution to CDLSPS(T). ε may be decomposed into a sum $\sum \varepsilon_\tau$, where

ε_τ finally serves to satisfy demand in some period $\tau > t$. Consider this policy as well as an alternative policy that has identical production decisions in periods $t+1$ to T, starts with cost value $K_t^*(X)$ at the end of period t and encounters additional stockouts of ε_τ in periods τ. The costs in period $t+1$ to T are identical except for additional holding costs $\sum \varepsilon_\tau (H_{\tau-1} - H_t)$ of the first policy and additional stockout costs of $\sum \varepsilon_\tau s_\tau$ of the second one. If it would be preferable to follow the first policy, then

$$K_t^*(X+\varepsilon) + \sum \varepsilon_\tau (H_{\tau-1} - H_t) < K_t^*(X) + \sum \varepsilon_\tau s_\tau$$

$$\begin{aligned} \Rightarrow K_t^*(X+\varepsilon) &< K_t^*(X) + \sum \varepsilon_\tau (s_\tau - \sum_{i=t+1}^{\tau-1} h_i) \\ &\overset{(8)}{<} K_t^*(X) + \varepsilon s_{t+1} \end{aligned}$$

which contradicts the assumption of the Lemma □

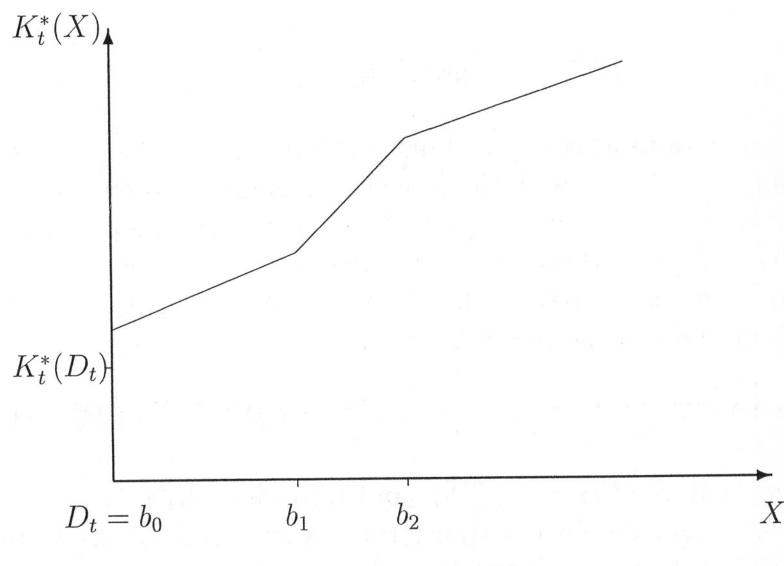

Figure 1: $K_t^*(X)$

Lemma 3 applies to points close to D_t as well as to linear segments of $K_t^*(X)$ with slope larger than s_{t+1}. It can be used as follows: consider the right boundaries b_i of the segments of $X \geq D_t$. If $K_t^*(b_i + \varepsilon) > K_t^*(b_i) + s_{t+1}\varepsilon$, initiate a ray with slope s_{t+1} in $(b_i, K_t^*(b_i))$, and compute the minimum of these rays and $K_t^*(X)$ to obtain a new function $K_t(X)$:

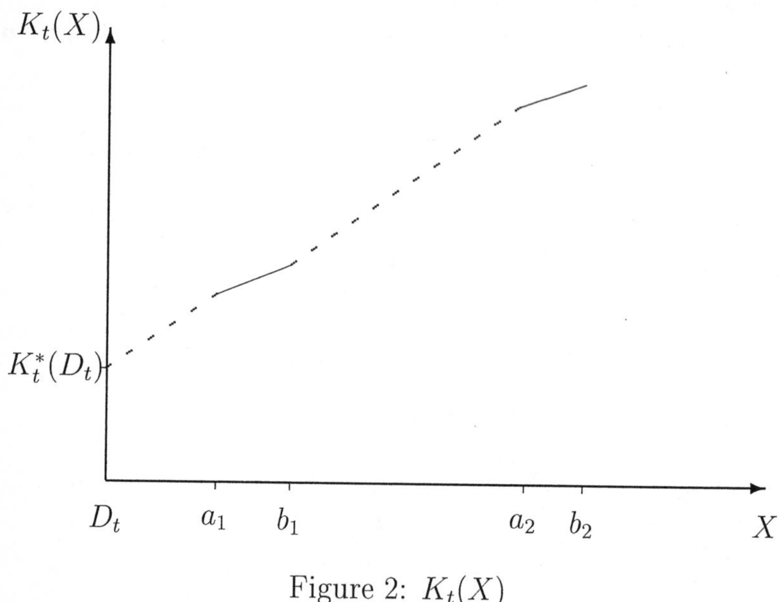

Figure 2: $K_t(X)$

One obtains a sequence of disjoint intervals $[D_t, D_t] = [a_0, b_0]$, $(a_i, b_i]$, $i = 1, 2, \ldots$, with the property that for X_t in one of these intervals this state may be part of a long-run optimal policy, and for $X_t \in (b_i, a_{i+1}]$ this state is to be avoided, because it is cheaper to go via b_i and to pay stockout costs for $X_t - b_i$ units.

A formal expression for $K_t(X)$ is

$$K_t(X) := \min_{D_t \leq Y \leq X} \{K_t^*(Y) + s_{t+1}(X - Y)\}, \quad \forall\, X \geq D_t \qquad (15)$$

Notice that $K_t(D_t) = K_t^*(D_t)$, and reconsider $K_t^+(X)$ as given by (14). Using an obvious transformation of variables, the second branch becomes

$$A_t + p_t C_t + \min_{\max\{D_{t-1}, X - d_t - C_t\} \leq Z \leq X - C_t} \{K_{t-1}^*(Z) + s_t(X - C_t - Z)\},$$

and by (15),(13) it may be rewritten in the form

$$A_t + p_tC_t + \begin{cases} K_{t-1}(X - C_t), & \max\{D_{t-1} + C_t, D_t\} \le X \le D_t + C_t \\ K_t^o(X - C_t), & D_t + C_t < X \le X_t^{max} \end{cases}$$

Thus, we obtain

$$K_t^+(X) = \begin{cases} K_{t-1}(D_{t-1}) + A_t + p_t(X - D_{t-1}), & X \in R_1 \\ K_{t-1}(X - C_t) + A_t + p_tC_t, & X \in R_2 \\ K_t^o(X - C_t) + A_t + p_tC_t, & X \in R_3 \end{cases} \tag{16}$$

$$\begin{aligned} \text{where} \quad R_1 &= \{X | D_t \le X \le D_{t-1} + C_t\} \\ R_2 &= \{X | \max\{D_{t-1} + C_t, D_t\} \le X \le D_t + C_t\} \\ R_3 &= \{X | D_t + C_t < X \le X_t^{max}\} \end{aligned}$$

For $X \ge D_t$ (15) and (12) yield:

$$\begin{aligned} K_t(X) &= \min_{D_t \le Y \le X} \{h_t(Y - D_t) \\ &\qquad + \min\{K_t^o(Y), K_t^+(Y)\} + s_{t+1}(X - Y)\} \\ &= h_t(X - D_t) \\ &\quad + \min \begin{cases} \min_{D_t \le Y \le X}\{K_t^o(Y) + (s_{t+1} - h_t)(X - Y)\} \\ \min_{D_t \le Y \le X}\{K_t^+(Y) + (s_{t+1} - h_t)(X - Y)\} \end{cases} \end{aligned}$$

We shall show that $K_t^o(Y)$ may be replaced by $K_{t-1}(Y)$ in the above recursion.

$$K_t^o(Y) \stackrel{(13)}{=} \min_{Y - d_t \le Z \le Y} \{s_t(Y - Z) + K_{t-1}^*(Z)\} \le K_{t-1}^*(Y),$$

and as $s_{t+1} - h_t < s_t$ by (8),

$$\begin{aligned} &\min_{D_t \le Y \le X} \{K_t^o(Y) + (s_{t+1} - h_t)(X - Y)\} \\ \le\ &\min_{D_{t-1} \le Y \le X} \{K_{t-1}^*(Y) + s_t(X - Y)\} \\ =\ &K_{t-1}(X). \end{aligned}$$

On the other hand, $K_t^o(Y) \ge K_{t-1}(Y)$ by definition, and therefore $K_t^o(Y)$ may be replaced by $K_{t-1}(Y)$ in the formula for $K_t(X)$

yielding:

$$\begin{aligned} K_t(X) = \; & h_t(X - D_t) \\ & + \min_{D_t \le Y \le X} \{\min\{K_t^1(Y), K_t^2(Y)\} + (s_{t+1} - h_t)(X - Y)\} \end{aligned} \tag{17}$$

where $K_t^1(Y) = K_{t-1}(Y)$ and

$$K_t^2(Y) = \begin{cases} K_{t-1}(D_{t-1}) + A_t + p_t(Y - D_{t-1}), & D_t \le Y \le D_{t-1} + C_t \\ K_{t-1}(Y - C_t) + A_t + p_t C_t, & Y \ge D_{t-1} + C_t \end{cases}$$

Consider now the linear transformation

$$V(X,t) := K_t(X) + E_t - XH_t, \qquad \forall\, X \ge D_t \tag{18}$$

Notice that $V(X,\bar{t})$, $\bar{t}$ fixed, is also a piecewise linear function which is continuous for $X > D_{\bar{t}}$. It will turn out later that computation of cost values is more comfortable using V instead of K.

Lemma 4 *Using the transformation*

$$K_t(X) = V(X,t) - E_t + XH_t, \qquad \forall\, X \ge D_t,$$

K_t may also be computed by the recursion:

$$V^+(X,t-1) = \begin{cases} \infty, & X = D_{t-1} \\ V(D_{t-1},t-1) + A_t + \tilde{p}_t(X - D_{t-1}), & \\ \qquad\qquad D_{t-1} < X \le D_{t-1} + C_t & \\ V(X - C_t, t-1) + A_t + \tilde{p}_t C_t, & D_{t-1} + C_t \le X \end{cases} \tag{19}$$

$$\begin{aligned} V(X,t) = \; & \min_{D_t \le Y \le X} \{\tilde{s}_{t+1}(X - Y) \\ & + \min\{V(Y,t-1), V^+(Y,t-1)\}\} \end{aligned} \tag{20}$$

Proof:

$$V(X,t-1) - E_t + XH_t \stackrel{(18)}{=} K_{t-1}'(X) - h_t D_t + X h_t \tag{21}$$

$$
\begin{aligned}
&V^+(X,t-1) - E_t + XH_t \\
&= \begin{cases}
\infty, & X = D_{t-1} \\
K_{t-1}(D_{t-1}) + E_{t-1} - D_{t-1}H_{t-1} & \\
\quad +A_t + \tilde{p}_t(X - D_{t-1}) - E_t + XH_t, & D_{t-1} < X \leq D_{t-1} + C_t \\
K_{t-1}(X - C_t) + E_{t-1} - (X - C_t)H_{t-1} & \\
\quad +A_t + \tilde{p}_t C_t - E_t + XH_t, & D_{t-1} + C_t \leq X
\end{cases}
\end{aligned}
$$

$$
= \begin{cases}
\infty, & X = D_{t-1} \\
K_{t-1}(D_{t-1}) + A_t + p_t(X - D_{t-1}) & \\
\qquad -h_t D_t + Xh_t, & D_{t-1} < X \leq D_{t-1} + C_t \\
K_{t-1}(X - C_t) + A_t + p_t C_t & \\
\qquad -h_t D_t + Xh_t, & D_{t-1} + C_t \leq X
\end{cases}
$$

$$
\Rightarrow V^+(X,t-1) - E_t + XH_t \;=\; h_t(X - D_t) + K_t^2(X) \tag{22}
$$

$$
\begin{aligned}
K_t(X) &= V(X,t) - E_t + XH_t \\
&\overset{(20)}{=} \min_{D_t \leq Y \leq X} \{\min\{V(Y,t-1), V^+(Y,t-1)\} + \tilde{s}_{t+1}(X-Y)\} \\
&\qquad - E_t + XH_t \\
&\overset{(21),(22)}{=} \min_{D_t \leq Y \leq X} \{\min\{K_t^1(Y), K_t^2(Y)\} + h_t(Y - D_t) \\
&\qquad + H_t(X - Y) + \tilde{s}_{t+1}(X - Y)\} \\
&= \min_{D_t \leq Y \leq X} \{\min\{K_t^1(Y), K_t^2(Y)\} + h_t(X - D_t) \\
&\qquad + (s_{t+1} - h_t)(X - Y)\} \qquad \Box
\end{aligned}
$$

4 Representing optimal subpolicies in a tree structure

The management of optimal subpolicies is nicely done in a dynamic tree structure that is closely related to the lot–trees which were introduced by Bastian [2, 3].

Each node α_i^ν refers to a production period ν and holds this index as well as a flag Y that indicates whether ν is the first period of a MPC or not.

$$Y := \begin{cases} Z, & \text{if } X_{\nu-1} = D_{\nu-1} \\ P, & \text{if } X_{\nu-1} > D_{\nu-1} \end{cases}$$

The path down to the root of the tree which originates from connecting consecutive production periods represents the optimal subpolicy leading to this node. If necessary, the first period $t_1 \leq \nu$ of the MPC to which ν belongs is added to the notation: α_i^{ν,t_1} . If ν is not the first period of a MPC (i. e. $\nu \neq t_1$), the production quantities on the path are uniquely determined, except for the lot–size in period t_1. The feasible range of x_{t_1} results in a range for the value of X_ν associated with this policy.

After iteration $t-1$ a certain state $\hat{X}_{t-1}$ may be reachable via different policies; some states can be excluded from being part of any long-run optimal policy. We shall keep track of this by a list

$$L_{t-1} = ([D_{t-1}, D_{t-1}], (a_1, b_1], (a_2, b_2], \ldots)$$

of nonempty disjoint intervals. Associated with each interval is a pointer to one of the nodes in the tree. If $\hat{X}_{t-1}$ is in one of the intervals of L_{t-1}, then it may still be a state in a long-run optimal policy, and the cheapest way to reach this state is via the subpolicy given by the associated node. In iteration t the tree will be enlarged by a set of new nodes corresponding to production decisions in period t. This leads to a modified list L_t of intervals. There may be some new intervals, and old ones may be shrunk or splitted for different reasons:

- a state may be reached via one of the new nodes at a lower cost,
- $a_i < D_t$ is not in the feasible range for X_t,
- a_i may be raised by Lemma 3.

In particular, an interval may become empty. In case this was the only interval of some node α_i, the node becomes **inactive**:

no long-run optimal policy will use this node. Inactive leaves are pruned (recursively). If, during this process, the tree gets a **stem**, i. e. the first decision is the same for all policies, and the root is inactive, then we know the period where the first production takes place. If, in addition, the stem consists of two arcs and the root as well as its only child are inactive, then the procedure may be stopped – the optimal initial decision is found, and it is known in which period the second production occurs.

Suppose, we have generated node α_i^{ν,t_1} in iteration ν and the policy leading to $X_\nu \in (a_i, b_i]$ was optimal via node α_i^{ν,t_1}. Then with $(a_i, b_i]$ the value $V_i := V(b_i, \nu)$ is stored as well as the slope σ_i of the linear segment of $V(X, \nu)$ over $(a_i, b_i]$. Thus, $V_i(b_i - \varepsilon, \nu) = V_i - \varepsilon\sigma_i$, for $0 \leq \varepsilon < b_i - a_i$. The pair of values (V_i, σ_i) remains valuable as long as node α_i^{ν,t_1} is not inactive.

Lemma 5 *Let $a \in (a_i, b_i]$, $a \geq D_{t-1}$, be a state that may be reached via node α_i^{ν,t_1} after iteration $t-1$. Then*

$$V((a, t-1) \mid \textit{last production in } \nu) = V(a, \nu) = V_i - (b_i - a)\sigma_i$$

Proof:

$$V((a, t-1) \mid \text{no production in } t-1) \stackrel{(20)}{=} \min_{D_{t-1} \leq Y \leq a} \{V(Y, t-2) + \tilde{s}_t(a - Y)\} = V(a, t-2),$$

because if the curbation by the rays with slope $\tilde{s}_t$ were relevant for state a, then it would not have remained in the interval $(a_i, b_i]$. Hence, $V(a, t-1 \mid \text{last production in } \nu) = V(a, \nu) = V_i - (b_i - a)\sigma_i$ □

5 Node generation and update operations in iteration t

Let $\Omega(t)$ denote the set of all generated but not yet inactivated nodes at the end of iteration t. Suppose, at the end of iteration $t-1$, we have at hand $\Omega(t-1)$, the pruned tree, as well as a representation of the function $V(X, t-1)$ by the list $L_{t-1} = ([D_{t-1}, b_{i_0} = D_{t-1}], (a_{i_1}, b_{i_1}], \ldots (a_{i_n}, b_{i_n}])$. Let the interval with right boundary b_{i_k} be associated with node $\alpha_{i_k}^{\nu_k} \in \Omega(t-1)$. Then we have to consider at most $n+1$ new nodes $\alpha_{j_k}^t$ corresponding to production decisions in period t: $\alpha_{j_k}^t$ becomes an offspring of $\alpha_{i_k}^{\nu_k}$.

The function $V^+(X,t-1)$ is given by the intervals $L_{t-1}^+ = ((a_{j_0}, b_{j_0}], (a_{j_1}, b_{j_1}], \ldots, (a_{j_n}, b_{j_n}])$ with $a_{j_0} = D_{t-1}$, $b_{j_0} = D_{t-1} + C_t$, $a_{j_k} = a_{i_k} + C_t$, $b_{j_k} = b_{i_k} + C_t$, $\forall k = 1, \ldots, n$ and

$$V_{j_0} = V_{i_0} + A_t + \tilde{p}_t C_t, \;\; \sigma_{j_o} = \tilde{p}_t$$

$$V_{j_k} = V_{i_k} + A_t + \tilde{p}_t C_t, \;\; \sigma_{j_k} = \sigma_{i_k}, \;\; 1 \le k \le n$$

$$V^+(X,t-1) = \begin{cases} \infty, & X = D_{t-1} \\ V_{j_k} - (b_{j_k} - X)\sigma_{j_k}, & a_{j_k} < X \le b_{j_k} \\ V_{j_k} + (X - b_{j_k})\tilde{s}_t, & b_{j_k} < X \le a_{j_{k+1}} \end{cases}$$

where $a_{j_{n+1}} := \infty$.

Notice that the necessity to generate node $\alpha_{j_k}^t$ depends on the value of D_t. Let $a_{i_{n+1}} := \infty$ and define

$$k^+ := \arg\min\{a_{j_k} \mid a_{j_k} > D_t\},$$
$$k^* := \arg\min\{a_{i_k} \mid a_{i_k} > D_t\}.$$

Then,

$V(X,t) = \min_{D_t \le Y \le X} \{\tilde{s}_{t+1}(X-Y) + \min\{V(Y,t-1), V^+(Y,t-1)\}\}$

does neither depend on V_{j_k}, $k < k^+ - 1$ nor on V_{i_k}, $k < k^* - 1$. Therefore, nodes α_{j_k} do not have to be generated for $k < k^+ - 1$, and for $k < k^* - 1$ nodes α_{i_k} become inactive at the end of iteration t.

It is now possible to formally describe iteration t, i. e. the generation of L_t, $\Omega(t)$ and the changes in the tree.

In the presentation we shall use a procedure INACTIVATE(i) which is called when we realise that in an optimal policy no state $X \in (a_i, b_i]$ will be reached via the associated node α_i.

INACTIVATE(i) performs the following operations:

delete the interval $(a_i, b_i]$ from the list it is a member of

IF $(a_i, b_i]$ is the only interval associated with node α_i
THEN mark node α_i as being inactive
    IF α_i is a leaf
    THEN prune the tree, i. e. delete α_i from the tree
        and iteratively proceed with the parent node,
        if it becomes a leaf and is inactive
    ENDIF
ENDIF

INITIALISE generates $[D_t, D_t]$, the first interval of L_t and the associated value $V(D_t, t)$.

$\Omega(t) := \emptyset$, $L_t = ([D_t, D_t])$
IF $D_t = D_{t-1}$
THEN $\Omega(t) = \Omega(t) \cup \{\alpha_{i_0}\}$;
    associate V_{i_0} with $[D_t, D_t]$;
    generate nodes α_{j_k}, $k = 0, \ldots n$ as well as
    the list L^+_{t-1} representing $V^+(X, t-1)$;
    delete $[D_{t-1}, D_{t-1}]$ from L_{t-1}
ELSE determine k^+ and k^*;
    generate nodes α_{j_k}, $k = k^+ - 1, \ldots, n$ as well as
    the list L^+_{t-1} starting from the interval
    with right boundary $b_{j_{k^+-1}}$
    FOR $k = 0, \ldots, k^* - 2$ INACTIVATE(i_k)
    ENDFOR;
    IF $b_{i_{k^*-1}} > D_t$
    THEN replace the corresponding interval
        by $(D_t, b_{i_{k^*-1}}]$ in L_{t-1};
        $W_1 := V_{i_{k^*-1}} - \sigma_{i_{k^*-1}}(b_{i_{k^*-1}} - D_t)$
    ELSE $W_1 := V_{i_{k^*-1}} + \tilde{s}_t(D_t - b_{i_{k^*-1}})$
        delete the corresponding interval from L_{t-1}
    ENDIF;
    IF $b_{j_{k^+-1}} > D_t$
    THEN replace the corresponding interval
        by $(D_t, b_{j_{k^+-1}}]$ in L^+_{t-1};
        $W_2 := V_{j_{k^+-1}} - \sigma_{j_{k^+-1}}(b_{j_{k^+-1}} - D_t)$

```
        ELSE   W_2 := V_{j_{k+-1}} + s̃_t(D_t - b_{j_{k+-1}})
               delete the corresponding interval from L^+_{t-1}
        ENDIF;
        IF     W_1 ≤ W_2
        THEN   associate W_1 with [D_t, D_t];
               Ω(t) := Ω(t) ∪ {α_{i_{k*-1}}};
               IF     b_{j_{k+-1}} ≤ D_t
               THEN   INACTIVATE(j_{k+-1})
               ENDIF
        ELSE   associate W_2 with [D_t, D_t]
               Ω(t) := Ω(t) ∪ {α_{j_{k+-1}}}
               IF     b_{i_{k*-1}} ≤ D_t
               THEN   INACTIVATE(i_{k*-1})
               ENDIF
        ENDIF
ENDIF
```

After initialisation we run through the lists L_{t-1} and L^+_{t-1} and make their intervals disjoint by cost considerations. If a complete interval is dominated, this gives rise to a call of the procedure INACTIVATE. Before these disjoint intervals enter L_t, they may be curbed from the left in case the slope of $V(X, t-1)$ or $V^+(X, t-1)$ is larger than $\tilde{s}_{t+1}$, because this is an upper bound for the slope of $V(X, t)$ (cf. (20)). This is done by the procedure SHRINKLEFT. In fact, we even apply this procedure before making the intervals disjoint, because it may happen that an interval vanishes which saves work.

The procedure SHRINKLEFT assumes that an interval with right boundary b and associated value W was just inserted into L_t. It is applied to L_{t-1} and L^+_{t-1}. Let $(a_i, b_i]$ be the first interval of one of the lists, i. e. $b \leq a_i$:

```
SHRINKLEFT

    IF        σ_i ≥ s̃_{t+1}
    THEN      INACTIVATE(i)
    ELSE      ε := [V_i − W − σ_i(b_i − b)]/(s̃_{t+1} − σ_i)
              {b + ε is the point where the ray W + s̃_{t+1}X cuts
              the line through the linear segment over (a_i, b_i] }
              IF        ε ≥ b_i − b
              THEN      INACTIVATE(i)
              ELSE      a_i := b + ε
              ENDIF
    ENDIF
```

In case INACTIVATE(i) was called, the routine is applied to the next interval. This is done for both lists.

SHRINKLEFT is called for the first time after the execution of INITIALISE with $b = D_t$ and $W = \min\{W_1, W_2\}$. The roles of the two lists are interchangeable; hence, let us simplify notation by talking of list1 and list2 and by denoting the first interval of list$i$ (after application of SHRINKLEFT) by $(a_i, b_i]$, $i = 1, 2$. Without loss of generality $b \leq a_1 \leq a_2$.

If $b < a_i$, SHRINKLEFT did change a_i which implies:

$$W + \tilde{s}_{t+1}(a_i - b) = V_i - (b_i - a_i)\sigma_i \tag{23}$$

If $b = a_i$, SHRINKLEFT did not change a_i, hence:

$$W \geq V_i - (b_i - a_i)\sigma_i \tag{24}$$

By definition, $V(X, t)$ is less or equal to the minimum of the functions $V(X, t-1)$ and $V^+(X, t-1)$. Therefore, equality holds in (24), and (23) is valid for $b \leq a_i$.

Consider now 5 different cases for the two intervals:

1. $a_1 < b_1 \leq a_2 < b_2$ (the intervals are disjoint)

   $(a_1, b_1]$ as well as $W := V_1$ are moved from list1 to L_t;
   $\Omega(t) := \Omega(t) \cup \{\alpha_1\}$;
   call SHRINKLEFT.

2. $a_1 < a_2 < b_1 < b_2$ (the intervals overlap)

   Consider the two curves given by list1 and list2 for $X = a_1 + \varepsilon \le b_1$. Curve 1 is the best choice, if

$$\begin{aligned} & W + \tilde{s}_{t+1}(a_1 - b) + \sigma_1 \varepsilon \\ & \le W + \tilde{s}_{t+1}(a_1 - b) + \tilde{s}_{t+1}\varepsilon, && \text{for } \varepsilon \le a_2 - a_1 \\ \text{and} \quad & W + \tilde{s}_{t+1}(a_1 - b) + \sigma_1 \varepsilon \\ & \le W + \tilde{s}_{t+1}(a_2 - b) + \sigma_2(\varepsilon - a_2 + a_1), && \text{for } \varepsilon > a_2 - a_1 \end{aligned}$$

   Obviously, curve 1 is best for $\varepsilon \le a_2 - a_1$, as $\tilde{s}_{t+1} > \sigma_1$. For $\varepsilon > a_2 - a_1$ the criterion becomes:

$$(\sigma_1 - \sigma_2)\varepsilon \le (\tilde{s}_{t+1} - \sigma_2)(a_2 - a_1)$$

   This allows us to make the intervals disjoint:

   IF $(\sigma_1 - \sigma_2)(b_1 - a_1) \le (\tilde{s}_{t+1} - \sigma_2)(a_2 - a_1)$
   THEN $a_2 := b_1$
   ELSE $\varepsilon^* := [(\tilde{s}_{t+1} - \sigma_2)(a_2 - a_1)]/(\sigma_1 - \sigma_2)$;
   $a_2 := a_1 + \varepsilon^*$;
   $V_1 := V_1 - (b_1 - a_1 - \varepsilon^*)\sigma_1$;
   $b_1 := a_1 + \varepsilon^*$;
   ENDIF;
   move $(a_1, b_1]$ and $W := V_1$ from list1 to L_t;
   $\Omega(t) := \Omega(t) \cup \{\alpha_1\}$;
   call SHRINKLEFT;

3. $a_1 < a_2 < b_2 \le b_1$ (one interval contains the other one)

   The optimality criterion is the same as in case 2.

   IF $(\sigma_1 - \sigma_2)(b_1 - a_1) \le (\tilde{s}_{t+1} - \sigma_2)(a_2 - a_1)$
   THEN INACTIVATE(2);
   apply SHRINKLEFT to list2
   ELSE compute ε^* as in case 2;
   $a_2 := a_1 + \varepsilon^*$;
   $L_t := L_t \cup (a_1, a_1 + \varepsilon^*]$;
   associate with this interval the value
   $W := V_1 - (b_1 - a_1 - \varepsilon^*)\sigma_1$;
   $\Omega(t) := \Omega(t) \cup \{\alpha_1\}$;

```
        IF      b_2 < b_1
        THEN    a_1 := b_2
        ELSE    delete (a_1, b_1] from list1;
                    apply SHRINKLEFT to list1
        ENDIF
    ENDIF;
```

4. $a_1 = a_2 < b_1 < b_2 \wedge \sigma_1 \leq \sigma_2$

   By (23), $V(a_1, t-1) = V^+(a_1, t-1)$; as $\sigma_1 \leq \sigma_2$, the interval of list1 is preferable on the whole range $(a_1, b_1]$:

   $a_2 := b_1$;
   move $(a_1, b_1]$ and $W = V_1$ from list1 to L_t;
   $\Omega(t) := \Omega(t) \cup \{\alpha_1\}$;
   call SHRINKLEFT;

5. $a_1 = a_2 < b_2 \leq b_1 \wedge \sigma_1 \leq \sigma_2$

   INACTIVATE (2);
   apply SHRINKLEFT to list2;

In case one of the lists becomes empty, SHRINKLEFT is just applied to the remaining list, and the curbed intervals are moved to L_t.

6 Example

Given are the following data:

Setup costs	$A_t = A = 7$	$\forall\ t$
Capacity constraints	$C_t = C = 5$	$\forall\ t$
Production costs	$p_t = 1$	$\forall\ t$
Inventory holding costs	$h_t = h = 1$	$\forall\ t$
Stockout costs	$s_t = 4$	$\forall\ t$

The demand structure can be taken from the following table:

t	1	2	3	4	5	6
d_t	2	3	1	4	6	2

Let the elements of the lists L and L^+ be given by quadrupels of the form $(I_i, \alpha_i, V_i, \sigma_i)$. The procedure starts with the root:

$$\boxed{\Omega(0) = \{\alpha_0\}, L_0 = (([0,0], \alpha_0, 0, *))}$$

- Iteration 1

  $D_1 = 2;$

  $k^+ = k^* = 1;$

  $L_0^+ = (((0,5), \alpha_1^1, 12, 1));$

  $\tilde{s}_1 = s_1 = 4; W_1 = 0 + 4 \cdot (2 - 0) = 8; L_0 = \emptyset;$

  $L_0^+ = (((2,5], \alpha_1^1, 12, 1)); W_2 = 12 - (5 - 2) = 9;$

  As $W_1 < W_2$, we associate the interval $[2,2]$ with node α_0 and insert $([2,2], \alpha_0, 8, *)$ into L_1.

  We now apply SHRINKLEFT to L_0^+ $(\tilde{s}_2 = 3 > \sigma_1 = 1)$:

  $\varepsilon = [12 - 8 - (5 - 2)]/(3 - 1) = 1/2$ (a_1 is raised to $5/2$).

  $L_0^+ = (((5/2, 5], \alpha_1^1, 12, 1)).$

  As L_0 is empty, iteration 1 stops with:

$$\boxed{\begin{array}{lcl} \Omega(1) & = & \{\alpha_0, \alpha_1^1\} \\ L_1 & = & (([2,2], \alpha_0, 8, *); ((5/2, 5], \alpha_1^1, 12, 1)) \end{array}}$$

- Iteration 2

  $D_2 = 5;\ \tilde{p}_2 = 0;$

  $k^+ = 1;\ k^* = 2.$

  Therefore, nodes α_2^2 and α_3^2 are generated, and node α_0 becomes inactive at the end of the iteration.

  $L_1^+ = (((2,7], \alpha_2^2, 15, 0);\ ((15/2, 10], \alpha_3^2, 19, 1));$

  $W_1 = 12 + (5 - 5) = 12;\ L_1 = \emptyset$ and $b_2 = 7 > D_2 = 5$ $\Rightarrow a_2$ is raised to 5:

  $L_1^+ = (((5,7], \alpha_2^2, 15, 0);\ ((15/2, 10], \alpha_3^2, 19, 1));$

$W_2 = 15 - 0 = 15$ implies $W_1 < W_2$, and the first quadrupel of L_2 becomes $([5, 5], \alpha_1^1, 12, 1)$.

We now apply SHRINKLEFT to L_1^+ ($\tilde{s}_3 = 2 > \sigma_2 = 0$):

$\varepsilon = (15 - 12 - 0)/(2 - 0) = 3/2$ and raises a_2 to $13/2$.

As $L_1 = \emptyset$, the second quadrupel of L_2 becomes $((13/2, 7], \alpha_2^2, 15, 0)$. Applying SHRINKLEFT again yields: $\varepsilon = [19 - 15 - (10 - 7)]/(2 - 1) = 1$ and a_3 is raised to 8. The result of this iteration is:

$$\begin{array}{lcl} \Omega(2) & = & \{\alpha_1^1, \alpha_2^2, \alpha_3^2\} \\ L_2 & = & (([5,5], \alpha_1^1, 12, 1); \\ & & ((13/2, 7], \alpha_2^2, 15, 0);\ ((8, 10], \alpha_3^2, 19, 1)) \end{array}$$

The current tree is shown in Figure 3 (Z–nodes are circled, P–nodes are boxed, inactive nodes are circled twice):

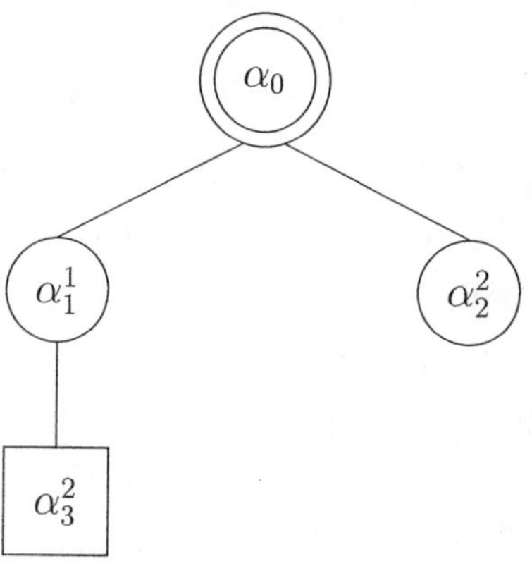

Figure 3 : tree after iteration 2

As the root is already inactive, we know at this point that initial production is in period 1 or in period 2 in an optimal policy.

- <u>Iteration 3</u>

  $D_3 = 6$; $\tilde{p}_3 = -1$;

  $k^+ = k^* = 1$.

  Three new nodes are generated as well as the list:

  $L_2^+ = (((5, 10], \alpha_4^3, 14, -1);\ ((23/2, 12], \alpha_5^3, 17, 0);$
  $\qquad ((13, 15], \alpha_6^3, 21, 1));$

  $W_1 = 12 + 2(6 - 5) = 14$ and the first quadrupel is dropped from L_2.

  As $b_4 = 10 > D_3 = 6$, the left boundary of the first interval in L_2^+ is raised to 6;

  $W_2 = 14 + (10 - 6) = 18 > W_1$, and the first quadrupel of L_3 becomes $([6, 6], \alpha_1^1, 14, 1)$.

  We now apply SHRINKLEFT to

  $L_2 = (((13/2, 7], \alpha_2^2, 15, 0);\ ((8, 10], \alpha_3^2, 19, 1))$ and

  $L_2^+ = (((6, 10], \alpha_4^3, 14, -1);\ ((23/2, 12], \alpha_5^3, 17, 0);$
  $\qquad ((13, 15], \alpha_6^3, 21, 1));$

  $\tilde{s}_4 = 1 \Rightarrow \varepsilon = [15 - 14 - 0]/1 = 1 \geq 7 - 6 = 1$.

  Hence, node α_2^2 is inactivated and eliminated from L_2.

  For the next interval of L_2, $\sigma_3 = 1 \geq \tilde{s}_4$ and α_3^2 becomes inactive, too ($L_2 = \emptyset$).

  For L_2^+ we obtain: $\varepsilon = [14 - 14 + (10 - 6)]/(1 + 1) = 2 < 10 - 6 = 4$ and a_4 is raised to 8.

  The quadrupel $((8, 10], \alpha_4^3, 14, -1)$ is moved to L_3, and SHRINKLEFT is applied again to L_2^+:

  $\varepsilon = [17 - 14 - 0]/(1 + 0) = 3 > 12 - 10 = 2$ and α_5^3 is inactivated.

  Finally, $\sigma_6 = 1 \geq \tilde{s}_4 = 1$, and α_6^3 becomes inactive.

  The result of iteration 3 is:

$$
\begin{array}{|lcl|}
\hline
\Omega(3) & = & \{\alpha_1^1, \alpha_4^3\} \\
L_3 & = & (([6, 6], \alpha_1^1, 14, 1);\ ((8, 10], \alpha_4^3, 14, -1)) \\
\hline
\end{array}
$$

The corresponding tree has a stem and is illustrated in Figure 4.

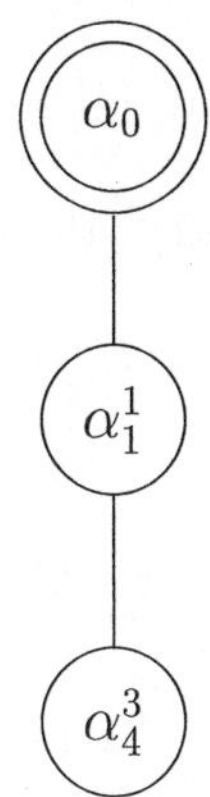

Figure 4: tree after iteration 3

The first lot size can now be fixed, because the root of the tree is inactive and has just one offspring α_1^1 that is a Z–node with interval [6,6]: $x_1 = 5$. All potential successors of α_1^1 will have production periods ≥ 3, i. e. $x_2 = 0$. The procedure may be stopped at this point. Continuing the algorithm until the given data file is exhausted leads to the following decisions in periods 3 to 5: $x_3 = 0$, $x_4 = 5$, $x_5 = 5$.

Generally, if the root α_0 of the tree is inactive and has just one offspring α_j^t at the end of iteration $\bar{t} \geq t$, the first decision can be determined, if the corresponding interval I_j equals $[D_{\bar{t}}, D_{\bar{t}}]$. In this case $x_1 = \cdots x_{t-1} = 0$ and $x_t = C_t$, because $D_{\bar{t}} \geq D_{t-1} + C_t$. If I_j does not consist of one point, only a range is known for x_t.

7 Proof that the algorithm is perfect

Let $V(X,t)$, $X \geq D_t$, be given by the list of intervals

$$L_t = (I_0 = [D_t, D_t],\ I_1 = (a_1, b_1], \ldots, I_n = (a_n, b_n]),$$

where I_j is associated with the values V_j and σ_j as well as with node $\alpha_j \in \Omega(t)$.

To show α_j may be part of a long-run optimal policy, we choose the data of the following period appropriately:

- $D_{t+1} \in I_j$,
- $C_{t+1} \geq D_{t+1} - D_t$,
- A_{t+1} such that $(s_{t+1} - p_{t+1})(D_{t+1} - D_t) < A_{t+1}$,
- h_{t+1} arbitrary and
- $s_{t+2} < p_{t+1} + h_{t+1}$.

$$\Rightarrow \quad (\tilde{s}_{t+1} - \tilde{p}_{t+1})(D_{t+1} - D_t) < A_{t+1}$$

$$\begin{aligned} \Rightarrow \quad & V(D_t, t) + \tilde{s}_{t+1}(D_{t+1} - D_t) \\ < \quad & V(D_t, t) + A_{t+1} + \tilde{p}_{t+1}(D_{t+1} - D_t) \\ \overset{(19)}{=} \quad & V^+(D_{t+1}, t) \end{aligned}$$

As the slope of each linear segment of $V(X, t)$ is less or equal $\tilde{s}_{t+1}$, the left side of the above inequality is an upper bound for $V(D_{t+1}, t)$. Hence,

$$V(D_{t+1}, t) < V^+(D_{t+1}, t)$$

$$\Rightarrow V(D_{t+1}, t + 1) = V(D_{t+1}, t)$$

But this means that node α_j is in $\Omega(t + 1)$, and all ranges I_i, $i < j$, are not in L_{t+1}. The slopes associated with all other nodes have the form $\tilde{p}_\tau$, $\tau \in \{1, 2 \ldots, t + 1\}$.

As $s_{t+2} < p_{t+1} + h_{t+1}$, (7) provides:

$$s_{t+2} < p_\tau + H_{t+1} - H_{\tau-1}, \; \forall \tau = 1, 2, \ldots, t + 1$$

$$\Rightarrow \tilde{s}_{t+2} < \tilde{p}_\tau, \; \forall \tau = 1, 2, \ldots, t + 1$$

As a consequence, $L_{t+1} = ([D_{t+1}, D_{t+1}])$, and each long-run optimal policy contains node α_j □

8 Complexity considerations

The worst case complexity of the algorithm heavily depends on the number of elements of the list L_t. This number doubles at the beginning of iteration $t+1$ (generation of L_t^+), and then the intervals are merged and made disjoint in order to obtain the function $V(X, t+1)$. When a segment enters L_{t+1}, the rest of that interval is eliminated in most cases. An exception may occur in case 3 of the algorithm, when a segment of list1 enters L_{t+1} and another piece of that interval remains in list1.

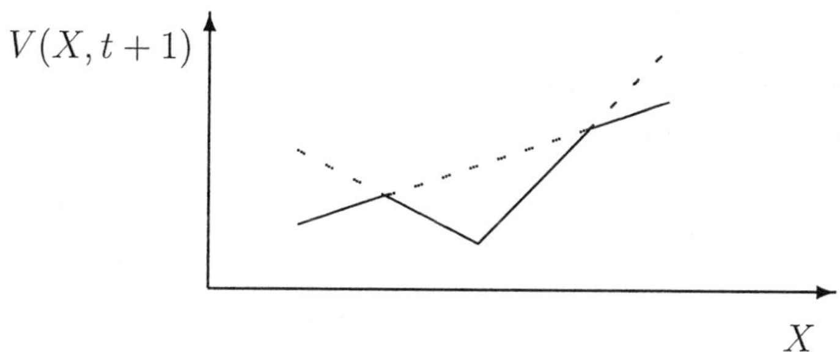

Figure 5: Splitting of an interval

If an interval associated with a large slope of the function follows in list2, the rest of the interval of list1 may be chosen again. The situation is sketched in Figure 5.

Thus, an additional interval may have to be encountered whenever one of the piecewise linear functions V or V^+ has a 'tooth', i.e. an increase of its slope. Fortunately, this does not occur too often. Consider equations (20) and (19), and notice that a tooth can not be the result of taking the minimum of two piecewise linear functions. It must have been present already in one of the functions. Teeth that are in $V(X, t)$ and have not been in $V(X, t-1)$ originate from the function $V^+(X, t-1)$, either from its last branch (which contains translated copies of the teeth of $V(X, t-1)$) or it is the tooth at the point $D_{t-1} + C_t$ where the left slope $\tilde{p}_t$ is the smallest slope encountered so far (cf. (7),(11)).

A tooth with left slope $\tilde{p}_i$, $i = 1, 2, \ldots, t$ is called a tooth of class i in the sequel. The first tooth may appear in iteration 2;

it is of class 2. In iteration $t > 2$ at most one tooth of class t is generated and the number of teeth of each existing class doubles. Therefore, an upper bound on the number of teeth of class i after iteration t is 2^{t-i}, $i = 2, \ldots, t$.

In each iteration τ, a tooth of class $i \leq \tau$ generates at most one additional interval, and the function associated with this interval has slope $\tilde{p}_j > \tilde{p}_i$ (i.e. $j < i$) on this region.

Up to the end of iteration t an upper bound for the number of intervals is obtained by summing up the $\sum_{\tau=i}^{t} 2^{\tau-i} = 2^{t-i+1} - 1$ additional intervals generated by teeth of classes $i = 2, \ldots, t$ and the 2^{t-i} 'regular' intervals with associated slopes $\tilde{p}_i$, $i = 1, 2, \ldots, t$.

Under a mild assumption, however, the number of intervals is bounded by a constant. Suppose, it never pays to produce and store m or more periods in advance, i.e. assume that the following **stockout assumption** holds:

$$\tilde{p}_r > \tilde{s}_{r+m}, \quad \forall r = 1, 2, \ldots \tag{25}$$

Theorem 3 *Under the stockout assumption (25) there will never be more than $2^m - (m+1)$ intervals in the lists L_t.*

Proof: The above considerations yield an upper bound of $2^{t+1} - (t+2)$ intervals at the end of iteration t. The Theorem obviously holds for iterations $t \leq m - 1$. For $t \geq m$ we get rid of the 'regular' intervals generated before period $t - m + 2$ as well as of the additional intervals generated by teeth of classes smaller than $t - m + 3$ by the stockout assumption. Hence, we are left with at most $\sum\limits_{i=t-m+2}^{t} 2^{t-i} + \sum\limits_{i=t-m+3}^{t} (2^{t-i+1} - 1) = 2^m - (m+1)$ intervals □

Theorem 4 *Under the stockout assumption (25) the proposed algorithm has a worst case complexity of $O(t^2)$*

Proof: Most operations are applied to the elements of the lists L_t and L_t^+. As their number is bounded by a constant, the effort is proportional to the number of iterations. Only the recursive procedure INACTIVATE may follow a path from a leaf down to the root of the tree. The length of the path is not longer than t and the number of calls during one iteration is bounded by a constant. This yields an effort of $O(t^2)$ for t iterations □

9 The case $C_t = C$ and $A_{t+1} \leq A_t$

In this section the special case of constant capacity bounds and non-increasing setup cost will be discussed. It will be shown that the algorithm remains perfect and that the number of active nodes is bounded by $m - 1$. Notice that the proof of section 7 has to be changed, because C can not always be chosen appropriately. If $C \geq D_{t+1} - D_t$ the proof remains correct. If $C < D_{t+1} - D_t$, the choice of A_{t+1} has to be revised as follows:

Let $l^* := \min\{l \mid b_l + C \geq D_{t+1} - D_t\}$.

i) If l^* does not exist $\Rightarrow b_n + C < D_{t+1} - D_t$.
   Choose $A_{t+1} > V(D_t, t) - V_n + \tilde{s}_{t+1}(b_n + C) - \tilde{p}_{t+1}C$

ii) If l^* does exist and $a_{l^*} + C \geq D_{t+1} - D_t$, then $l^* - 1$ takes the place of n in case i).

iii) If l^* does exist and $a_{l^*} + C < D_{t+1} - D_t$: Choose $A_{t+1} > V(D_t, t) - V_{l^*} - \tilde{p}_{t+1}C + \sigma_{l^*}(b_{l^*} + C - (D_{t+1} - D_t))$

Hence, the algorithm remains perfect. The upper bound on the number of active nodes is much smaller in this case.

Lemma 6 *Assume* $C_t = C \wedge A_{t+1} \leq A_t \; \forall t = 1, 2, \ldots,$ *and let* α_j^{t-1} *be a P-node, i. e.* $x_{t-1} = C$. *If* $a_j \geq D_{t-1} + C$, *then node* α_j^{t-1} *becomes inactive at the end of period* t.

Proof: If α_j^{t-1} were active at the end of period t, then there would exist a long-run optimal policy via node α_j^{t-1} with $x_t = 0$. The cost of this policy in period $t-1$ is $A_{t-1} + p_{t-1}C + h_{t-1}(X_{t-1} - D_{t-1})$. Consider now an alternative policy that produces quantity C in period t instead of period $t-1$. The cost in period $t-1$ plus production costs at the beginning of period t amount to:

$$h_{t-1}(X_{t-1} - C - D_{t-1}) + A_t + p_t C$$

(Notice that $X_{t-1} > C + D_{t-1},\ X_{t-1} \in (a_j, b_j])$.
By (7) and $A_t < A_{t-1}$, we have

$$A_t + p_t C < A_{t-1} + (p_{t-1} + h_{t-1})C$$

which implies

$$A_t + p_t C + h_{t-1}(X_{t-1} - C - D_{t-1}) < A_{t-1} + p_{t-1}C + h_{t-1}(X_{t-1} - D_{t-1})$$

i. e. the alternative policy is cheaper □

According to Lemma 6 the number of active nodes increases at most by one in each iteration. If the stockout assumption (25) is satisfied then for all $t \geq m$ at least one node becomes inactive, yielding at most $m-1$ active nodes.

Also, the handling of the list of intervals may be simplified in this special case. In iteration t it is sufficient to compute L_t the way we have discussed in the algorithm (i.e. by making overlapping intervals disjoint) just up to $X = D_t + C$. From this point it is optimal to confine the curbation of intervals to the list L^+_{t-1}, i.e. equation (20) changes to:

$$V(X,t) = \begin{cases} \min\limits_{D_t \leq Y \leq X} \{\tilde{s}_{t+1}(X-Y) + \min\{V(Y,t-1), V^+(Y,t-1)\}\} \\ \hfill D_t \leq X \leq D_t + C \\ \\ \min\limits_{D_t + C \leq Y \leq X} \{\tilde{s}_{t+1}(X-Y) \\ \qquad + \min\{V(D_t + C, t), V^+(Y, t-1)\}\} \quad X > D_t + C \end{cases}$$

This is easily seen by the following consideration. For $X \geq D_t + C$ we have:

$$\begin{aligned} V(X,t-1) &\overset{(20)}{\geq} V^+(X,t-2) \\ &\overset{(19)}{=} V(X-C,t-2) + A_{t-1} + \tilde{p}_{t-1}C \\ &\overset{(20),ass.}{\geq} V(X-C,t-1) + A_t + \tilde{p}_t C \\ &\overset{(19)}{=} V^+(X,t-1) \end{aligned}$$

Notice, that we do not generate any teeth, and therefore there is just one interval associated with each active node in the tree: on the interval $[D_t, D_t+C]$, the new line segment has the smallest slope encountered so far, and beyond $D_t + C$ we do not have any intersections of intervals.

References

[1] K.R. BAKER; P. DIXON; M.J. MAGAZINE; E.A. SILVER (1978), An algorithm for the dynamic lot-size problem with time-varying production capacity constraints, *Management Science*, Vol 24., pp. 1710-1720.

[2] M. BASTIAN (1990), Lot-trees: a unifying view and efficient implementation of forward procedures for the dynamic lot-size problem, *Computers Opns. Res.*, Vol. 17, pp. 255-263.

[3] M. BASTIAN (1992), A perfect lot–tree procedure for the discounted dynamic lot–size problem with speculation, *Naval Research Logistics*, Vol. 39, pp. 651-668.

[4] M. BASTIAN; M. VOLKMER (1992), A perfect forward procedure for a single facility dynamic location/relocation problem, *Operations Research Letters*, Vol. 12, pp. 11-16.

[5] G. R. BITRAN; H. MATSUO (1986), Approximation formulations for the single–product capacitated lot size problem, *Operations Research*, Vol. 34, pp. 63-74.

[6] G. R. BITRAN; H. H. YANASSE (1982), Computational complexity of the capacitated lot size problem, *Management Science*, Vol. 28, pp. 1174-1186.

[7] S. CHAND; T.E. MORTON (1986), Minimal forecast horizon procedures for dynamic lot size models, *Naval Research Logistics Quarterly*, Vol. 33, pp. 111-122.

[8] G. D. EPPEN; R. K. MARTIN (1987), Solving multi-item capacitated lot-sizing problems using variable redefinition, *Operations Research*, Vol. 35, pp. 832-848.

[9] A. FEDERGRUEN; M. TZUR (1994), Minimal forecast horizons and a new planning procedure for the general dynamic lot size model: nervousness revisited, *Operations Research*, Vol. 42, pp. 456-468.

[10] B. FLEISCHMANN (1990), The discrete lot–sizing and scheduling problem, *EJOR*, Vol. 44, pp. 337-348.

[11] R. LUNDIN; T.E. MORTON (1975), Planning horizons for the dynamic lot size model: Zabel vs. protective procedures and computational results, *Operations Research*, Vol. 23, pp. 711-734.

[12] M. SALOMON; L.G. KROON, R. KUIK; L.N. VAN WASSENHOVE (1991), Some extensions of the discrete lotsizing and scheduling problem, *Management Science*, Vol. 37, pp. 801-812.

[13] R.A. SANDBOTHE; G.L. THOMPSON (1990), A forward algorithm for the capacitated lot size model with stockouts, *Operations Research*, Vol. 38, pp. 474-486.

[14] H.M. WAGNER; T.M. WHITIN (1958), Dynamic version of the economic lot size model, *Management Science*, Vol. 5, pp. 89-96.

Capacitated Lot-Sizing with Linked Production Quantities of Adjacent Periods

Knut Haase
University of Kiel

Abstract: One of the most important tasks operations manager are confronted with is to determine production quantities over a medium-size planning horizon such that demand is met, scarce production facilities are not overloaded and that the sum of holding and setup costs is minimized. For the single machine case the well-known Capacitated Lot-Sizing Problem (CLSP) has been proposed to determine minimum cost solutions. The CLSP is based on the assumption that for each lot produced in a period setup cost is incurred. But in practice the machine setup can be preserved over idle time very often. In such cases the setup cost of a CLSP solution can be reduced by linking the production quantities of an item which is scheduled in two adjacent periods. Therefore we propose the CLSP with linked lot-sizes of adjacent periods. The problem is formulated as a mixed-integer programming model. For the heuristic solution we present a priority rule based scheduling procedure which is backward-oriented, i.e. at first lot-sizes are fixed in the last period, then in the last but one period, and so on. The priority rule consists of a convex combination of estimated holding and setup cost savings. Since the solution quality depends on realisation of the convex combination we perform a simple local search method on the parameter space to obtain low cost solutions. We show by a computational study that our procedure is more efficient than a two stage approach which first solves the CLSP with the Dixon-Silver or the Kirca-Kökten heuristic and performs linking of lots afterwards.

1 Introduction

The capacitated lot-sizing problem is characterized as follows: A number $j = 1, \ldots, J$ of different items is to be manufactured on one machine (corresponding to a single capacity constraint). The planning horizon is segmented into a finite number $t = 1, \ldots, T$ of time periods. In period t the machine is available with C_t capacity units. Producing one unit of item j uses p_j capacity units (finite production speed). The demand for item j in period t, $d_{j,t}$, has to be satisfied without delay, i.e. shortages are not allowed. Setting up the machine for item j causes setup cost of s_j. Inventory cost per unit of h_j (holding cost coefficient) are incurred for the inventory of item j at the end of a period. The objective is to minimize the sum of setup and holding costs.

Basically assuming that

Setup cost occur for each lot produced in a period.

and defining the decision variables

- $I_{j,t}$ the inventory of item j at the end of period t (w.l.o.g. $I_{j,0} = 0$ for all $j = 1, \ldots, J$),
- $q_{j,t}$ the quantity (lot-size) of item j to be produced in period t, and
- $x_{j,t}$ a binary variable indicating whether setup occurs for item j in period t ($x_{j,t} = 1$) or not ($x_{j,t} = 0$),

we can state the well-known capacitated lot-sizing problem (CLSP) as follows:

Problem CLSP

$$Minimize \sum_{j=1}^{J}\sum_{t=1}^{T}(s_j x_{j,t} + h_j I_{j,t}) \quad (1)$$

subject to

$$I_{j,t-1} + q_{j,t} - I_{j,t} = d_{j,t} \qquad j = 1, \ldots, J; t = 1, \ldots, T \quad (2)$$

$$\sum_{j=1}^{J} p_j q_{j,t} \leq C_t \qquad t = 1, \ldots, T \quad (3)$$

$$C_t x_{j,t} - p_j q_{j,t} \geq 0 \qquad j = 1, \ldots, J; t = 1, \ldots, T \quad (4)$$

$$x_{j,t} \in \{0,1\} \quad j = 1, \ldots, J; t = 1, \ldots, T \quad (5)$$

$$I_{j,t}, q_{j,t} \geq 0 \qquad j = 1, \ldots, J; t = 1, \ldots, T \quad (6)$$

The objective to minimize the sum of setup and inventory holding costs is expressed by (1). Equations (2) determine the inventory of an item at the end of each period. Constraints (3) make sure that the total production in each period does not exceed the capacity. For each produced lot in a period constraints (4) force a setup, i.e. the corresponding binary setup variable must equal one, thus increasing the sum of setup costs. The suitable domains of the variables are determined by the restrictions (5) and (6). The non-negativity condition for the inventory ensures that shortages do not occur.

In the literature for the CLSP a multitude of heuristics (cf. [5], [6], [8], [13], [15], [16], [18], [14]) and exact methods (cf. [1], [7]) have been proposed.

Let us now look at the following example which reveals the drawback associated with the basic assumption.

Example 1: Let $J = 2$, $T = 3$, $s_1 = s_2 = 100$, $p_1 = p_2 = 1$, $h_1 = h_2 = 1$, $C_t = 10$ for $t = 1, \ldots, 3$, and

$$(d_{j,t}) = \begin{pmatrix} 5 & 0 & 6 \\ 0 & 3 & 0 \end{pmatrix}.$$

The optimal solution of the CLSP with an objective function value $Z^*_{CLSP} = 300$ is determined by equalizing lot-sizes and demands, i.e. $q_{j,t} = d_{j,t}$ for all $j = 1, \ldots, J$, and $t = 1, \ldots, T$. Let us now consider the solution

$$(q_{j,t}) = \begin{pmatrix} 5\cup & 6 & 0 \\ 0 & 3 & 0 \end{pmatrix}$$

where $\cup$ denotes a linking of the production quantities of adjacent periods. That is, in the second period we first produce item $j = 1$. Hence, it is not necessary to change the setup state of the machine because the machine has been already setup for item $j = 1$ in period $t = 1$. This way we reduce total cost by $-s_1 + 6 \times h_1 = -100 + 6 \times 1 = 106$ which gives us a new minimum cost solution of 206.

The example shows, that the CLSP solution quality can be poor in the case where the setup state can be preserved between adjacent periods. Furthermore, a solution approach for lot–sizing and semi–sequencing, i.e. determining the last and the first item in a period, may provide a better solution quality than an algorithm for the CLSP.

The remainder of the paper is structured as follows: In Section 2 we extend the CLSP with respect to the possibility of linking lot–sizes of adjacent periods. An efficient straight–forward heuristic which is backward oriented and which relies on a priority rule is presented in Section 3. How a given CLSP solution can be improved by performing links subsequently is shown in Section 4. The results of a computational study are covered by Section 5. A summary and conclusions are given in Section 6.

2 The CLSP with Linked Lot-Sizes of Adjacent Periods

Due to the basic assumption of the CLSP the setup state at the beginning of a period is ignored. As shown in Example 1, in practical cases, cf. e.g. [4], where the setup state of a production facility can be preserved from one period to the succeeding period the CLSP may be not well-suited. To overcome the drawback of the CLSP we propose the following extension of the CLSP (cf. [11]).

Let us define the additional binary variable

$z_{j,t}$ indicating whether the quantities of item j in period $t-1$ and period t are linked ($z_{j,t}=1$) or not ($z_{j,t}=0$),

we state the CLSP with linked lot-sizes of adjacent periods, denoted by CLSPL, as follows:

Problem CLSPL

$$Minimize \sum_{j=1}^{J}\sum_{t=1}^{T}(s_j(x_{j,t}-z_{j,t})+h_jI_{j,t}) \qquad (7)$$

subject to

$$I_{j,t-1}+q_{j,t}-I_{j,t} \;=\; d_{j,t} \qquad j=1,\ldots,J;t=1,\ldots,T \quad (8)$$

$$\sum_{j=1}^{J} p_j q_{j,t} \leq C_t \qquad t = 1, \ldots, T \tag{9}$$

$$C_t x_{j,t} - p_j q_{j,t} \geq 0 \qquad j = 1, \ldots, J; t = 1, \ldots, T \tag{10}$$

$$\sum_{j=1}^{J} z_{j,t} \leq 1 \qquad t = 1, \ldots, T \tag{11}$$

$$2z_{j,t} - x_{j,t} - x_{j,t-1} + z_{j,t-1} \leq 0 \qquad j = 1, \ldots, J; t = 1, \ldots, T \tag{12}$$

$$x_{j,t}, z_{j,t} \in \{0,1\} \quad j = 1, \ldots, J; t = 1, \ldots, T \tag{13}$$

$$I_{j,t}, q_{j,t} \geq 0 \qquad j = 1, \ldots, J; t = 1, \ldots, T \tag{14}$$

where w.l.o.g. $I_{j,0} = 0$, $x_{j,0} = 0$, $z_{j,0} = z_{j,1} = 0$ for all $j = 1, \ldots, J$.

The objective function (7) charges no setup cost for a production quantity in period t which is linked with a production quantity of the preceeding period $t-1$. (8) corresponds to ordinary inventory balance constraints. (9) secures feasibility with respect to the machine capacity. (10) couples the production decisions with the setup state of the machine. (11) secures that only one product can be produced at the end of a period. By (12) linking of an item in one of two adjacent periods is allowed only, if the machine is setup for the item in both periods, i.e. $z_{j,t}$ can be "1" only if $x_{j,t} = x_{j,t-1} = 1$ and $z_{j,t-1} = 0$. Note that, the CLSPL is equivalent to the CLSP if we set $z_{j,t} = 0$ for $j = 1, \ldots, J$, and $t = 1, \ldots, T$.

As a large time scale assumed in the CLSP, we expect that generally more than one item will be scheduled in a period. That is the required capacity of a lot will be strictly less than the available period capacity. Let us now consider the (unusual) case where in period t only item j is scheduled and that item j is also scheduled in periods $t-1$ and $t+1$. Thus, the setup state may be preserved for item j from the end of period $t-1$ up to the beginning of period $t+1$. To preserve the setup state two links are necessary for item j which is not allowed due to (12). More precisely, (12) requires that $z_{j,t} + z_{j,t-1} \leq 1$.

Note, we allow lot-splitting. For example, in spite of producing the quantities $q_{j,t-1}$ and $q_{j,t}$ in one lot, i.e. $z_{j,t} = 1$, $q_{j,t-1}$ may be used to satisfy the demand $d_{j,t-1}$.

3 Improving a CLSP Solution

If a CLSP is solved, then the provided schedule can be improved by performing the links afterwards, which reduces setup cost. Moreover, holding cost is saved, if a quantity of a linked lot can be "right-shifted" . This will be clarified by the following example:

Example 2: Let $J = 4$, $T = 4$, $C_t = 100$ for $t = 1, \dots, 4$, $(p_j) = (1,1,1,1)$, $(h_j) = (1,1,1,1)$, $(s_j) = (200,150,100,150)$, and

$$(d_{j,t}) = \begin{pmatrix} 20 & 10 & 30 & 20 \\ 30 & 10 & 30 & 30 \\ 0 & 30 & 10 & 60 \\ 20 & 20 & 0 & 10 \end{pmatrix}.$$

Solving this instance as CLSP to optimality we receive the solution

$$(q_{j,t}) = \begin{pmatrix} 40 & 0 & 40 & 0 \\ 40 & 0 & 60 & 0 \\ 0 & 40 & 0 & 60 \\ 20 & 30 & 0 & 0 \end{pmatrix}$$

which is $Z^*_{CLSP} = 1,320$ costly.

This solution can be improved by linking and "right-shifting" . For example, a right-shifting of 30 units is feasible for $q_{2,3} = 60$. The cost of the respective modified CLSP solution

$$(q_{j,t}) = \begin{pmatrix} 40 & 0 & 40 & 0 \\ 40 & 0 & 30\cup & 30 \\ 0 & 30\cup & 10 & 60 \\ 20\cup & 30 & 0 & 0 \end{pmatrix}$$

is $Z^*_{modifiedCLSP} = 1,130$.

However, the objective function value of the optimal CLSPL solution

$$(q_{j,t}) = \begin{pmatrix} 20\cup & 60 & 0 & 0 \\ 30 & 10\cup & 60 & 0 \\ 0 & 30 & 10\cup & 60 \\ 50 & 0 & 0 & 0 \end{pmatrix}$$

is $Z^*_{CLSPL} = 1,000$.

Comparing the objective function values we see the modified CLSP solution is 11.3% more costly than the CLSPL solution. This demonstrates that the solution quality can be substantially improved by integrating semi–sequencing. Due to the constraints (2)–(6) and (8)–(14), respectively, we can state that the CLSP and the CLSPL have the same set of feasible solutions. But on account of the objective functions (1) and (7), the following inequality holds:

$$Z^*_{CLSP} \leq Z^*_{modifiedCLSP} \leq Z^*_{CLSPL} \tag{15}$$

That is, solutions associated with minimum cost are always CLSPL-based. Since the CLSPL contains more binary variables than the CLSP it may be more difficult to determine an optimal solution in reasonable time. However, in practical applications fast heuristics may be applied. So a CLSPL approach which computes a solution in reasonable time may be more attractive than an exact or heuristic method for the CLSP.

Note that a similar model with machine state preserving has been presented in [9] which is a special case of the multi-machine case with setup times introduced in [4].

4 A Priority Rule Based Heuristic

In the following we describe a simple heuristic for the CLSPL which starts with scheduling in the period $t = T$ and steps backward to the first period $t = 1$. The lot–size and semi–sequencing decisions are based on a simple priority rule which consists of a convex combination of holding and setup cost savings.

At first we derive a simple feasibility check: Let initially $q_{j,t} = 0$. Consider now a period t, $1 \leq t \leq T$, where we have already made production decisions for the periods $\tau = t$ to T by fixing $q_{j,t}$. Then the cumulative demand $D_{j,t}$ of item j from period t to the horizon T which is not satisfied is calculated by

$$D_{j,t} = \sum_{\tau=t}^{T} (d_{j,\tau} - q_{j,\tau})$$

Note, $D_{j,\tau-1} = D_{j,\tau} + d_{j,\tau-1}$ for $\tau = 2, \ldots, t$. Thus, the total still required capacity can easily be specified by

$$TRC = \sum_{j=1}^{J} p_j D_{j,1}$$

By

$$AC_t = C_t - \sum_{j=1}^{J} p_j d_{j,t}$$

and

$$CC_t = \sum_{\tau=1}^{t} C_t$$

we compute the available capacity in period t and the cumulated available capacity from period 1 to period t, respectively.

Obviously, a feasible solution exists only if the inequality

$$\sum_{\tau=1}^{t} \sum_{j=1}^{J} p_j d_{j,t} \leq CC_t$$

is satisfied for all t. From another point of view

$$TAC_t = CC_{t-1} - \sum_{\tau=1}^{t-1} \sum_{j=1}^{J} p_j d_{j,t}$$

counts the capacity units available in periods earlier than t. Thus TAC_t can be used to produce demands from the periods $\tau = t, \ldots, T$. If we step backwards from period t to period $t-1$ there may be unsatisfied demand in period t, i.e. $\sum_{j=1}^{J} D_{j,t} > 0$, which can only be satisfied if

$$TAC_t \geq \sum_{j=1}^{J} p_j D_{j,t}$$

or

$$TRC \leq CC_{t-1}$$

which follows by simple substitutions.

Such a *feasibilty check* is only necessary if capacity in period t is available and a backwards step to period $t-1$ is intended. This may occur if we preserve the setup state (linking) from period t to $t-1$ for an item which is already scheduled in period t.

If a feasible solution exists for the problem under consideration we will always derive a feasible solution due to the feasibility check. It should be noted here that backward lot-sizing and scheduling methods in the capacitated case are from a conceptual point of view superior to forward-oriented ones like, e.g., the Dixon-Silver heuristic, since there is no need for complicated and time-consuming look ahead procedures in order to secure resource feasibility.

To decide on the item to be produced in period t we use a "savings"-based priority rule. (The quotation marks emphasize that the savings are roughly estimated.) Four cases are distinguished:

1. There is unsatisfied demand of item j in period t and the available capacity in period t is greater or equal than the capacity which will be required if item j is scheduled in period t. Thus, to schedule item j in period t rather than in period $t-1$ saves holding cost $h_j D_{j,t}$ and incurs setup cost s_j.

2. Item j is not already scheduled in period t and the unsatisfied demand of item j in period t requires more capacity than still available in period t. Thus to schedule the unsatisfied demand of item j in one lot requires a linking from period $t-1$ to t. This decision causes setup cost s_j in period $t-1$ and "saves" holding cost $h_j D_{j,t-1}$ $(= h_j D_{j,t} + d_{j,t-1})$.

3. Item j is already scheduled in period t, there exists positive unsatisfied demand of item j in period $t-1$, and there is no linking performed for item j in period $t+1$. Thus a link in period t, avoids setup cost s_j and holding cost $h_j d_{j,t-1}$. Moreover, linking is feasible (feasibilty check).

4. Linking does not improve the solution quality. This case occurs if there is no unsatisfied demand in period t and for

all items scheduled in period t no demand exists in period $t-1$.

Formally we define the priority value $r_{j,t}$ for item j in period t by

$$r_{j,t} = \begin{cases} \alpha_j D_{j,t} - \beta_j & : \quad AC_t \geq p_j D_{j,t} > 0 \\ \alpha_j D_{j,t-1} - \beta_j & : \quad AC_t < p_j D_{j,t} \wedge t > 0 \\ \alpha_j d_{j,t-1} + \beta_j & : \quad z_{j,t+1} = 0 \wedge q_{j,t} > 0 \wedge t > 0 \wedge \\ & \qquad D_{j,t-1} > 0 \wedge TRC \leq CC_{t-1} \\ -\infty & : \quad \text{otherwise} \end{cases}$$

where $\alpha_j = (1-\gamma)h_j$ and $\beta_j = \gamma s_j$ with $0 \leq \gamma \leq 1$. For initialization we set $z_{j,T+1} = 0$. The larger $r_{j,t}$, the more preferable it is to schedule item j in period t. Thus, the item with the largest priority value will be produced. With the parameter γ we control sizes of lots, e.g., if $\gamma = 1$ we expect large lot-sizes for items with high setup cost.

The backward oriented scheduling procedure, denoted by BA, is sketched out in Table 1: In *step 0* we initialize the variables and parameters. The priority values are obtained in *step 1*. In *step 2* we select the item i with maximum priority. In *step 3* "case" corresponds to the case which was true for the calculation of $r_{i,t}$. In the fourth case only a backward step is required, i.e. we have to reduce the period counter t by 1. If one of the other cases is true item i is scheduled which forces some updates of the variables and parameters. Note, a link in period t induces always a backward step to period $t-1$ but only in case 3 a feasibilty check is required.

To give more insights, we apply BA for the instance in Example 2 with $\gamma = 0.5$: Before starting the scheduling phase in period $T = 4$, we set $q_{j,t} = 0$, and hence derive

$$(D_{j,t}) = \begin{pmatrix} 80 & 60 & 50 & 20 \\ 100 & 70 & 60 & 30 \\ 100 & 100 & 70 & 60 \\ 50 & 30 & 10 & 10 \end{pmatrix},$$

Table 1: Outline of the Heuristic BA

---

step 0: Set $\forall j\ z_{j,t} = x_{j,t} = q_{j,t} = 0, D_{j,t} = \sum_{\tau=1}^{\tau=t} d_{j,\tau}$;
$\forall j\ z_{j,T+1} = z_{j,0} = 0$;
$\forall t\ AC_t = C_t,\ CC_t = \sum_{\tau=1}^{\tau=t} C_t$;
$TRC = \sum_{j,t} p_j d_{j,t},\ t = T$.
while $(t > 0)$
{
step 1: Calculate $r_{j,t}\ \forall j$.
step 2: Select $i | r_{i,t} \geq r_{j,t}\ \forall j$.
step 3: If *case 1* then
set $q_{i,t} = D_{i,t},\ x_{i,t} = 1,\ AC_t = AC_t - p_i q_{i,t}$,
$D_{i,\tau} = D_{i,\tau} - q_{i,t}\ (\tau = 1, \ldots, t)$
else if *case 2* then
set $q_{i,t} = AC_t/p_i,\ AC_t = 0,\ t = t-1,\ z_{i,t} = 1$,
$q_{i,t} = min\{D_{i,t} - q_{i,t+1}, AC_t/p_i\}$,
$AC_t = AC_t - p_i q_{i,t},\ x_{i,t} = 1$,
$D_{i,\tau} = D_{i,\tau} - q_{i,t} - q_{i,t+1}\ (\tau = 1, \ldots, t)$
else if *case 3* then
$q_{i,t} = min\{D_{i,t}, AC_t/p_i\},\ AC_t = AC_t - p_i q_{i,t}$,
$z_{i,t} = 1,\ x_{i,t} = 1$,
$D_{i,\tau} = D_{i,\tau} - q_{i,t}\ (\tau = 1, \ldots, t)$
else $t = t - 1$.
}
step 4: if $\exists D_{j,1} > 0$ then
message ("No feasible solution exists!")
else evaluate solution.

Table 2: Objective Function Values Depending on γ

γ	0	0.25	0.50	0.75	1
Z_{CLSPL}	1250	1000	1150	1150	

$TRC = 330$, $(\alpha_j) = (0.5, 0.5, 0.5, 0.5)$, and $(\beta_j) = (100, 75, 50, 75)$. For all items the first case occurs, hence:

$$\begin{aligned} r_{1,4} &= 0.5 \times 20 - 100 = -90 \\ r_{2,4} &= 0.5 \times 30 - 75 = -60 \\ r_{3,4} &= 0.5 \times 60 - 50 = -20 \\ r_{4,4} &= 0.5 \times 10 - 75 = -70 \end{aligned}$$

The item associated with the largest savings is scheduled in period $t = 4$, i.e. $j = 3$. The corresponding production is $q_{34} = 60$, i.e. we schedule the complete quantity $D_{j,t}$. Now we perform some updates: $TRC = 330 - 60 = 270$, $AC_4 = 100 - 60 = 40$, and $D_{3,t} = (40, 40, 30, 0)$. As long as no linking is performed and a $D_{j,t} > 0$ exists we try to schedule other items, i.e. we compute again priority values. Only for item 3 the priority value changes, i.e. we derive (third case)

$$r_{3,4} = 0.5 \times 10 + 50 = 55$$

Note that $TRC = 270 \leq CC_3 = 300$. Thus item $j = 3$ is selected again, i.e. we set $z_{3,4} = 1$ and $q_{3,3} = 10$. Afterwards we schedule period $T - 1 = 3$. Continuing this way, we end up with a (non-optimal) schedule with overall cost of 1,150. Trying to get a better solution, we have also tried alternative values for γ. Table 2 reports the corresponding objective function values.

As can be seen, the solution quality depends heavily on the choice of the parameter value. For our instance, the optimal solution was computed with the parameter $\gamma = 0.25$.

Figure 1: Local Search in the Parameter Space

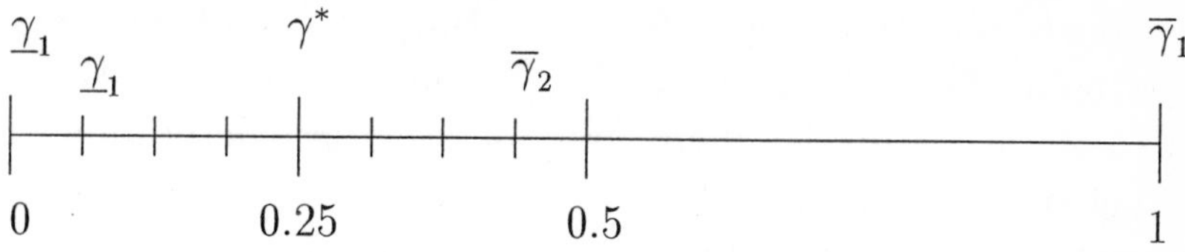

Since the heuristic is a mapping from a problem to a solution, each pair (BA, γ) is an encoding of a specific solution. A subset of the solution space can then be generated by different values for $\gamma \in [0, 1]$. We therefore propose a local search method where the search space equals the parameter space and the neighbourhood is a subset of the parameter space as follows: First, we determine a promising region $[\underline{\gamma}, \overline{\gamma}] \subseteq [0, 1]$. Then, we intensify the search in that region. Let γ^* be the parameter value where BA has computed the best solution and b an integer greater than 2. We start with $\gamma = \underline{\gamma} = 0$ and $\overline{\gamma} = 1$. Then γ is incremented by $\delta = (\overline{\gamma} - \underline{\gamma})/b$ and a new solution is computed. This is repeated as long as the objective function value can be improved or $\gamma = \overline{\gamma}$. If an improvement has been achieved a more detailed search will be started with $\delta = \delta/b$, $\underline{\gamma} = max\{\delta, \gamma^* - \delta/(b-1)\}$, and $\overline{\gamma} = min\{1 - \delta, \gamma^* + \delta(b-1)\}$; otherwise the search procedure stops.

Now, let $b = 4$, $\underline{\gamma} = 0$, $\overline{\gamma} = 1$, thus $\delta = (1-0)/4 = 0.25$. Furthermore, for our instance, γ will be increased up to 0.5, and γ^* will be 0.25. Then a new (and last) search will be started with $\delta = 0.0625$, $\underline{\gamma} = 0.0625$, and $\overline{\gamma} = 0.4375$. This is illustrated in Figure 1 where the index values 1 and 2 denote the first and second search phase, respectively. Note, no solution will be computed for $\gamma = 0.75$ and $\gamma = 1$.

5 Computational Study

The purpose of our computational study is twofold: First, we investigate the performance of BA. Second, we compare BA with a two stage-approach where first lot sizes are determined and linking of lot-sizes is performed afterwards. Before presenting the computational results we will describe an algorithm which generates the test instances used: Given the number of items, J, the number of periods, T, and the expected capacity utilization, U, the other problem parameters are determined as follows:

- In all periods the capacity is constant, i.e. $C_t = J \times 50$.
- The production time to produce one unit of item j, p_j, is chosen at random from the set $\{1, 2\}$.
- The demand of item j in period t, $d_{j,t}$, is chosen at random from the set $\{0, 1, \ldots, 100\}$ with respect to capacity constraints and the expected capacity utilization, U.
- The holding cost coefficients are constant, i.e. $h_j = 1$.
- The setup cost, s_j, are chosen randomly from the set {50, 51,..., 250}.

We use the uniform distribution to choose a parameter at random. For each of the a priori given parameter combinations provided in Table 3, we generated 10 instances and solved them to optimality by using the general mixed integer programming solver LINDO [17]. Each entry reflects the average percentage deviation from the optimum objective function value for the parameter combination under investigation.

From this small investigation we can conclude that the overall performance of BA is with an overall deviation of 5.28 % quite promising. Furthermore, it can be noticed that neither the capacity utilization nor the relation between the number of items and periods systematically influences the quality of the solutions.

Now we are going to explore the computational performance of BA for larger and thus realistic data sets. Since these instances

Table 3: Average Percentage Deviation from Optimum

U	$J=4, T=8$	$J=8, T=4$
0.75	7.8	3.7
0.95	4.4	5.2

cannot be solved to optimality, the main focus is on the comparison of BA with a two-stage approach which works as follows: First we apply a CLSP heuristic, then we improve the derived schedules by linking. We employed the well-know benchmark-instances of [2]. These 120 instances may be briefly introduced as follows. They are divided in three sets which differ with respect to the number of items and periods, (J, T), as given in the first line of Table 8. The instances of a set differ with respect to the three factors capacity utilization (U), capacity requirements (C), and demand variation (S). For each factor two levels exits, which are low (L) and high (H) for

- capacity utilization (LU, HU), i.e. in the average $U = 75\%$, or $U = 95\%$, and
- demand variation (LS, HS), where given an average demand of 100 the standard deviation of the demand for the instances with low and high demand variation is 6 and 35, respectively.

Then two factors constant (C) and varying (V) are respected for

- the capacity requirements (CC, VC), i.e. $p_j = 1$ or $p_j \in \{1, 2, 3, 4\}$ chosen at random.

Based on the results in [15] it can be sated that the heuristics introduced in [15] and [18] are outperformed by the Kirca-Kökten heuristc (KK). The same is true for the Dixon-Silver heuristic (DS), but for instances with a large number of items, i.e. $(J, T) = (50, 8)$, DS has computed very often the best solution. Therefore

Table 4: Comparison of DS, KK, and BA

J, T	alg.	HU	LU	HS	LS	CC	VC	av.
50,8	DS	0.13	2.91	0.91	2.13	1.28	2.09	1.52
	KK	1.25	3.06	2.44	1.86	2.18	2.09	2.15
	BA	1.39	0.09	0.82	0.66	0.36	1.63	0.74
20,20	DS	12.15	4.39	3.75	12.79	9.68	4.98	8.27
	KK	5.36	2.63	3.57	4.42	4.17	3.57	3.99
	BA	0.67	0.55	1.19	0.02	0.58	0.67	0.61
8,50	DS	16.30	8.34	8.91	15.73	13.15	10.13	12.32
	KK	10.91	9.18	8.68	11.41	10.09	9.94	10.05
	BA	0.41	0.00	0.41	0.00	0.11	0.45	0.20
av.	DS	9.53	5.22	4.52	10.22	8.04	5.73	7.37
	KK	5.84	4.96	4.90	5.90	5.48	5.20	5.40
	BA	0.82	0.21	0.81	0.23	0.35	0.91	0.52

we choose for a comparision with BA the heuristics DS and KK. For a given DS or KK solution setups are reduced a posteriori, by combining two lots of adjacent periods (for the item with the highest setup cost, if there is a choice). The average percentage deviation between the reduced cost of DS and the cost of BA are reported in Table 4. As can be seen for $J = 50$ and $T = 8$, the solution quality of DS, KK and BA is almost identical if we have a large number of items. This may be reasoned by the fact that for a large number of items it is "easy" to find in the second stage a "good" linking between production quantities of adjacent periods. But for medium (20) and small numbers (8) of items the solution quality of BA is substantially better than the solution quality of DS and KK, i.e. integrated sequencing is very important. This is especially valid for instances with high capacity utilization, low

Table 5: Average Cumputation Times of BA in Seconds

$J = 50, T = 8$	$J = 8, T = 20$	$J = 8, T = 50$
4.5	1.2	0.73

Table 6: Average Number of BA Executions

	$J = 50, T = 8$	$J = 8, T = 20$	$J = 8, T = 50$
best solution	7.48	6.93	7.90
total	14.48	13.85	14.90

demand variation and/or constant capacity.

We have coded BA in Pascal and implemented it on a personal computer with 486DX2 processor and 50 MHz. Table 5 reports the average computation times of the 40 instances per entry.

It is demonstrated that BA solves even large instance sizes within a few seconds. The computational effort increases rapidly with the number of items. The average number of executions and the average number of executions until the best solution has been reached are reported in Table 6, respectively.

6 Summary and Conclusions

Integrating sequencing in lot-sizing can be very attractive. A mixed-integer formulation has been presented where lot-sizes of adjacent periods can be linked (semi-scheduling), denoted by CLSPL, which is an extension of the well-known capacitated lot-sizing problem (CLSP). The CLSPL can be solved efficiently by a backward-oriented approach where lot-sizing and linking depends on a priority rule. Especially for instances with a small to medium number of items the method solves the CLSPL more efficient than a two stage approach which first solves the CLSP (heuristically) and performs linking of lots afterwards.

We have shown that in the case of small to medium number of items production managers can save a lot of expense if they apply production planning and control methods which perform lot-sizing and sequencing, simultaneously. Hence, integrating sequencing in lot-sizing methods has a great potential for improving the solution quality.

The proposed priority rule depends on a parameter which affects the solution quality. Instead of using a fixed value for the parameter a higher solution quality is achieved by applying a simple local search method on the parameter space. The search space and neighbourhood are defined in general terms. An extension for more than one parameter is straightforward (cf. [10], [12]). Thus the local search method may be applied for any parameterized heuristic.

Acknowledgement

The author is grateful to Andreas Drexl for his helpful support, to Jörg Latteier for his assistance in the computational study, and to Rainer Kolisch and Andreas Schirmer for the improvement of the readability of the paper.

References

[1] Barany, I., van Roy, T.J., and Wolsey, L.A. (1984), Strong Formulations for Multi–Item Capacitated Lot–Sizing, *Management Science*, Vol. 30, pp. 1255–1261

[2] Cattrysse, D., Maes, J., and Van Wassenhove, L.N. (1990), Set Partitioning and Column Generation

Heuristics for Capacitated Lotsizing, *European Journal of Operational Research*, Vol. 46, pp. 38–47

[3] DIABY, M., BAHL, H.C., KARWAN, M.H., AND ZIONTS, S. (1992), A Lagrangean Relaxation Approach for Very–Large–Scale Capacitated Lot–Sizing, *Management Science*, Vol. 38, pp. 1329–1340

[4] DILLENBERGER, C., ESCUDERO, L.F., WOLLENSAK, A., AND ZHANG, W. (1992), On Solving a Large–Scale Resource Allocation Problem in Production Planning, in: Operation Research in Production Planning and Control, FANDEL, G. et al. (eds.), Berlin, Springer, pp. 105–119

[5] DIXON, P. S. AND SILVER, E.A. (1981), A Heuristic Solution Procedure for Multi–Item Single–Level, Limited Capacity, Lot–Sizing Problem, *Journal of Operations Management*, Vol. 2/1, pp. 23–39

[6] DOGRAMACI, A., PANAYIOTOPOULOS, J.C., AND ADAM, N.R. (1981), The Dynamic Lot–Sizing Problem for Multiple Items under Limited Capacity, *AIIE Transactions*, Vol. 13, pp. 294–303

[7] GELDERS, L.F., MAES, J., AND VAN WASSENHOVE, L.N. (1986), A Branch and Bound Algorithm for the Multi–Item Single Level Capacitated Dynamic Lot–Sizing Problem, in: Multi-stage Production Planning and Inventory Control, AXSÄTER, S. et al. (eds.), Berlin, Springer, pp. 92–108

[8] GÜNTHER, H.O. (1987), Planning Lot Sizes and Capacity Requirements in a Single Stage Production System, *European Journal of Operational Research*, Vol. 31, pp. 223–231

[9] HAASE, K. (1994), *Lotsizing and Scheduling for Production Planning*, Berlin, Springer

[10] HAASE, K. (1996), Capacitated lot-sizing with sequence dependent setup costs, *OR Spektrum*, Vol. 18, pp. 51–59

[11] HAASE, K., DREXL, A. (1994), Capacitated Lot–Sizing with Linked Production Quantities of Adjacent Periods, in: Operations Research '93, BACHEM, A. et al. (eds.), pp. 212–215

[12] HAASE, K. AND LATTEIER, J. (1997), Deckungsbeitragsorientierte Lehrgangsplanung bei der Lufthansa Technical Training GmbH, *OR Spektrum*, to appear

[13] KIRCA, Ö. AND KÖKTEN, M. (1994), A New Heuristic Approach for the Multi–Item Dynamic Lot Sizing Problem, *European Journal of Operational Research*, Vol. 75, pp. 332–341

[14] KOHLMORGEN, U., SCHMECK, H. AND HAASE, K. (1996), Experiences with Fine-Grained Parallel Genetic Algorithms, *Annals of OR*, to appear

[15] LAMBRECHT, M.R. AND VANDERVEKEN, H. (1979), Heuristic Procedures for the Single Operations, Multi–Item Loading Problem, *AIIE Transactions*, Vol. 11, pp. 319–326

[16] MAES, J. AND VAN WASSENHOVE, L.N. (1986), A Simple Heuristic for the Multi–Item Single Level Capacitated Lotsizing Problem, *Operations Research Letters*, Vol. 4, pp. 265–274

[17] SCHRAGE, L. (1986), *Linear, Integer and Quadratic Programming with LINDO*, Redwood City

[18] THIZY, J.M. AND VAN WASSENHOVE, L.N. (1985), Lagrangean Relaxation for the Multi–Item Capacitated Lot–Sizing Problem: A Heuristic Implementation, *IIE Transactions*, Vol. 17, pp. 308–313

Cash-Flow Oriented Lot Sizing in MRP II Systems

Stefan Helber
Ludwig-Maximilians-Universität München

Abstract: This paper treats the multi-level dynamic lot sizing problem under multiple capacity constraints and setup times. In the framework of Manufacturing Resource Planning (MRP II), lot sizing is an intermediate planning step based on given or estimated product demand which gives rise to short-term and detailed scheduling problems. We analyze the cash-flow impact of lot sizing decisions within the MRP II context and derive cash-flow oriented setup and holding cost parameters. The analysis shows that the economic impact of lot sizing decisions in MRP II systems may be smaller than the use of conventional accounting-based setup and holding cost parameters suggest. We formulate a version of the Multi-Level Capacitated Lot Sizing Problem. In the objective function of this model, the net present value of future cash flows is minimized. Several solution procedures are available to solve an approximation of this model. Two heuristic decomposition approaches as well as two local search heuristics are described and evaluated in a numerical study. A Lagrangean procedure developed by Tempelmeier and Derstroff shows a superior performance.

1 Introduction

Production planning and control (PPC) systems are used to plan and control physical transformation processes in production systems. Many PPC systems used in practice follow the well-known MRP II logic [29, 41]. MRP II essentially is a successive planning concept that begins with a (not necessarily feasible) master production schedule derived from end product demand. The bill of material is used together with inventory records, lot sizing rules, and expected lead times to calculate the time-phased material requirements at lower production stages. Capacity limits of work centers are not explicitly taken into account, therefore the planned process may be infeasible. Vendors of PPC systems propose rough cut capacity planning [41] to "close the loop" and to determine feasible plans. However, since this requires highly complex manual changes by trial-and-error, rough cut capacity planning will usually lie "...somewhere between the difficult and the impossible" [4, p. 657]. The well-known results are problems like long and unpredictable lead times as well as high work-in-process inventories and poor customer service [12, 22, 36]. The defining element of this approach is the separation of material requirements and capacity planning. It can lead to feasible plans only by coincidence.

Even if plans generated in this way were feasible with respect to the capacity constraints of the production system, they might still not be efficient in the sense that they use resources in the most productive way. The reason is that lot sizing decisions may in practice be driven by setup and holding cost parameters that do not properly reflect the cash-flow impact of setup and inventory holding decisions. The economic impact of a decision in a firm operating in a market system is the difference of the net present value (NPV) of the cash-flow that is due to this decision. If there is no such difference, the decision does not have an economic impact. If it is a short-term decision that is based on a long-term plan, this difference is called a *cost* ([34, p. 215], [26]). That is, costs are based on higher-level plans and they can only calculated given this higher-level plan.

In practice, however, setup cost may be calculated as the product of the setup time and some hourly rate of the machinery and required personal plus possibly some direct expenses, e.g. for energy or material consumed during the setup ([43, p. 11.4.8]). Setup cost that are calculated this way do not properly reflect that the expenses for machinery and personal are sunk by the time the setup decision is made as they are based on a higher-level plan to buy machinery and hire personal.

A similar argument holds for inventory holding costs parameters. They should reflect the difference in the NPV of the future cash flow that is due to the decision to hold one product unit for one time unit. In the case of given end-product demand as in the MRP II framework, holding costs are basically the interest on the expenses for raw material that are due to the inventory holding decision. If - as in the MRP II context - decisions to buy equipment and hire personal are made less frequently and on a higher planning level, then holding cost parameters should not include these components as they do not depend on lot sizing decisions. In practice, however, holding cost parameters may be determined on the basis of an interest rate and of manufacturing costs that include these components ([43, p. 11.4.8],[36, p. 230.]). If manufacturing costs include in addition some factory overhead, the economic impact of an inventory holding decision is even more overestimated.

If both setup and holding cost parameters are overestimated, so is the importance of obtaining cost-minimal lot sizes. However, it *is* important to have plans that are *feasible* and that can be *adjusted* to some extent to unforeseeable events such as machine failures or rush jobs. This is a difficult task in the presence of multiple capacity constraints, multiple production stages, multiple products, dynamic demand, and setup times even if a deterministic environment is assumed. However, several procedures can be used to determine production schedules that meet these restrictions. All these procedures are heuristics that have to be evaluated with respect to accuracy and computational effort.

The remainder of the paper is organized as follows: In Section 2, a cash-flow oriented model of lot sizing in the MRP II context is developed and related to a well-known cost-oriented lot sizing model. Heuristic solution procedures are described in Section 3 and tested in a numerical study in Section 4. Section 5 contains concluding remarks.

2 A Lot Sizing Model Based on Cash Flows

Lot sizing models for the MRP II context usually *assume* the existence of setup and holding cost parameters which enter the objective function of the model. However, these parameters are nothing to be found somewhere in the accounting department. They reflect assumptions about the future economic impact of current decisions that should be stated explicitly as they direct the planning process. Furthermore, knowing the cash-flow impact of a decision field allows to assess the *relative* importance of the decision field under study, compared to other decision fields. For this reason we first derive a cash-flow oriented lot sizing model that is later approximated by a cost-oriented model. The setup and holding cost parameters of this second model are based on explicit assumptions about cash flows.

The cash-flow oriented lot sizing model is based on the well-known multi-level, multi-item capacitated lot sizing problem (MLCLSP) with multiple resources [3]. The following assumptions reflect the way how lot sizing could be embedded into the MRP II context:

- An item is the result of one particular operation that requires one resource.
- Each item may have several predecessors and successors, i.e. the product structure is of the general type.
- The capacities of the different resources are given.
- Processing (operation) times for each item are given.
- A setup occurs whenever a positive quantity of an item is produced during a period.
- Setup times are sequence-independent.
- The time scale consists of a finite number of periods.
- The demand for each end item and period is given and must be met.
- There is a lead time of one period between successive production stages.
- There is no initial or ending inventory at any production stage.

- A setup in a period may lead to direct expenses in this period.
- Processing an item in a period may lead to direct expenses in this period.

A resource may be a single machine or a group of machines, and an operation may consist of one or several steps as long as they can be related to a resource. The given demand for each end item is derived from the master production schedule. The sequence of the lots within periods is not taken into account. For this reason it is necessary to have a lead time of at least one period between successive production stages. Otherwise it may not be possible to disaggregate the plan into a feasible schedule for each resource and period. An alternative approach is to solve the lot sizing and the scheduling problem simultaneously [8, 17, 23]. However, this approach leads to large and detailed models which are difficult to solve and to plans which are vulnerable to any change in the data [20, 21].

The direct expenses related to setups and operations do not include accounting based costs such as hourly rates for machinery and personal as these are sunk by the time the lot sizing decision is made. If some raw material is used in an operation, there will usually be some payment related to the use of this raw material. If the material is used earlier, the payment will tend to occur earlier. As an approximation, we assume that the payment occurs in the same period as the operation. There may be setups and operations to which no such direct expenses can be assigned. Setup times may also be zero.

The cash-flow oriented lot-sizing model uses the following notation:

a_{ij}	number of units of Item i required to produce one unit of Item j
b_{mt}	available capacity of Resource m in Period t
c	interest rate on a perfect capital market
B	big number

d_{it} external demand of Item i in Period t
I number of items i
K_m set of items that require Resource m
M number of resources m
P_i set of predecessors of Item i (immediate or other)
ps_i direct expense (payment) related to a setup for Item i
po_i direct expense (payment) related to an operation for Item i
Q_{it} production quantity (lot size) of Item i in Period t
S_i set of immediate successors of Item i
T number of periods t
to_i operation (processing) time per unit of Item i
ts_i setup time of Item i
X_{it} binary setup variable of Item i in Period t
Y_{it} planned end-of-period inventory of Item i in Period t

Given the above notation, the mathematical programming formulation is as follows:

$$\text{Minimize } Z = \sum_{t=1}^{T}\sum_{i=1}^{I}(ps_i X_{it} + po_i Q_{it})\frac{1}{(1+c)^t} \tag{1}$$

subject to

$$Y_{i,t-1} + Q_{it} - \sum_{\substack{j \in S_i \\ t+1 \leq T}} a_{ij} Q_{j,t+1} - Y_{it} = d_{it}, \qquad \forall i,t \tag{2}$$

$$\sum_{i \in K_m}(ts_i X_{it} + to_i Q_{it}) \leq b_{mt}, \quad \forall m,t \tag{3}$$

$$Q_{it} - B \cdot X_{it} \leq 0, \quad \forall i,t \tag{4}$$

$$Q_{it} \geq 0, \quad \forall i,t \tag{5}$$

$$Q_{i,1} = 0, \quad \forall i \text{ with } P_i \neq \{\} \tag{6}$$

$$Y_{it} \geq 0, \quad \forall i,t \tag{7}$$

$$X_{it} \in \{0,1\}, \quad \forall i,t \tag{8}$$

The objective (1) is to minimize the NPV of payments due to setups and operations. The constraints (6) prohibit the production of any item in the first period if it requires any preceding item, i.e. that has a non-empty set of predecessors P_i. These equations reflect the assumption that there is no initial inventory of any item $j \in P_i$ that could be used to produce Item i in Period 1. Equations (2) connect adjacent periods and production stages. A quantity Q_{it} of Item i produced in Period t must not enter Item j before Period $t+1$, i.e. after the lead time of one period. The capacity constraints are modeled in restriction (3). Equation (4) enforces a setup for Item i whenever a positive quantity is produced in Period t. Production quantities and inventory levels are non-negative (5),(7) and setup variables are binary (8).

The objective function (1) is formulated in terms of setups X_{it} and production quantities Q_{it}. Objective functions of lot sizing models such as the MLCLSP, however, are usually formulated in terms of setups and end-of-period inventory levels Y_{it}. Production quantities Q_{it} are tied to inventory levels Y_{it} through the balance equation (2). It is possible to express all production quantities Q_{it} in (1) in terms of inventory levels Y_{it}. Production quantities can be successively eliminated from the objective function, starting with the first production stage, and working all the way to the last production stage [20, p. 170-177].

The following additional notation is required to express all production quantities in terms of inventory levels:

A_{ki}	set of all possible paths in the product structure along which Item k enters Item i
p	path in the product structure from one to another item
g_{kip}	number of units of Item k required to produce one unit of Item i that enter Item i along Path p
w_{kip}	number of production stages between Items k and i along Path p

To illustrate the notation, we use the product structure in Figure 1. The production coefficients a_{ij} are given next to the arcs depicting input-output relations between the items.

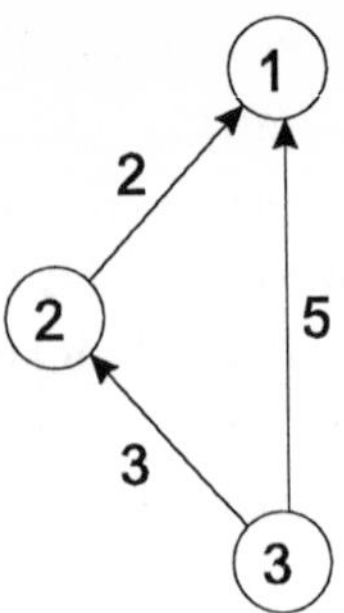

Figure 1: Example of a Product Structure

There is only one path along which Item 2 enters Item 1:

$$A_{2,1} = \{[2 \to 1]\} \tag{9}$$

Along this path $[2 \to 1]$, we have

$$g_{2,1,[2\to 1]} = 2 \tag{10}$$

units of Item 2 which enter each unit of Item 1 with a delay of

$$w_{2,1,[2\to 1]} = 1 \tag{11}$$

periods as the items are on succeeding production stages.

Item 3, however, enters Item 1 along two paths:

$$A_{3,1} = \{[3 \to 2 \to 1], [3 \to 1]\} \tag{12}$$

Six units of Item 3 are required with a delay of 2 periods along Path $[3 \to 2 \to 1]$, i.e.

$$g_{3,1,[3\to 2\to 1]} = 3 \cdot 2 = 6$$
$$w_{3,1,[3\to 2\to 1]} = 1 + 1 = 2$$

and 5 units are needed with a delay of 1 period along Path $[3 \to 1]$, i.e.

$$g_{3,1,[3\to 1]} = 5$$
$$w_{3,1,[3\to 1]} = 1.$$

This takes the different paths and time lags into account for each unit of Item 3 which enters Item 1.

Using this notation, substitution of production quantities Q_{it} by inventory levels Y_{it} leads to the following equivalent objective function:

$$
\begin{aligned}
Z &= \sum_{i=1}^{I}\sum_{t=1}^{T} ps_i \frac{1}{(1+c)^t} X_{it} \\
&+ \sum_{i=1}^{I}\sum_{t=1}^{T} \left[po_i + \sum_{k\in P_i}\sum_{p\in A_{ki}} po_k\, g_{kip}\,(1+c)^{w_{kip}} \right] \frac{1}{(1+c)^t} d_{it} \\
&+ \sum_{i=1}^{I}\sum_{t=1}^{T} \left[po_i + \sum_{k\in P_i}\sum_{p\in A_{ki}} po_k\, g_{kip}\,(1+c)^{w_{kip}} \right] \frac{\left(1-\frac{1}{1+c}\right)}{(1+c)^t} Y_{it}
\end{aligned}
\tag{13}
$$

The first term on the right hand side of (13) is the net present value of the payments due to setups. The coefficient of the binary setup variable X_{it} can be interpreted as a time-variant setup cost parameter s_{it} with

$$
s_{it} = ps_i \frac{1}{(1+c)^t} \tag{14}
$$

which reflects the current value of the cash flow that is due to a setup for Item i in Period t. The coefficient

$$
\left[po_i + \sum_{k\in P_i}\sum_{p\in A_{ki}} po_k\, g_{kip}\,(1+c)^{w_{kip}} \right] \frac{1}{(1+c)^t} \tag{15}
$$

of the demand parameter d_{it} is the NPV of time-phased payments for operations that are due to the *given* demand if everything is produced as late as possible in the context of the model, i.e. if there is no planned end-of-period inventory.

If one unit of Item i is produced at time t due to the demand d_{it}, this might lead to a direct expense po_i in Period t. The NPV of this expense is $po_i(1+c)^{-t}$.

In addition, there might be some expense po_k for each of the g_{kip} units of the preceding item k that enters Item i along Path

p in the product structure. Since we assume a lead time of one time period between any two adjacent production stages, these payments occur w_{kip} periods earlier than the direct payments for Item i. This requires the factor $(1+c)^{w_{kip}}$ to determine the NPV of the payments for these items k. Since the demand d_{it} is a given parameter, this second term on the right hand side of (13) is a constant with respect to lot sizing decisions. Thus, it can be omitted from the objective function.

The third term on the right hand side of (13) is the NPV of the cash flow that is due to inventory holding decisions. The coefficient of the inventory level variables Y_{it} in the third term can be interpreted as a time-variant holding cost parameter h_{it} with

$$h_{it} = \left[po_i + \sum_{k \in P_i} \sum_{p \in A_{ki}} po_k \, g_{kip} \, (1+c)^{w_{kip}} \right] \frac{\left(1 - \frac{1}{1+c}\right)}{(1+c)^t} \tag{16}$$

Holding a unit of end-of-period inventory of Item i requires that this unit is produced one period earlier, compared to the zero-inventory situation. The expenses for Item i and all its predecessors k occur one period earlier. Producing one period earlier leads to the factor $\left(1 - (1+c)^{-1}\right)$ by which the coefficients of d_{it} in (15) and X_{it} in (16) differ.

Setup and holding cost parameters s_{it} and h_{it} as given in (14) and (16) are time-variant as the NPV of a payment depends on when it occurs. Many solution procedures for lot sizing models are designed for time-invariant parameters. Time variant parameters can be approximated by time-invariant parameters by omitting the factor $\frac{1}{(1+c)^t}$. This approximation gets worse as the interest rate c and the length of the planning horizon increase.

However, in the MRP II context, lot sizes are determined in a rolling fashion and only the results for the first few periods of each instance are actually implemented. Approximating time-variant setup and holding cost parameters s_{it} and h_{it} by time-invariant parameters s_i and h_i with

$$s_i = ps_i \tag{17}$$

$$h_i = \left[po_i + \sum_{k \in P_i} \sum_{p \in A_{ki}} po_k \, g_{kip} \, (1+c)^{w_{kip}} \right] \left(1 - \frac{1}{1+c}\right) \quad (18)$$

in a cost-oriented objective function with

$$Z = \sum_{t=1}^{T} \sum_{i=1}^{I} (s_i X_{it} + h_i Y_{it}) \quad (19)$$

may therefore lead to very similar results unless production plans differ strongly during the first periods and these differences have a major cash-flow impact. Given the relatively short planning horizon, this does not appear to be very likely.

Objective function (19) and restrictions (6) to (8) constitute the Multi-Level Capacitated Lot Sizing Problem (MLCLSP) with setup times. If setup times are non-zero, the feasibility problem ("Does a solution exist?") is already NP-complete [28]. For this reason, several researchers have developed heuristic solution procedures. Surveys are given by [2, 3, 5, 9, 24, 28, 31, 36].

In [20, 39, 35] the shortest-route formulation [1, 10] of the dynamic lot sizing problem is extended to the MLCLSP. This shortest-route formulation can be used to calculate lower bounds on the MLCLSP via an LP-relaxation using some commercially available MIP solver for problems that do not have too many binary variables. It can also be used to solve very small problems exactly. For larger problems, however, only heuristic solution procedures have been developed yet.

3 Solution Procedures

Several heuristic solution procedures for the MLCLSP have been developed during the last years. One procedure is a greedy heuristic that extends the Dixon and Silver [7] approach to multi-level problems with setup times. Others use stochastic local search techniques such as tabu search and evolution strategies. A third approach developed by Tempelmeier and Derstroff [6, 36, 38] is based on Lagrangean techniques. The following outlines the main features of most of these heuristics.

3.1 A Decomposition Approach Based on the Dixon and Silver Heuristic

One way to solve the MLCLSP is to decompose it into a series of single-level capacitated lot sizing problems (CLSPs) subject to a side constraint. Starting at the end item level, these CLSPs are solved one after the other, resulting in dependent demand rates at preceding production stages. The problem with this decomposition approach is that each lot sizing decision must not only be feasible with respect to the current resource, but also with respect to those resources at preceding production stages that will be considered later in the planning process.

The Dixon and Silver heuristic [7] is a single-level forward-planning procedure for the single-level CLSP that adds future demands to current production quantities as long as average costs per period decrease and/or future infeasibilities are reduced. Multiple production stages and setup times are not considered in the original paper by Dixon and Silver. During the determination of production quantities for Period t, the feasibility of a partial solution with respect to the remaining Periods $t+1$ to T it is checked via the following cumulative capacity constraint. It says that the cumulated required capacity in Periods $t+1$ to T must not exceed the cumulated available capacity, i.e.

$$\sum_{\tau=t+1}^{T^*} \sum_{i \in K_m} to_i \, d_{i\tau} \leq \sum_{\tau=t+1}^{T^*} b_{m\tau}, \qquad \forall m,\ T^* = t+1, \ldots, T. \qquad (20)$$

It is not possible to use this criterion in the presence of setup times as in this case the single-level feasibility problem is already NP-complete [27, 28]. Quite often, however it is possible to use a multi-level, finite capacity, backward loading procedure to determine a *tentative schedule* that meets capacity constraints over all production stages. This tentative production schedule is subsequently improved in a level-by-level manner similar to an MRP run. This approach has been proposed in [39] for multi-level problems without setup times and extended in [20] for problems with

Step 1:
Assign a low level code $L(i)$ to each item i. Determine cost parameters for each item i. Use a backward loading procedure to determine a tentative feasible production schedule Q^F that meets capacity constraints at all production stages.

Step 2:
Consider sequentially all levels $l = 0, 1, ...L$. Consider all resources M_l required at Level l. Determine the set of items I_{ml} at Level l that use Resource m. Determine the remaining capacity for the current resource. Use a modified Dixon-Silver heuristic to solve a single-level CLSP for the current resource in order to improve the tentative feasible production schedule Q^F.

Figure 2: Outline of the Decomposition Approach for the ML-CLSP

setup times. The basic structure of the decomposition approach is outlined in Figure 2. Some details are discussed below:

Assigning low level codes: Low level codes can be assigned based on

- the bill-of-material ('gozinto') structure (end items = level 0, immediate predecessor items = level 1, etc.)
- the precedence relations between resources in the production process (last resource = level 0, second to last resource = level 1, etc.)

Items on different levels of the bill-of-material structure that use the same resource will be treated in separate single level CLSPs in Step 2 of the procedure if low-level codes are assigned based on the bill-of-material structure. However, assigning low level codes based on the structure of resources tends to reduce the number of single level CLSPs as items on different levels of the bill-of-material structure are treated within one single-level CLSP if possible. This may improve the solution quality (see [20, 36, 39] for details).

Assigning cost parameters: In each of the single-level CLSPs in Step 2, either

- the original cost parameters s_i and h_i from the MLCLSP or
- modified cost parameters

can be used. The modification approximates the cost impact of lot sizing decisions at the current level l on preceding production stages [18, 19, 36]. Using modified cost parameters may improve the solution quality.

Feasibility check: In Step 1, a backward scheduling procedure is used to heuristically determine a feasible schedule for all items and periods. Each possible lot sizing decision considered in Step 2 gives rise to a new feasiblity problem with respect to the current and preceding production stages. A decision is considered to be feasible if the backward scheduling procedure finds another feasible schedule for all items and periods.

Starting at level 0 (in Step 1) or the current level l (in Step 2), the independent or dependent demand for the yet unscheduled items is determined. Starting with the last period T, all production quantities are scheduled as late as possible within the given limits of the remaining capacity. It may not be possible to satisfy the demand for an item in Period t by production in Period t if capacity limits are too tight. In this case, the necessary quantities are shifted to earlier Periods $\tau < t$.

This effort is not successful if for some resource the capacity limit for the first period is exceeded. In this case, the problem as a whole (in Step 1) or the possible decision (in Step 2) is considered to be infeasible.

Due to the heurstic nature of the feasibility check, this does not mean that a feasible solution does not exist. However, in the presence of setup times the single level feasibility problem is already NP-complete and fast exact procedures for the feasibility problem are not available.

Shifting production quantities to earlier periods in order to generate a tentative feasible schedule has an impact on the amount of capacity required for setups. In the backward scheduling procedure, reducing an overload $O_{mt} > 0$ defined as

$$O_{mt} = \sum_{i \in K_m} (ts_i X_{it} + to_i Q_{it}) - b_{mt} \tag{21}$$

of Resource m in Period t by shifting production quantities to Period $t-1$ should create as little overloading in Period $t-1$ as possible. In order to achieve this goal, the procedure outlined in Figure 3 can be used.

For all periods $t = T, \ldots, 2$
  Determine all items I_{ml} for the current resource and level
  Determine demand d_{it} and set $Q_{it} = d_{it}, \forall i \in I_{ml}$
  Determine overload O_{mt} of Resource m in Period t
  If $O_{mt} > 0$ then
    Phase I: Reduce setups in t and $t-1$ and adjust O_{mt}
  If $O_{mt} > 0$ then
    Phase II: Shift setups from t to $t-1$ and adjust O_{mt}
  If $O_{mt} > 0$ then
    Phase III: Shift one item i from t to $t-1$ and adjust O_{mt}
  If $O_{mt} < 0$ then
    Phase IV: Try to undo the steps from Phases I to III

Figure 3: Outline of the Feasibility Check with Setup Times

The example in Figures 4 and 5 illustrates the four phases of the feasibility check in Figure 3. In the example, four items have to be scheduled backwards over two periods in order to assess the feasibility of the initial situation. The bright parts of the bars

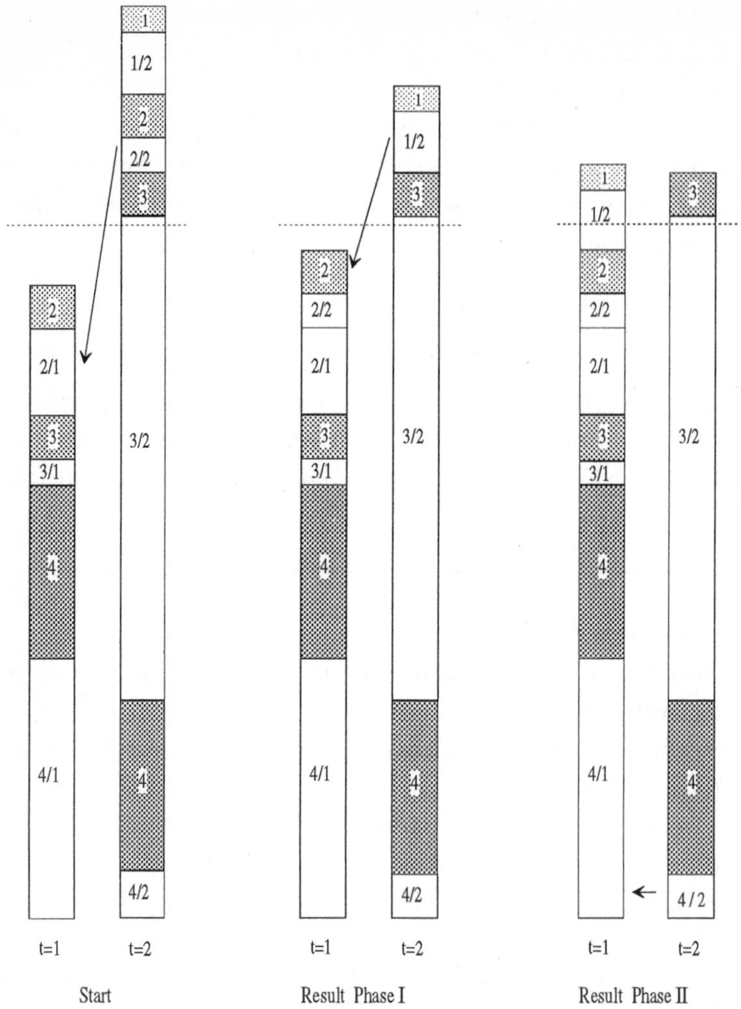

Figure 4: Capacity Check with Setup Times. Part 1

depict processing times for the scheduled lots i/t of Item i in Period t and the dark parts depict setup times of Item i.

In the initial situation depicted in the leftmost part of Figure 4, the currently scheduled setups and operations exceed the capacity limit in Period 2 while there is excess capacity in Period 1.

During *Phase I*, an attempt is made to reduce the number of setups for those items that are currently scheduled in Period $t = 2$

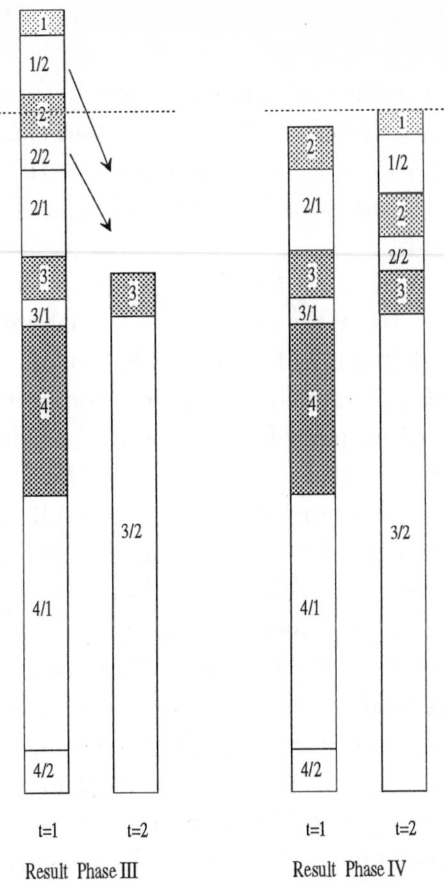

Figure 5: Capacity Check with Setup Times. Part 2

and $t-1=1$ that meet the condition

$$ts_i + to_i Q_{it} \leq O_{mt} \tag{22}$$

Shifting the complete production quantity from Period t to $t-1$ cannot lead to excess capacity in Period t. Items are considered according to ascending numbers. In Figure 4, Item 1 is not yet scheduled in Period 1. It is therefore not possible to reduce the number of setup for Item 1 in Periods 1 and 2 by shift lot Q_{12} to Period 1. However, the whole production quantity of Item 2 can be shifted to Period 1, thereby eliminating one setup for Item 2. After this step, it is not possible to eliminate another setup in

Period 2 without creating excess capacity, i.e. $O_{mt} < 0$ in Period $t = 2$.

In *Phase II*, production quantities and setups for those items that are not yet scheduled in Period $t-1$ and that meet condition (22) are shifted to Period $t-1$. In Figure 4, this applies to Item 1 which can be shifted from Period 2 to 1 without creating excess capacity in Period 2.

After this step, it is no longer possible to shift complete production quantities from Period t to $t-1$ without creating excess capacity $O_{mt} < 0$ in Period t. For this reason it is sufficient to shift in *Phase III* a production quantity (complete or incomplete) of *one* additional item from Period t to $t-1$ in order to eliminate the remaining overload $O_{mt} > 0, t \geq 2$. The procedure chooses the item and quantity that leads to the smallest increase of required capacity in Period $t-1$. In Figure 4, shifting the complete production quantity of Item 4 from Period 2 to 1 leads to the smallest increase of the overload O_{m1} in Period 1. The result of this step is depicted on the left side of Figure 5.

However, this leads to excess capacity in Period t if an additional setup is eliminated in Period t. It may hence be possible to *undo* some of the steps from Phases I to II in *Phase IV*. This is the case in the example in Figure 5. Due to the large setup time of Item 4, it is possible to undo both the steps from Phase I and II. The final schedule is depicted in the rightmost part of Figure 5. Note that the capacity limit in the first period is not exceeded. For this reason, the initial situation in the leftmost part of Figure 4 is considered to be *feasible*.

Solving the Single-Level CLSPs: In Step 2 of the decomposition approach, a modified version of the heuristic by Dixon and Silver [7] is used to solve the single-level CLSPs. The modifications are due to the setup times that were not considered in the original paper by Dixon and Silver. The procedure is outlined in Figure 6.

In *Phase I* of the procedure in Figure 6, complete future production quantities $Q_{i\tau}$ are shifted from Period τ to Period t in

Set $Q_{it} = d_{it}, i \in I_{ml}, \forall t$
For all periods $t = 1, \ldots, T-1$
Phase I: Shift complete future quantities $Q_{i\tau}$ to Period t, $t < \tau$
Phase II: Ensure feasibility for Periods $t+1$ to T.

Figure 6: Outline of the Dixon and Silver Heuristic

order to reduce costs or to reduce the overload of Periods $t+1$ to T.

Items are selected according to the Silver-Meal criterion which is based on the average cost per period $AC_{it\tau}$ of a lot $Q_{it\tau}$ with

$$AC_{it\tau} = \frac{s_i + \sum_{\theta=t+1}^{\tau} h_i\,(\theta - t) d_{i\theta}}{\tau - t + 1} \tag{23}$$

that includes the demand until Period τ. If the current production quantity Q_{it} is increased by $Q_{i,\tau+1}$, this may lead to a decrease $AC_{it\tau} - AC_{it,\tau+1}$ of the average cost per period. On the other hand, it does lead to an increase $CP_{it,\tau+1}$ of the required capacity in Period t. The increase

$$CP_{it,\tau+1} = \begin{cases} to_i\, Q_{i,\tau+1} + ts_i & \text{if } Q_{it} = 0 \\ to_i\, Q_{i,\tau+1} & \text{if } Q_{it} > 0 \end{cases} \tag{24}$$

depends on whether a production quantity for Item i has already been scheduled in Period t. The priority index $\rho_{it\tau}$ with

$$\rho_{it\tau} = \frac{AC_{it\tau} - AC_{it,\tau+1}}{CP_{it,\tau+1}} \tag{25}$$

relates the cost decrease to the additional required capacity in Period t. In *Phase I* of the modified Dixon and Silver heuristic, items are sorted according to descending values of $\rho_{it\tau}$ and complete future production quantities are shifted from Period $\tau+1$ to t as long as this reduces costs or an overload $O_{mt} > 0$ in Periods

$t+1$ to T. Each of the considered shifts gives rise to a feasibility check and the one that is eventually performed leads to an updated tentative feasible schedule Q^F.

In *Phase II*, we have to ensure that no overload with respect to Periods $\tau+1$ to T remains. In this phase, feasibility can always be achieved in one of the following three ways:

- Shifting an additional complete production quantity $Q_{i,t+1}$ to Period t
- Shifting a fraction of a future production quantity $Q_{i,t+1}$ to Period t such that the overload of Periods $t+1$ to T is eliminated.
- Implementing the tentative schedule Q^F in Period t.

The choice between the first two options is based on the absolute cost increase in Period t. Each of these possible steps has to pass the feasibility check. It is possible that none of the two options leads to a feasible solution. This is due to the heuristic feasibility check. In this situation the production quantities for Period t from the tentative schedule Q^F are implemented. Now the remaining problem for Periods $t+1$ to T is solvable and the next Period $t+1$ can be considered. In what follows this decomposition approach will be named DA.

3.2 Local Search Heuristics

Local search procedures for the MLCLSP operate on one or a group of setup patterns simultaneously in an iterative manner. Current setup patterns are first slightly modified and then transformed into new production schedules. The cost associated with a new production schedule will include some penalty if the schedule is not feasible. The penalty should reflect the degree of infeasibility. The new schedules replace the current solution if they are better with respect to the corresponding objective function value. This iterative procedure will reach some (possibly local) optimum. Techniques like simulated annealing [40], tabu search [13, 14, 15], genetic algorithms [16] and evolution strategies [30, 33] have been proposed to leave local optima during a local search and they have been applied to the MLCLSP in [20, 21]. In what follows, the presentation is limited to tabu search and evolution strategy algorithms. Both simulated annealing and genetic algorithms were too slow to solve larger problems with several hundred binary setup variables. Even the faster tabu search and evolution strategy algorithms were not able to solve the largest problem class with 80 items and 16 periods.

3.2.1 Tabu Search

Salomon [32] has proposed tabu search techniques for multi-level lot sizing problems with one capacity constraint (see also [25]). In [20] his procedure has been extended to multiple capacity constraints and setup times. The tabu search heuristic named TS uses a tabu list of fixed size. During each iteration only a limited number of stochastically selected neighborhood solutions is generated. A neighborhood solution differs from the current solution in the value of exactly one setup variable. The TS heuristic starts from a setup pattern with all setup variables set to one. An alternative is to start with the setup pattern of a (hopefully good) solution determined by the decomposition approach DA. If the tabu search heuristic is initialized with the setup pattern derived from the decomposition approach it is named TS-DA.

3.2.2 Evolution Strategies

Rechenberg [30] has proposed an approach that operates on a population of m solutions at the same time. From these μ (old) solutions, λ new solutions are derived. They differ from old setup patterns in one or more setup variables. The result is a temporary population of $(\mu+\lambda)$ solutions. In a $(\mu+\lambda)$ evolution strategy, the best μ solutions are selected [33]. Thus, old solutions can usually be replaced only by better solutions. However, if the algorithm finds a possibly local optimum, inferior solutions that meet some aspiration level are also accepted in order to leave the (local?)

optimum and continue the search. In [20] an evolution strategy with $\mu = 1$ is described. It is named ES when initialized by the setup pattern and ES-DA when initialized by the setup pattern derived from the decomposition approach DA.

3.3 A Lagrangean Based Procedure

Tempelmeier and Derstroff [6, 37, 38] describe a solution procedure for the MLCLSP that is based on Lagrangean relaxation of both the inventory balance equations (2) and the capacity constraints (3). In the first step, the inventory quantities Y_{it} are expressed in terms of production quantities Q_{it}. Assuming, for the purpose of explanation, that all lead times are zero, this yields:

$$Y_{it} = \sum_{\tau=1}^{t} Q_{i\tau} - \sum_{\tau=1}^{t} \left(d_{i\tau} + \sum_{j \in N_i} a_{ij}\, Q_{j\tau} \right), \forall i, t \tag{26}$$

Substituting the variables Y_{it} in the objective function results in

$$Z = \sum_{i=1}^{I} \sum_{t=1}^{T} \left(e_i \, (T - t + 1)\, Q_{it} + s_i \, X_{it} \right) - F \tag{27}$$

where the e_i denote echelon holding costs and F is a constant term that does not depend on the decision variables. In a second step they add both the non-negativity constraints on physical inventory ($Y_{it} \geq 0$) and the capacity constraints (3) using Lagrangean multipliers v_{it} and u_{mt}, respectively. This yields the lower bound:

$$\begin{aligned} LB \;=\; & \sum_{i=1}^{I} \sum_{t=1}^{T} \left(e_i \, (T - t + 1)\, Q_{it} + s_i \, X_{it} \right) \\ - \; & \sum_{i=1}^{I} \sum_{t=1}^{T} v_{it} \sum_{\tau=1}^{t} \left(Q_{i\tau} + \sum_{j \in N_i} a_{ij}\, Q_{j\tau} - d_{i\tau} \right) \\ + \; & \sum_{m=1}^{M} \sum_{t=1}^{T} u_{mt} \left(\sum_{i \in K_m} \left(to_i \, Q_{it} + ts_i \, X_{it} \right) - b_{mt} \right) - F \end{aligned} \tag{28}$$

Step 0:
Initialize the Lagrangean multipliers v_{it} and u_{mt}.

Step 1:
Update the lower bound by solving I uncapacitated lot sizing problems of the Wagner-Whitin type using the current set of Lagrangean multipliers.

Step 2:
Update the new set of Lagrangean multipliers reflecting the current workload and backorders from Step 1.

Step 3:
Update the new upper bound based on the current set of Lagrangean multipliers by eliminating backorders and work overload using a finite capacity scheduling heuristic. Unless a stopping criterion is met, go to Step 1.

Figure 7: Outline of the Lagrangean Based Heuristic by Tempelmeier and Derstroff

After some reformulation, the relaxed MLCLSP can be separated into I uncapacitated, single level dynamic lot sizing problems of the Wagner-Whitin type. They can be solved efficiently using the algorithms proposed by Federgruen and Tzur [11] and Wagelmans et al. [42]. The cost coefficients of these single-level problems are time-dependent as they include time-dependent Lagrangean multipliers. Subgradient optimization is used to update the Lagrangean multipliers. Solving the I uncapacitated, single level lot sizing problems of the Wagner-Whitin type may result in workloads and backorders that are infeasible with respect to the original MLCLSP formulation. Therefore, an heuristic upper bound is computed that meets both inventory balance and capacity constraints. The heuristic upper bounding procedure uses the cost parameters determined by the subgradient optimization of the Lagrangean multipliers. The basic structure of the Tempelmeier and Derstroff heuristic is outlined in Figure 7.

The Tempelmeier and Derstroff heuristic yields an upper bound on the relative deviation from the optimum ((Upper Bound-Lower Bound)/Lower Bound). Unfortunately, the lower bound of the solution may be rather weak. A useful feature of this heuristic is the trade-off between computation time and solution quality as the quality of the feasible upper bounds tends to get better from iteration to iteration. It should be noted that Tempelmeier and Derstroff's procedure can be extended to cover overtime whereas this does not seem to be a straightforward extension of the decomposition approach DA. Details of the procedure are given in [6, 36, 38].

4 Numerical Results

The above described solution procedures have been tested against two classes of problems in a computational study [20, 21]. The problem classes differ with respect to the problems size, i.e. the number of products I and periods T. The parameter sets used for the local search procedures (number of iterations, length of the tabu list etc.) are reported in [20]. All computations were performed on a 486/50 PC. Derstroff's procedure TD and the decomposition approach DA were programmed in Microsoft FORTRAN 5.0, the local search procedures were coded in Turbo Pascal 6.0.

4.1 Problem Class I ($I = 40$, $T = 16$)

Problem class I consists of 18 different problems. These problems do not include setup times. They cover two setup cost profiles, three capacity profiles, and three demand time series (see Appendix A.1). The product structure is depicted in Figure 8. Table 1 lists the resources and the corresponding items/operations.

It has not been possible to compute optimal solutions through a MIP-solver even though the tight shortest-route based formulation of the MLCLSP [10, 20, 39] was used. It is, however, possible to compute lower bounds (LB) using either a Lagrangean relax-

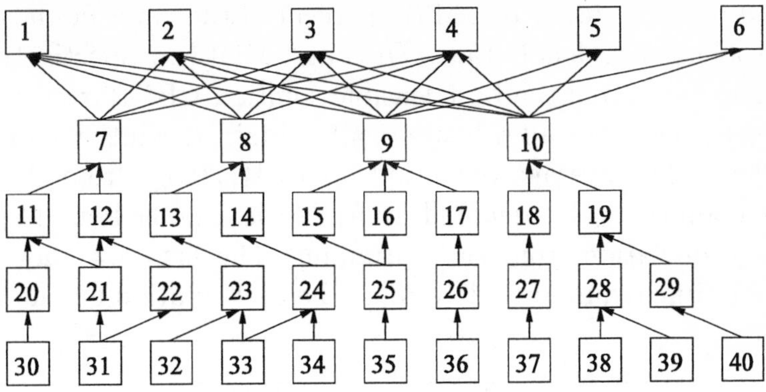

Figure 8: Product Structure of Problem Class I

Resource	Items	Resource	Items
A	1 ... 3, 5	D	11 ... 15, 25 ... 29
B	4, 6	E	16 ... 24
C	7 ... 10	F	30 ... 40

Table 1: Resources and Corresponding Items of Problem Class I

ation of the MLCLSP or the LP-relaxation of the shortest route formulation [20, 39]. Lower bounds for Problem Class I were determined via the LP-relaxation of the shortest-route based formulation of the MLCLSP. For Problem Class II, the Lagrangean relaxation together with a subgradient optimization has been used to compute the Lagrangean dual as the LB.

We tried to solve the resulting 18 problems by each of the heuristics DA, TS, TS-DA, ES, ES-DA and TD. Procedures ES-DA and TS-DA did not solve all the problems with the highest capacity utilization (90 %). Therefore, the 90 % utilization profile was omitted for these procedures in Problem Class I. The results are summarized in Table 2.

Each of the solutions generated by one of the heuristics is an upper bound on the objective function value of the underlying MLCLSP. The average and the maximum deviation of the upper bound from the lower bound (in percent of the lower bound) over 18 problems are denoted as ADLB and MDLB, respectively. We did also determine the best feasible solution UB* (lowest known upper bound) over all 6 heuristically obtained solutions for each problem. The average deviation (UB-UB*)/UB* from the best known solution UB* denoted as ADUB is a lower bound on the average deviation from the optimum. The last row shows the average time required to determine the best solution for each procedure.

	ADLB [%]	MDLB [%]	ADUB [%]	Time [s]
DA	13.92	31.82	2.73	42
TS	14.94	34.02	3.69	547
TS-DA	6.16	13.13	0.68	165
ES	16.48	42.17	4.90	449
ES-DA	5.67	12.59	0.23	158
TD	14.29	37.72	3.01	19

Table 2: Results for Problem Class I

The results for Problem Class I show a slightly superior solution quality of the DA and TD heuristics compared to TS and ES. The computational requirements of the DA and TD heuristics are rather modest. The local search procedures either require long computation time (TS and ES) or are not able to solve all the problems (TS-DA and ES-DA). In some cases they may, however, yield high quality solutions at the expense of considerable additional computational effort when starting with a "good" initial setup pattern.

4.2 Problem Class II ($I = 80$, $T = 16$)

Problem class II consists of 75 different problems. It covers five setup cost profiles, five capacity/setup time profiles and three demand time series (see Appendix A.2). The product structure is depicted in Figure 9. Table 3 lists the resources and the corresponding items/operations. Due to computational requirements, the local search procedures TS, TS-DA, ES, and ES-DA have not been applicable in this problem class. Hence, Derstroff's Lagrangean procedure competed against the decomposition approach only.

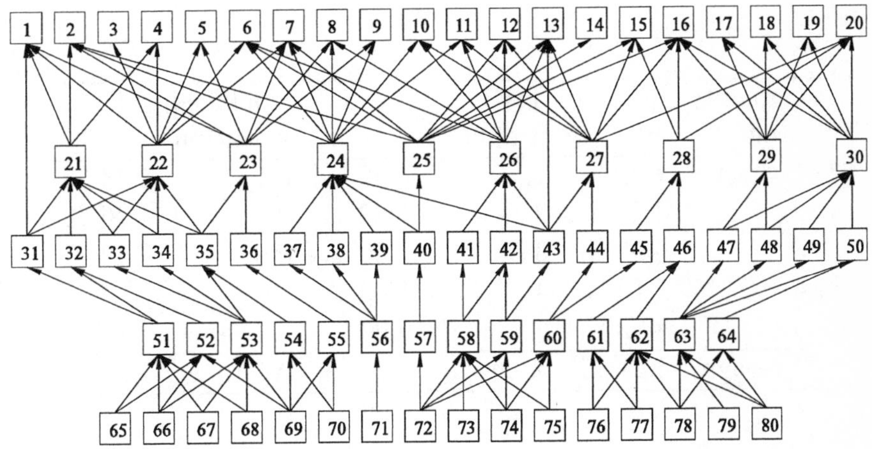

Figure 9: Product Structure of Problem Class II

Resource	Items
A	1 ... 5, 24 ... 28
B	6 ... 12, 21 ... 23, 29, 30
C	13 ... 20, 31 ... 33
D	34 ... 37, 51 ... 53
E	38 ... 43
F	44 ... 48, 60
G	49, 50, 54 ... 59, 61 ... 64
H	65 ... 80

Table 3: Resources and Corresponding Items of Problem Class II

	ADLB [%]	MDLB [%]	ADUB [%]	Time [s]
DA	38.73	111.88	4.54	415
TD	33.42	99.09	0.36	81

Table 4: Results for Problem Class II

Capacity profile		TBO profile		Coefficient of variation	
1	1.22 %	1	4.46 %	0.1	2.44 %
2	3.65 %	2	5.82 %	0.5	4.76 %
3	6.95 %	3	3.98 %	0.9	5.42 %
4	5.10 %	4	5.82 %		
5	4.12 %	5	0.96 %		
Average	4.21 %	Average	4.21 %	Average	4.21 %

Table 5: Average Improvement of the TD Solution over the DA Solution for Problem Class II

The results for Problem Class II are presented in Table 4. This problem class demonstrates again the superior solution quality of the TD heuristic. Table 5 presents the average improvement of the TD solutions with respect to the DA solutions in percent of the DA solution. The results suggest that one can expect a strong improvement whenever the capacity constraints are not too tight. Tight capacity constraints may prevent batching even though this might seem "economically worthwhile". The deviation per TBO profile (see Appendices A.1 and A.2) shows that Derstroff's procedure will yield superior solutions unless high setup costs arise at late production stages only (TBO profile 5). Late production stages (i.e. end products) are considered first by the DA heuristic. Low setup costs on early production stages that are considered later will not give rise to any additional batching of production quantities in this case. The last column suggests that the TD heuristic is especially strong whenever there is a high variability in the demand time series, i.e. when the coefficient of variation of the demand time series is high. The average computation times reported in Table 4 should not be overrated as it is within both procedures possible to trade in solution time for solution quality.

5 Conclusions

Lot sizing in the MRP II context has to consider capacity limits of the production system in order to generate feasible production schedules. We considered the case of multiple capacity constraints, setup times and time-varying demand in a deterministic environment. In order to assess the economic importance of optimizing lot sizes in the MRP II context, we asked for the cash-flow impact of these decisions. The economic impact of lot sizing decisions is overrated if setup cost parameters based on hourly rates for machinery and holding costs parameters based on manufacturing costs are used. Starting with a cash-flow oriented model, we derived setup and holding cost parameters for the given decision field that are based on explicitly stated assumptions about how payments are related to decision variables. The well-known cost-oriented MLCLSP approximates the cash-flow oriented model. In the second part of the paper, several heuristic approaches for the MLCLSP under multiple capacity constraints were outlined and tested in a numerical study. A Lagrangean procedure developed by Tempelmeier and Derstroff shows a superior performance for problems with up to 80 items, 5 productions stages, 8 restricted resources, and 16 periods. It is therefore possible to solve the multi-level dynamic lot sizing problem subject to multiple capacity constraints. An open problem is the lead time that is induced by the separation of lot sizing and scheduling in this approach.

Acknowledgement

I thank Christian Hofmann and an anonymous referee for their helpful comments.

References

[1] AFENTAKIS, P., GAVISH, B., AND KARMARKAR, U. S. (1984), Computationally efficient optimal solutions to the lot-sizing problem in multi-stage assembly systems, *Management Science*, Vol. 30, pp. 222–239

[2] BAHL, H. C., RITZMAN, L. P., AND GUPTA, J. N. D. (1987), Determining lot sizes and resource requirements: A review, *Operations Research*, Vol. 35, pp. 329–345

[3] BILLINGTON, P. J., MCCLAIN, J. O., AND THOMAS, L. J. (1983), Mathematical programming approaches to capacity-constrained MRP systems: Review, formulation and problem reduction, *Management Science*, Vol. 29, pp. 1126–1141

[4] CHASE, R., AND AQUILANO, N. (1989), *Production and Operations Management: A Life Cycle Approach*, 5 ed., Homewood (Illinois), Irwin

[5] DE BODT, M. A., GELDERS, L. F., AND VAN WASSENHOVE, L. N. (1984), Lot sizing under dynamic demand conditions: A review, *Engineering Costs and Production Economics*, Vol. 8, pp. 165–187

[6] DERSTROFF, M. (1995), *Mehrstufige Losgrößenplanung mit Kapazitätsbeschränkungen*, Heidelberg, Physica

[7] DIXON, P. S., AND SILVER, E. A. (1981), A heuristic solution procedure for the multi-item, single-level, limited capacity lot-sizing problem, *Journal of Operations Management*, Vol. 2, pp. 23–39

[8] DREXL, A. AND HAASE, K. (1995), Proportional lotsizing and scheduling, *International Journal of Production Economics*, Vol. 40, pp. 73–87

[9] DREXL, A. AND KIMMS, A. (1997), Lot sizing and scheduling — Survey and extensions, *European Journal of Operational Research*, Vol. 99, pp. 221-235

[10] EPPEN, G. D., AND MARTIN, R. K. (1987), Solving multi-item capacitated lot-sizing problems using variable redefinition, *Operations Research*, Vol. 35, pp. 832–848

[11] FEDERGRUEN, A., AND TZUR, M. (1991), A simple forward algorithm to solve general dynamic lot sizing models with n periods in 0(n log n) or 0(n) time, *Management Science*, Vol. 37, pp. 909–925

[12] FLEISCHMANN, B. (1988), Operations-Research-Modelle und -Verfahren in der Produktionsplanung, *Zeitschrift für Betriebswirtschaft*, Vol. 58, pp. 347–372

[13] GLOVER, F. (1989), Tabu search - Part I, *ORSA Journal of Computing*, Vol. 1, pp. 190–206

[14] GLOVER, F. (1990), Tabu search - Part II, *ORSA Journal of Computing*, Vol. 2, pp. 4–42

[15] GLOVER, F., AND GREENBERG, H. J. (1989), New approaches for heuristic search: A bilateral linkage with artificial intelligence, *European Journal of Operational Research*, Vol. 39, pp. 119–130

[16] GOLDBERG, D. E. (1989), *Genetic Algorithms in Search, Optimiziation, and Maschine Learning*, Reading (Massachusetts), Addison-Wesley

[17] HAASE, K. (1994), *Lot-Sizing and Scheduling for Production Planning*, Berlin, Springer

[18] HEINRICH, C. E. (1987), *Mehrstufige Losgrößenplanung in hierarchisch strukturierten Produktionplanungssystemen*, Berlin, Springer

[19] HEINRICH, C. E., AND SCHNEEWEISS, C. (1986), Multi-stage lot-sizing for general production systems, In *Multi-Stage Production Planning and Inventory Control*, S. Axsäter, C. Schneeweiß, and E. Silver, Eds., Berlin, Springer, pp. 150–181

[20] HELBER, S. (1994) *Kapazitätsorientierte Losgrößenplanung in PPS-Systemen*, Stuttgart, Metzler & Poeschel

[21] HELBER, S. (1995), Lot sizing in capacitated production planning and control systems, *Operations Research Spectrum*, Vol. 17, pp. 5–18

[22] HOPP, W. J., AND SPEARMAN, M. L. (1996), *Factory Physics. Foundation of Manufacturing Management*, Chicago, Irwin

[23] KIMMS, A. (1997), *Multi-Level Lot Sizing and Scheduling — Methods for Capacitated, Dynamic, and Deterministic Models*, Heidelberg, Physica

[24] KUIK, R., SALOMON, M., AND VAN WASSENHOVE, L. (1994), Batching decisions. Structure and models, *European Journal of Operational Research*, Vol. 75, pp. 243–263

[25] KUIK, R., SALOMON, M., VAN WASSENHOVE, L. N., AND MAES, J. (1993), Liniear programming, simulated annealing and tabu search heuristics for lotsizing in bottleneck assembly systems, *IIE Transactions*, Vol. 25, pp. 62–72

[26] KÜPPER, H.-U., (1991), Multi-period production planning and managerial accounting, In *Modern Production Concepts. Theory and Applications*, G. Fandel and G. Zäpfel, Eds., Berlin, Springer, pp. 46–62

[27] MAES, J. (1987), *Capacitated Lotsizing Techniques in Manufacturing Resource Planning*, PhD thesis, Katholieke Universiteit Leuven

[28] MAES, J., MCCLAIN, J. O., AND VAN WASSENHOVE, L. N. (1991), Multi-level capacitated lotsizing complexity and LP-based heuristics, *European Journal of Operational Research*, Vol. 53, pp. 131–148

[29] ORLICKY, J. (1975), *Material Requirement Planning*, New York, McGraw Hill

[30] Rechenberg, I. (1973), *Evolutionsstrategie - Optimierung technischer Systeme nach Prinzipien der biologischen Evolution*, Stuttgart, Frommann

[31] Roll, Y., and Karni, R. (1991), Multi-item, multi-level lot sizing with an aggregate capacity constraint, *European Journal of Operational Research*, Vol. 51, pp. 73–87

[32] Salomon, M. (1991), *Deterministic Lotsizing Models for Production Planning*, Berlin, Springer

[33] Schwefel, H.-P., and Bäck, T. (1992), Künstliche Evolution - eine intelligente Problemlösungsstrategie? *Künstliche Intelligenz*, pp. 20–27

[34] Schweitzer, M., and Küpper, H.-U. (1995), *Systeme der Kosten- und Erlösrechnung*, 6. Aufl., München, Vahlen

[35] Stadtler, H. (1996), Mixed integer programming model formulation for dynamic multi-item multi-level capacitated lotsizing, *European Journal of Operational Research*, Vol. 94, pp. 561–581

[36] Tempelmeier, H. (1995), *Material-Logistik. Grundlagen der Bedarfs- und Losgrößenplanung in PPS-Systemen*, 3. Aufl., Berlin, Springer

[37] Tempelmeier, H., and Derstroff, M. (1993), Mehrstufige Mehrprodukt-Losgrößenplanung bei beschränkten Ressourcen und genereller Erzeugnisstruktur, *Operations Research Spectrum*, Vol. 15, pp. 63–73

[38] Tempelmeier, H., and Derstroff, M. (1996), A lagrangean-based heuristic for dynamic multilevel multiitem constrained lotsizing with setup times, *Management Science*, Vol. 42, pp. 738–757

[39] Tempelmeier, H., and Helber, S. (1994), A heuristic for dynamic multi-item multi-level capacitated lotsizing for general product structures, *European Journal of Operational Research*, Vol. 75, pp. 296–311

[40] VAN LAARHOVEN, P. J. M., AND AARTS, E. H. L. (1987), *Simulated Annealing: Theory and Applications*, Dordrecht

[41] VOLLMANN, T. E., BERRY, W. L., AND WHYBARK, D. C. (1997), *Manufacturing Planning and Control Systems*, 4 ed., Homewood (Illinois), Irwin

[42] WAGELMANS, A., VAN HOESEL, S., AND KOLEN, A. (1992), Economic lot sizing: An 0(n log n) algorithm that runs in linear time in the Wagner-Whitin case, *Operations Research*, Vol. 40, pp. 145–156

[43] WEMMERLÖV, U. (1982), Inventory management and control, In *Handbook of Industrial Engineering*, G. Salvendy, Ed., New York, Wiley, pp. 11.4.1–11.4.28

A Appendix

A.1 Description of Problem Class I

Problem class I consists of three time series of end product demand with a coefficient of variation (CV) of 0.1, 0.5, and 0.9, respectively. The average demand D_i of the six end products is 40, 20, 30, 60, 20, and 30. In each time series, the cumulated demand from Period 1 to t, $t \leq T$, never exceeds the cumulated average demand from Period 1 to t. The demand time series follows a normal distribution which is truncated to avoid negative demand. These time series are available on request. The echelon holding cost e_i of each item is 1. The setup cost s_i corresponding to Item i is derived from the time between orders TBO_i, echelon holding cost e_i, and average demand D_i via $s_i = 0.5\, e_i\, D_i\, TBO_i^2$ [32, p. 113] . All processing times to_i are 1.

Setup times ts_i of different items requiring one particular resource are assumed to be identical ($ts_i = ts_{(m)}, \forall m \in M, \forall i \in K_m$). Capacity utilization of Resource m is expressed as an interval $[\varrho_{1m}, \varrho_{2m}]$ where ϱ_{1m} denotes the utilization that is due to processing times alone, i.e.

$$\varrho_{1m} = \frac{\sum_{i \in K_m} \sum_{t=1}^{T} to_i \, Q_{it}}{\sum_{t=1}^{T} b_{mt}}, \qquad \forall m \tag{29}$$

whereas ϱ_{2m} denotes the utilization that would result if a setup were performed for each item during each period, i.e.

$$\varrho_{1m} = \frac{\sum_{i \in K_m} \left(T \, ts_{(m)} + \sum_{t=1}^{T} to_i \, Q_{it} \right)}{\sum_{t=1}^{T} b_{mt}}, \qquad \forall m \tag{30}$$

As there is no way to guarantee a priori that a problem with $\varrho_{2m} > 1$ is feasible, we only study problems for test purposes with a maximum ϱ_{2m} of 95%.

The TBO and the capacity utilization profiles are presented in Tables 6 and 7, respectively. Note that the problems in Problem Class I do not include setup times ($\varrho_{1m} = \varrho_{2m}, \forall m$).

TBO profile	$TBO_i = 1$	$TBO_i = 2$	$TBO_i = 4$
1	$i = 1 \ldots 6$	$i = 7 \ldots 40$	-
2	$i = 1 \ldots 7$	-	$i = 8 \ldots 20$

Table 6: TBO Profiles for Problem Class I

Capacity profile	[90%, 90%]	[70%, 70%]	[50%, 50%]
1	$m =$ A … F	-	-
2	-	$m =$ A … F	-
3	-	-	$m =$ A … F

Table 7: Capacity Profiles for Problem Class I

A.2 Description of Problem Class II

Problem class II consists of three time series of end product demand with a CV of 0.1, 0.5 and 0.9. The average demand of the twenty end products is given in Table 8. The TBO and the capacity utilization/setup time profiles are presented in Table 9 and 10, respectively.

i	D_i	i	D_i	i	D_i	i	D_i
1	30	6	30	11	10	16	17
2	10	7	20	12	30	17	35
3	15	8	25	13	35	18	40
4	20	9	15	14	8	19	22
5	5	10	20	15	12	20	11

Table 8: Average End-Product Demand in Problem Class II

TBO profile	$TBO_i = 1$	$TBO_i = 3$	$TBO_i = 5$
1	$i= 1 \ldots 80$	-	-
2	-	$i= 1 \ldots 80$	-
3	-	-	$i= 1 \ldots 80$
4	$i= 1 \ldots 25$	$i= 26 \ldots 50$	$i= 51 \ldots 80$
5	$i= 51 \ldots 80$	$i= 26 \ldots 50$	$i= 1 \ldots 25$

Table 9: TBO Profiles for Problem Class II

Capacity profile	[80%, 95%]	[65%, 75%]	[45%, 55%]
1	m= A ... H	-	-
2	-	m= A ... H	-
3	-	-	m= A ... H
4	m= A ... C	m= D ... E	m= F ... H
5	m= F ... H	m= D ... E	m= A ... C

Table 10: Capacity Profiles for Problem Class II

Multi–Level Lot Sizing — An Annotated Bibliography

Alf Kimms and Andreas Drexl
University of Kiel

Abstract: Lot sizing certainly belongs to the most established production planning problems. First scientific reports of this subject date from the beginning of the 20th century and at least one chapter about lot sizing can be found in almost every good textbook about production research issues. But, as we will show, some topics of practical importance such as multi–level lot sizing where capacity is scarce and demand is time variant have first been successfully attacked in the recent past. Furthermore, lot size and sequence decisions are usually not integrated as it ought to be for the short–term planning. This paper reviews the history of multi–level lot sizing from the very early work to the state–of–the–art.

1 Scope of the Review

The huge amount of literature can be classified into several categories. We discriminate three important criteria.

First, publications may assume stochastic or deterministic data. The focus here is on deterministic cases. We refer to [12, 33, 62, 66, 82, 147] for an introduction into the problems associated with stochastic data.

Second, authors assume a single– or a multi–level production. Approaches of the former type do not fit for multi–level structures by definition. Hence, we do not give a comprehensive overview of single–level lot sizing here. [7, 14, 15, 54, 55, 57, 60, 93, 132] contain reviews. However, modifications of single–level heuristics may be used to attack multi–level problems, e.g. on a level–by–level basis. Such methods are called improved heuristics. Subsequent sections deal with these.

Last, demand may be assumed to be dynamic or stationary. In a stationary situation, the demand occurs continuously and with a constant rate. Usually, an infinite planning horizon comes along with stationary demand. This is appropriate for mass production for instance. In most cases, the common basis is an economic order quantity (EOQ) policy [1, 73] with no capacity considerations, or the economic lot sizing problem (ELSP) [57] where capacity limits are taken into account. There is a vast amount of literature about multi–level lot sizing for stationary demand. Most of it is based on EOQ–like assumptions. But, these approaches are of minor help in the dynamic case.

In the subsequent sections we should have a closer look at the multi–level lot sizing literature where data are deterministic and demand is dynamic. Though our interest lies in capacitated situations primarily, uncapacitated cases are closely related and shall therefore be reviewed, too.

2 Contributions of General Interest

2.1 Reviews

This is of course not the first review of (multi–level) lot sizing. So, let us start with a look at the history of surveys. A very brief survey was given by Jacobs and Khumawala [80] with seven references only. Research issues for multi–level lot sizing are discussed by Collier in [45]. DeBoth et al. [51] review dynamic lot sizing. The review of Bahl et al. [14] probably is the one that is most often referenced. Fleischmann [60] gives insight how operations research methods were applied to different production planning problems. So he does for lot sizing. Gupta and Keung [68] also provide a review. Goyal and Gunasekaran [64] provide a collection of literature about multi–level production/inventory systems. Recently, Kuik et al. [93] report on different assumptions for lot sizing, among them multi–level structures. Simpson and Erenguc [124] compiled literature related to multi–level lot sizing, too, and conclude: "Clearly, work in this area has only begun." Drexl and Kimms [55] provide a survey on lot sizing and scheduling. In the textbooks written by Domschke et al. [54] and Tempelmeier [132] we also find notable subsections explaining multi–level lot sizing research. The book by Kimms [89] also gives a review which is updated in this paper.

2.2 Issues about Complex Product Structures

The most basic problem in material requirements planning (MRP) introduced by multi–level structures is the parts requirement (or explosion) problem which is that of computing the total number of parts needed to fulfill demand. In this context, the low level code (or explosion level) of an item is defined as the number of arcs on the longest path from that particular item to an end item. Especially in the presence of positive component inventory levels, the determination of the so–called net requirement has attracted early researchers. Vazsonyi [138] gives a mathematical statement. Elmaghraby [56] is among the first who present an algorithm. Thompson [136] follows up and coins the notion of a technological order which is a partial order on the basis of the low level codes of the items. Both of them use matrix operations to compute net requirements. Nowadays, this problem does no longer exist since efficient methods using the low level codes of the items are available.

The bill of material (BOM) is a formal statement of the successor–predecessor relationship of items and thus defines the gozinto–structure. Special cases of acyclic structures are those where each item has at most one successor (so–called assembly structures), and those where each item has at most one predecessor (so–called divergent, distribution, or arborescence structures, respectively). Structures which belong to the assembly as well as to the distribution type are called serial (or linear). A gozinto–structure that is neither assembly nor divergent is general. Collier [43, 44] has long been the only who tries to define a measure of complexity for gozinto–structures by making use of the information in the BOM. He introduces the degree of commonality index which reflects the average of the number of successors per component part. Recently, Kimms [89] has defined an alternative complexity measure.

3 Unlimited Capacity

Clark and Scarf [37] provide an early analysis of optimal lot sizing policies in a distribution network. They also invent the notion echelon stock for what is in inventory plus what is in transit but not sold. Veinott [139] formulates lot sizing as minimizing a concave function over the solution set of a Leontief substitution system. His work is a generalization of Zangwill's paper [151]. For serial gozinto–structures Zangwill [152] presents a dynamic programming approach based on a reinterpretation of the constraints as flow constraints in a single–source network. Love [100] uses these results and considers serial structures under certain assumptions about the costs related to different production levels. He presents a dynamic programming algorithm, too, to find so–called nested extreme optimal schedules where the attribute nested means that production for an item in a certain period implies production for all successor items in the same period. Crowston and Wagner [46] extend Love's model and develop dynamic programming and B&B–methods for assembly systems. On the basis of Veinott's and Zangwill's results, Kalymon [85] decomposes problems with divergent gozinto–structures into a series of single–item problems. He uses an enumeration procedure to determine the sequence in which these single–item problems are then solved.

Steinberg and Napier [128] determine optimal solutions for general gozinto–structures with standard MIP–solvers on the basis of a network model. The production quantities may be constrained by an item–specific upper bound. An alternative formulation with a less cumbersome notation is given by McClain et al. [105]. In response, Steinberg and Napier [129] make clear that the alternative formulation is more compact, but that there is no evidence which proves greater computational efficiency.

Rao [114] uses Benders' decomposition to compute optimal solutions for problems with general gozinto–structures.

Krajewski et al. [90] perform a computational study and by using its outcome they determine the parameters of three regression models with least squares estimators. For instance, they specify the average size of unreleased orders for each item as a function of the number of product levels above the item, the total number of components for the item, the number of immediate components, and the general lot size for the system (which is a binary variable indicating small or large lots).

Bitran et al. [26] suggest a planning approach where the gozinto–structure is aggregated into just two levels. Then, the two–level problem is solved and the solution is eventually disaggregated. However, they do not consider setups which makes both, the two–level planning problem as well as the disaggregation, become linear with continuous decision variables only.

Bahl and Ritzman [13] formulate a non–linear mixed–integer program for simultaneous master production scheduling, lot sizing, and capacity requirements planning. They also develop a heuristic. The traditional level–by–level approach is retained. Furthermore, they assume a lot–for–lot policy for all component parts. In fact, they reduce the multi–level problem to a single–level one under these assumptions.

By means of simulation studies Biggs [19] examines the interaction effect of different lot sizing and sequencing rules when used conjunctively. In a first step, he solves a multi–level lot sizing problem using a single–level lot sizing rule on a level–by–level basis. Note that the same rule is applied to all levels. For each period he computes a sequence of different items with a simple priority rule, afterwards. The test–bed consists of six lot sizing rules combined with five sequencing rules. The performance is analyzed using several criteria, among them the total number of setups for instance. As an outcome of this study, it turns out that interaction effects between lot size and sequence decisions do exist, but, "... a total explanation will have to wait for future research...".

Billington et al. [24] discuss how to reduce the size of problem instances by means of gozinto–structure compression. A similar idea is published by Axsäter and Nuttle [10, 11] where assembly structures are taken into account only. Zangwill [153] analyzes problems with serial gozinto–structures to find out at which levels inventory cannot occur. Though his motivation is to give advice where to invest for setup cost reduction in order to enable just–

in–time production and to achieve zero inventories irrespective of demand, the information obtained can also be used to compress the gozinto–structure.

When testing six single–level heuristics on a level–by–level basis using the same lot sizing rule on each level, Benton and Srivastava [16] find out that neither the depth nor the breadth of a gozinto–structure has a significant effect on the performance of lot sizing. Afentakis [3] conducts a computational study which contradicts this result. The reason might be that Benton and Srivastava use gozinto–structures with up to five levels and six items while Afentakis uses structures with up to 45 levels and 200 items. Sum et al. [131] provide a study which confirms Afentakis. They employ 11 lot sizing rules on a level–by–level basis tested on 1980 instances. By the way, a heuristic that is introduced by Bookbinder and Koch [31] performs best. An analysis of variance reveals the impact of the number of items, the number of levels, and the average number of immediate successors per item [43]. It is interesting to note that the ranking of the rules remains stable.

Simplistic applications of single–level lot sizing rules are also described by Berry [17] and Biggs et al. [18, 21, 22]. Another evaluation of the level–by–level technique is done by Veral and LaForge [140]. Choi et al. [36], Collier [41, 42], Ho [79], LaForge [94, 95], and Lambrecht et al. [97] provide computational studies, too.

A study in which single–level lot sizing rules are applied to a multi–level structure in a level–by–level manner is done by Yelle [148]. He uses four different simple lot sizing rules and a two–level test–bed. His basic idea is to apply different rules to different levels. This study is extended by Jacobs and Khumawala [81] who add three more single–level methods plus a multi–level algorithm proposed by McLaren [107] which in turn bases on the single–level Wagner–Whitin procedure [6, 59, 144, 145]. Tests with a three–level test–bed are done by Choi et al. [35].

Blackburn and Millen [27] apply single–level heuristics level–by–level to assembly structures using the same rule for all levels. Three heuristics are tested this way. Following the ideas of New [108] and McLaren [107], they modify the cost parameters of the items in order to take the multi–level structure indirectly into account. Five types of modifications are tested. Closely related to their lot sizing approach is the one of Rehmani and Steinberg [116] who keep the cost parameters, but, modify the computation of some derived values that guide a single–level heuristic.

Gupta et al. [67] compare the cost modification approach with others. Studies of Wemmerlöv [146] also examine the impact of cost modifications.

Raturi and Hill [115] introduce something similar to capacity constraints into the model of Blackburn and Millen. But, capacity usage is only estimated on the basis of average demand. They heuristically compute shadow prices for the capacities and derive modified setup costs from that. Eventually, the lot sizing problem is solved using the Wagner–Whitin procedure level by level.

An extension of Blackburn and Millen's approach for general gozinto–structures is presented by Dagli and Meral [47]. The combination of lot sizing and capacity planning is treated in [29] where the subject is confined to assembly structures. The impact of a rolling horizon is discussed in [28].

Combining the idea of using different single–level heuristics at different levels with the idea of modifying the cost parameters is evaluated by Blackburn and Millen [30]. Seven algorithms are used in six combinations. In all tests they choose one algorithm for the end item level and another algorithm for the remaining levels.

While the above level–by–level algorithms construct a solution in one pass, Graves [65] presents a multi–pass procedure. Assuming that external demand occurs for end items only, the basic idea is to perform a single pass, and then to modify the cost parameters (how this is done is of no relevance here). Afterwards, he solves the single–item problems for those items again for which external demand exists. If nothing changes for these items the procedure terminates. If there are any changes in the production plan then the whole procedure repeats using the modified cost parameters this time. A proof of convergence is given.

Another multi–pass heuristic is invented by Peng [110] who uses the Wagner–Whitin procedure to generate an initial solution.

This solution is then repeatedly reviewed to combine lots until no further cost savings can be achieved.

Moving away from level–by–level lot sizing, Lambrecht et al. [98] compare level–by–level approaches with a period–by–period approach. The latter one computes production quantities for the first period and for all items before the second period is concerned. This is going on period by period. All computational tests are restricted to assembly structures. As a result, the period–by–period approach seems to be competitive. A similar method proposes Afentakis [3]. Starting with solving the lot sizing problem consisting of the first period only, he generates a solution which includes the next period by augmenting the intermediate result. This is done until all periods are covered. While Afentakis uses the term sequential approach instead of level–by–level, he now calls his method a parallel one instead of period–by–period. The proposed method decidedly outperforms all (five) single–level methods that are used level–by–level in his study.

Rosling [119] gives an uncapacitated plant location reformulation of the problem with assembly structures. On the basis of this new MIP–model, he develops an optimal procedure.

Heinrich [76, 75] presents a heuristic for general gozinto–structures which operates in two phases. First, the dynamic multi–level lot sizing problem is reduced into a problem with constant demand (using the average demand for each item for instance). Using modified cost parameters, the resulting problem is then solved with a deterministic search in the set of feasible reorder points. In the second phase, the resulting production amounts are modified to meet the demand of the original problem instance.

Coleman and McKnew [40] keep the idea of level–by–level methods alive and design a new heuristic based on an earlier work [39]. Although their computational study gives promising results, they fail to compete with sophisticated, established methods (e.g. [65]). A study by Simpson and Erenguc [125] indicates that, if this is done, the new heuristic appears rather poor.

Gupta and Brennan [67] evaluate level–by–level approaches if backorders are allowed. Other improved heuristics for models with backlogging are discussed by Vörös and Chand [142].

Afentakis et al. [5] efficiently compute optimal solutions for assembly structures. The proposed method is of the B&B–type. Based on a MIP–reformulation, lower bounds are determined with a Lagrangean relaxation of those constraints which represent the multi–level structure. As a result, a set of single–item, uncapacitated lot sizing problems is to be solved. This is done with a shortest path algorithm. Optimal solutions for general gozinto–structures are determined by Afentakis and Gavish in [4] by transforming complex structures into assembly structures where some items occur more than once. Chiu and Lin [34] also compute optimal solutions for assembly structures with dynamic programming. Their idea is based on a graphical interpretation of Afentakis' reformulation. A level–by–level heuristic with postprocessing operations is described as well.

Kuik and Salomon [91] apply a stochastic local search method, namely simulated annealing, to multi–level lot sizing. This approach can handle general gozinto–structures. A comparison with other approaches is not done.

Salomon [122] suggests a decomposition of the multi–level problem into several single–item problems. Roughly speaking, this is achieved by eliminating inventory variables by substitution, doing a Lagrangean–like relaxation of inventory balances, and adding some new constraints. He then solves the single–item problems with some lot sizing algorithm sequentially, and upon termination updates the Lagrangean multipliers to repeat the process.

Joneja [83] considers assembly gozinto–structures. By adapting results from lot sizing with stationary demand [120], he develops a so–called cluster algorithm and proves worst case performance bounds. The projection of this work to serial structures closely relates to Zangwill's paper [153]. Roundy [121] presents two cluster algorithms for general structures and worst case bounds for these. In a computational study, one of them turns out to be slightly better than Afentakis' method [3].

McKnew et al. [106] present a linear zero–one–model for multi–level lot sizing. Assembly gozinto–structures are assumed. They claim that the LP–relaxation of the model always yields integer solutions. Rajagopalan [112] arguments that they are wrong. But, his counter–example is false.[1]

Pochet and Wolsey [111] consider general gozinto–structures and derive cutting planes to ease the computational effort when standard solvers are running.

Atkins [9] suggests a way to compute lower bounds for problems with assembly structures. His basic idea is to convert an assembly structure into a set of serial structures by finding all paths from the lowest level to end items. Some items may now appear in more than one serial structure having their own identity. In such a case the original setup costs are split up and assigned to the (new) items. Eventually, the approach in [100] can be used to solve the resulting instances which, in summary, gives a lower bound.

Simpson and Erenguc [125] invent a neighborhood search procedure and slightly outperform [65], [40], and three other improved heuristics. Lower bounds are obtained with a Lagrangean relaxation.

Richter and Vörös [117] are trying to find a so–called setup cost stability region which is defined as the set of all setup cost values for which a given solution remains optimal. Recently, Vörös [143] analyzes the sensitivity of the setup cost parameters for facilities in series and computes stability regions with a dynamic programming procedure.

Arkin et al. [8] give a classification of the complexity of uncapacitated, multi–level lot sizing problems. Optimization problems with general gozinto–structures and no setup times are proven to be NP–hard.

4 Scarce Capacity

Lambrecht and Vander Eecken [96] consider a serial structure and formulate an LP–model on the basis of a network representation.

[1]See [112], page 1025: The determinant of the submatrix in the example evaluates to 1 and not to 2 as stated in the paper. Hence, there is no contradiction to total unimodularity.

Setups are not part of their focus which is the reason for having continuous variables only (such approaches are used for master production scheduling). Furthermore, capacity constraints exclusively restrict the production amounts of the end item. Optimal solutions are computed by means of extreme flow considerations. A similar problem is concerned by Gabbay [61] who has several serial structures being processed in parallel in mind. In contrast to Lambrecht and Vander Eecken he assumes capacity limits on all levels. Zahorik et al. [150] generalize Gabbay's work, but still assume serial structures. Assembly structures are investigated by Afentakis [2] who transforms the problem into a job shop scheduling problem.

A first MIP–model is formulated by Haehling von Lanzenauer [70]. Remarkably to note that he considers lot sizing and scheduling simultaneously. Another early multi–level lot sizing model is that of Elmaghraby and Ginsberg [58]. An LP–model for master production scheduling in a multi–level system is given in [71].

Gorenstein [63] provides a case description for planning tire production with three levels. He formulates a MIP–model, but does not present any methods. In a short note he points out that LP–based rounding methods could work fine. Vickery and Markland [141] report about experiences in using MIP–model formulations being solved with standard solvers in a real–world situation. Their field of application is a pharmaceutical company.

Zäpfel and Attmann [149] present some kind of a tutorial for multi–level lot sizing. They assume several serial gozinto–structures being processed in parallel. A MIP–model is given on the basis of which a fixed charge problem reformulation and examples are discussed.

Ramsey and Rardin [113] consider serial structures and propose several heuristics. The basic assumption in this work is that items do not share common resources. Two of the heuristics are LP–based methods where a particular choice of the arc weights in a network flow representation makes the solution of the LP–network–problem become a feasible solution. One out of the two LP–based procedures uses the optimal solution of the uncapacitated problem [152] to set the arc weights. Two other heuristics are ad hoc approaches. One makes the multi–level problem collapse into a single–level problem by making additional assumptions. The other one is a greedy heuristic. A computational study shows that the greedy heuristic gives the best results, although the idea of collapsing the gozinto–structure leads to much shorter execution times. Both LP–based approaches are worse with respect to the run–time as well as with respect to the deviation from a lower bound.

Biggs [20] informally discusses the problem of scheduling component parts in order to meet the demand for the end items. In a computational study he uses, among others, the total number of setups to evaluate several rules. As a result, he states that "...the use of the various rules did cause the system to respond differently...".

Harl and Ritzman [72] use modifications of uncapacitated, single–level heuristics to develop a capacity–sensitive multi–level procedure.

Billington et al. [23, 25] assume a general gozinto–structure. Some (potentially all) items share a single common resource with scarce capacity. Beside a MIP–formulation they also present a heuristic to solve the problem. The basic working principle of the heuristic is a B&B–strategy. The B&B–method enumerates over the setup variables. At each node lower bounds are computed by relaxing capacity and multi–level constraints. The uncapacitated, single–level subproblem is then heuristically solved being embedded into a Lagrangean–like iteration.

Hechtfischer [74] heuristically solves problems with an assembly structure. Similar to Billington et al. he assumes that (some) items share a single common resource. But additionally, all items which require the resource must have the same low level code. Basically, his approach follows traditional MRP II ideas and operates level–by–level. If production amounts exceed the available capacity, some items are shifted into earlier periods. Guided with a Lagrangean–like penalty expression, the whole procedure iterates until a maximum number of iterations is performed.

Salomon [122] introduces a simulated annealing and a tabu search heuristic for the problem earlier defined by Billington et al. [25]. Furthermore, he discusses LP–based heuristics for assembly structures. Among traditional rounding approaches, he also presents combinations of simulated annealing and tabu search, respectively, with LP–based rounding. Reprints of parts of this outlet are [92] and [123].

Roll and Karni [118] attack the multi–level lot sizing problem with a single resource as well. In contrast to Billington et al. they do not consider setup times. Starting with a first feasible solution, they pass multiple phases making changes to intermediate solutions to gain improvements.

Toklu and Wilson [137] assume assembly gozinto–structures. Furthermore, only end items require a single commonly used scarce resource. They develop a heuristic for this type of problem.

Maes et al. [101, 102, 103] suggest LP–based rounding heuristics for assembly type problems based on the solution of the LP–relaxation of a reformulation as a plant location model [119]. Their ideas cover simple single–pass heuristics fixing binary variables to 1, or fixing them to 1 or to 0, respectively, meanwhile solving a new LP–relaxation in–between. Another suggestion is that of curtailed B&B–procedures.

Mathes [104] extends the MIP–model given by Afentakis et al. [5] and introduces capacity constraints whereby items do not share common resources (so–called dedicated machines). Several valid constraints are derived to improve the run–time performance of standard solvers.

Stadtler [126, 127] compares the impact of different MIP–model formulations on the performance of standard solvers. His assumptions are quite unrestrictive: general gozinto–structures, multiple resources which are shared in common, and positive setup times. Due to his assumption that an unlimited amount of overtime per period is allowed, there exists no feasibility problem which would make the development of heuristics really hard.

Helber [77, 78] develops heuristics for the multi–level lot sizing problem. He assumes general gozinto–structures. Items may share common resources, but no item requires more than one resource. Positive setup times are allowed. Among the proposed procedures are simulated annealing, tabu search, and genetic algorithms which guide a search in the space of setup patterns on the basis of which production plans are derived. Tempelmeier and Helber [135] test a modification of the Dixon–Silver heuristic [53]. Derstroff and Tempelmeier [52, 133, 134] use a Lagrangean relaxation of capacity constraints and inventory balances to provide a lower bound. Within each iteration an upper bound can be computed by solving an uncapacitated, multi–level problem on a level–by–level basis using the objective function including the penalty expressions for evaluation. A postprocessing stage is used to smooth capacity violations.

Clark and Armentano [38] present a heuristic for general gozinto–structures which operates in two phases. First, the Wagner–Whitin procedure is used to solve the uncapacitated problem on a level–by–level basis. Second, production in overloaded periods is shifted into earlier periods to respect the capacity constraints. Interesting to note is that items may require more than just one resource.

Following the successive planning idea, Sum and Hill [130] perform lot sizing and scheduling. They discriminate orders and operations where each order is fulfilled by doing a plenty of operations. They have both, a network (i.e. gozinto–structure in our terminology) of orders and for each order a subnetwork of operations. Operations are scheduled with a modified version of a resource constrained project scheduling algorithm [50, 109]. After this is done, they merge and/or break some orders which results in a new network and the whole process repeats until the recent solution is not improved. The key element of lot sizing is the merging and breaking part in this approach. The decision what to merge or to break, respectively, relies on a simple priority rule.

Dauzère–Péres and Lasserre [48, 49, 99] take an integrated approach in lot sizing and scheduling into account. In a first step they make lot sizing decision regarding capacity constraints. Then, they determine a sequence given fixed lot sizes. Afterwards, a new lot sizing problem is solved with new precedence constraints among items which share the same resource. These precedence relations stem from the sequence decisions just made. The whole procedure is repeated until a stopping criterion is met. For solving the lot sizing and the scheduling subproblems, respectively, they employ methods from the literature.

Haase [69] gives a MIP–model for multi–level lot sizing and scheduling with a single bottleneck facility. Solution methods are, however, not provided.

Brüggemann and Jahnke [32] employ simulated annealing to find suboptimal solutions for lot sizing and scheduling problems with batch production. However, their heuristic works only for two–level gozinto–structures. The underlying idea is to proceed level by level again.

Jordan [84] considers batching and scheduling. Due to restrictive assumptions, he is able to compute optimum results for small instances.

Kimms [89] tackles the problem of multi–level lot sizing and scheduling under quite general assumptions. The gozinto–structures may be of the general type and several scarce resources may impose capacity restrictions (single–machine cases are discussed in [86, 87, 88]). Several heuristics are developed, among them regret based random sampling, a cellular automaton, a genetic algorithm, disjunctive arc based tabu search, and a so–called demand shuffle procedure which, by the way, performs best. The basic idea of demand shuffle is to combine random sampling with a problem specific data structure manipulation. Roughly speaking, this data structure contains the information until when certain (external or internal) demand is to be met. Manipulating this data structure basically equals modifying the deadline of some demand artificially.

5 Conclusion

Under unconstrained capacities most authors favorite level–by–level approaches which are easy to implement and which can make use of many single–level heuristics as well as efficient exact methods. However, recent research indicates that other ideas are superior [121, 125]. Due to its complexity [8], optimal solution procedures are unlikely to be able to solve medium– to large–sized problem instances. Hence, heuristics will dominate for practical purposes. Definitely, this holds for problems under capacity constraints.

Reviewing the work on capacitated, multi–level lot sizing reveals that most researchers consider very restrictive cases. Most of them assume a single resource only. Also, a lot of work does not deal with general gozinto–structures. Exceptions can be found in very recent outlets only [52, 77, 78, 133, 134, 135].

Although the importance of considering multi–level lot sizing and scheduling simultaneously was recognized rather early [70], research in this field has just begun. A first step away from pure successive MRP II concepts towards taking interaction effects of lot sizing and scheduling into account is done with iterative procedures [48, 49, 99, 130]. Models and methods which solve the combined problems in one step are presented in [89].

Acknowledgement

This work was done with partial support from the DFG–project Dr 170/4–1.

References

[1] ANDLER, K., (1929), Rationalisierung der Fabrikation und optimale Losgrösse, München, Oldenbourg

[2] AFENTAKIS, P., (1985), Simultaneous Lot Sizing and Sequencing for Multistage Production Systems, IIE Transactions, Vol. 17, pp. 327–331

[3] AFENTAKIS, P., (1987), A Parallel Heuristic Algorithm for Lot–Sizing in Multi–Stage Production Systems, IIE Transactions, Vol. 19, pp. 34–42

[4] AFENTAKIS, P., GAVISH, B., (1986), Optimal Lot–Sizing Algorithms for Complex Product Structures, Operations Research, Vol. 34, pp. 237–249

[5] AFENTAKIS, P., GAVISH, B., KARMARKAR, U., (1984), Computationally Efficient Optimal Solutions to the Lot–Sizing Problem in Multistage Assembly Systems, Management Science, Vol. 30, pp. 222–239

[6] AGGARWAL, A., PARK, J.K., (1993), Improved Algorithms for Economic Lot–Size Problems, Operations Research, Vol. 41, pp. 549–571

[7] AKSOY, Y., ERENGUC, S.S., (1988), Multi–Item Inventory Models with Coordinated Replenishments: A Survey, International Journal of Production Research, Vol. 22, pp. 923–935

[8] ARKIN, E., JONEJA, D., ROUNDY, R., (1989), Computational Complexity of Uncapacitated Multi–Echelon Production Planning Problems, Operations Research Letters, Vol. 8, pp. 61–66

[9] ATKINS, D.R., (1994), A Simple Lower Bound to the Dynamic Assembly Problem, European Journal of Operational Research, Vol. 75, pp. 462–466

[10] AXSÄTER, S., NUTTLE, H.L.W., (1986), Aggregating Items in Multi–Level Lot Sizing, in: Axsäter, S., Schneeweiss, C, Silver, E., (eds.), Multi–Stage Production Planning and Inventory Control, Lecture Notes in Economics and Mathematical Systems, Vol. 266, Springer, Berlin, pp. 109–118

[11] AXSÄTER, S., NUTTLE, H.L.W., (1987), Combining Items for Lot Sizing in Multi–Level Assembly Systems, International Journal of Production Research, Vol. 25, pp. 795–807

[12] AXSÄTER, S., ROSLING, K., (1994), Multi–Level Production–Inventory Control: Material Requirements Planning or Reorder Point Policies?, European Journal of Operational Research, Vol. 75, pp. 405–412

[13] BAHL, H.C., RITZMAN, L.P., (1984), An Integrated Model for Master Scheduling, Lot Sizing and Capacity Requirements Planning, Journal of the Operational Research Society, Vol. 35, pp. 389–399

[14] BAHL, H.C., RITZMAN, L.P., GUPTA, J.N.D., (1987), Determining Lot Sizes and Resource Requirements: A Review, Operations Research, Vol. 35, pp. 329–345

[15] BAKER, K.R., (1990), Lot–Sizing Procedures and a Standard Data Set: A Reconsiliation of the Literature, Journal of Manufacturing and Operations Management, Vol. 2, pp. 199–221

[16] BENTON, W.C., SRIVASTAVA, R., (1985), Product Structure Complexity and Multilevel Lot Sizing Using Alternative Costing Policies, Decision Sciences, Vol. 16, pp. 357–369

[17] BERRY, W.L., (1972), Lot Sizing Procedures for Material Requirements Planning Systems: A Framework for Analysis, Production and Inventory Management Journal, Vol. 13, No. 2, pp. 19–35

[18] BIGGS, J.R., (1975), An Analysis of Heuristic Lot Sizing and Sequencing Rules on the Performance of a Hierarchical Multi–Product, Multi–Stage Production Inventory System Utilizing Material Requirements Planning, Ph.D. dissertation, Ohio State University

[19] BIGGS, J.R., (1979), Heuristic Lot–Sizing and Sequencing Rules in a Multistage Production–Inventory System, Decision Sciences, Vol. 10, pp. 96–115

[20] BIGGS, J.R., (1985), Priority Rules for Shop Floor Control in a Material Requirements Planning System under Various Levels of Capacity, International Journal of Production Research, Vol. 23, pp. 33–46

[21] BIGGS, J.R., GOODMAN, S.H., HARDY, S.T., (1977), Lot Sizing Rules in a Hierarchical Multi–Stage Inventory

System, Production and Inventory Management Journal, Vol. 18, No. 1, pp. 104–115

[22] Biggs, J.R., Hahn, C.K., Pinto, P.A., (1980), Performance of Lot–Sizing Rules in an MRP System with Different Operating Conditions, Academy of Management Review, Vol. 5, pp. 89–96

[23] Billington, P.J., (1983), Multi–Level Lot–Sizing with a Bottleneck Work Center, Ph.D. dissertation, Cornell University

[24] Billington, P.J., McClain, J.O., Thomas, L.J., (1983), Mathematical Programming Approaches to Capacity–Constrained MRP Systems: Review, Formulation and Problem Reduction, Management Science, Vol. 29, pp. 1126–1141

[25] Billington, P.J., McClain, J.O., Thomas, L.J., (1986), Heuristics for Multilevel Lot–Sizing with a Bottleneck, Management Science, Vol. 32, pp. 989–1006

[26] Bitran, G.R., Haas, E.A., Hax, A.C., (1983), Hierarchical Production Planning: A Two–Stage System, Operations Research, Vol. 30, pp. 232–251

[27] Blackburn, J.D., Millen, R.A., (1982), Improved Heuristics for Multi–Stage Requirements Planning Systems, Management Science, Vol. 28, pp. 44–56

[28] Blackburn, J.D., Millen, R.A., (1982), The Impact of a Rolling Schedule in Multi–Level MRP Systems, Journal of Operations Management, Vol. 2, pp. 125–135

[29] Blackburn, J.D., Millen, R.A., (1984), Simultaneous Lot–Sizing and Capacity Planning in Multi–Stage Assembly Processes, European Journal of Operational Research, Vol. 16, pp. 84–93

[30] Blackburn, J.D., Millen, R.A., (1985), An Evaluation of Heuristic Performance in Multi–Stage Lot–Sizing Systems, International Journal of Production Research, Vol. 23, pp. 857–866

[31] Bookbinder, J.H., Koch, L.A., (1990), Production Planning for Mixed Assembly / Arborescent Systems, Journal of Operations Management, Vol. 9, pp. 7–23

[32] Brüggemann, W., Jahnke, H., (1994), DLSP for 2–Stage Multi–Item Batch Production, International Journal of Production Research, Vol. 32, pp. 755–768

[33] Chen, F., Zheng, Y.S., (1994), Lower Bounds for Multi–Echelon Stochastic Inventory Systems, Management Science, Vol. 40, pp. 1426–1443

[34] Chiu, H.N., Lin, T.M., (1989), An Optimal Model and a Heuristic Technique for Multi–Stage Lot–Sizing Problems: Algorithms and Performance Tests, Engineering Costs and Production Economics, Vol. 16, pp. 151–160

[35] Choi, H.G., Malstrom, E.M., Classen, R. J., (1984), Computer Simulation of Lot–Sizing Algorithms in Three–Stage Multi–Echelon Inventory Systems, Journal of Operations Management, Vol. 4, No. 4, pp. 259–278

[36] Choi, H.G., Malstrom, E.M., Tsai, R.D., (1988), Evaluating Lot–Sizing Methods in Multilevel Inventory Systems by Simulation, Production and Inventory Management Journal, Vol. 29, pp. 4–10

[37] Clark, A.J., Scarf, H., (1960), Optimal Policies for a Multi–Echelon Inventory Problem, Management Science, Vol. 6, pp. 475–490

[38] Clark, A.R., Armentano, V.A., (1995), A Heuristic for a Resource–Capacitated Multi–Stage Lot Sizing Problem with Lead Times, Journal of the Operational Research Society, Vol. 46, pp. 1208–1222

[39] Coleman, B.J., McKnew, M.A., (1990), A Technique for Order Placement and Sizing, Journal of Purchasing and Materials Management, Vol. 26, pp. 32–40

[40] Coleman, B.J., McKnew, M.A., (1991), An Improved Heuristic for Multilevel Lot Sizing in Material Requirements Planning, Decision Sciences, Vol. 22, pp. 136–156

[41] Collier, D.A., (1980), The Interaction of Single–Stage Lot Size Models in a Material Requirements Planning System, Production and Inventory Management, Vol. 4, pp. 11–20

[42] Collier, D.A., (1980), A Comparison of MRP Lot–Sizing Methods Considering Capacity Change Costs, Journal of Operations Management, Vol. 1, pp. 23–29

[43] Collier, D.A., (1981), The Measurement and Operating Benefits of Component Part Commonality, Decision Sciences, Vol. 12, pp. 85–96

[44] Collier, D.A., (1982), Aggregate Safety Stock Levels and Component Part Commonality, Management Science, Vol. 28, pp. 1296–1303

[45] Collier, D.A., (1982), Research Issues for Multi–Level Lot Sizing MRP Systems, Journal of Operations Management, Vol. 2, pp. 113–123

[46] Crowston, W.B., Wagner, M.H., (1973), Dynamic Lot Size Models for Multi–Stage Assembly Systems, Management Science, Vol. 20, pp. 14–21

[47] Dagli, C.H., Meral, F.S., (1985), Multi–Level Lot–Sizing Heuristics, in: Bullinger, H.J., Warnecke, H.J., (eds.), Towards the Factory of the Future, Berlin, Springer

[48] Dauzère–Péres, S., Lasserre, J.B., (1994), Integration of Lotsizing and Scheduling Decisions in a Job-Shop, European Journal of Operational Research, Vol. 75, pp. 413–426

[49] DAUZÈRE–PÉRES, S., LASSERRE, J.B., (1994), An Integrated Approach in Production Planning and Scheduling, Lecture Notes in Economics and Mathematical Systems, Berlin, Springer, Vol. 411

[50] DAVIS, E.W., PATTERSON, J.H., (1975), A Comparison of Heuristic and Optimum Solutions in Resource–Constrained Project Scheduling, Management Science, Vol. 21, pp. 944–955

[51] DEBOTH, M.A., GELDERS, L.F., VAN WASSENHOVE, L.N., (1984), Lot Sizing Under Dynamic Demand Conditions: A Review, Engineering Costs and Production Economics, Vol. 8, pp. 165–187

[52] DERSTROFF, M., (1995), Mehrstufige Losgrössenplanung mit Kapazitätsbeschränkungen, Heidelberg, Physica

[53] DIXON, P.S., SILVER, E.A., (1981), A Heuristic Solution Procedure for the Multi–Item, Single–Level, Limited Capacity Lot–Sizing Problem, Journal of Operations Management, Vol. 2, pp. 23–39

[54] DOMSCHKE, W., SCHOLL, A., VOSS, S., (1998), Produktionsplanung – Ablauforganisatorische Aspekte, Berlin, Springer, 2nd edition

[55] DREXL, A., KIMMS, A., (1997), Lot Sizing and Scheduling — Survey and Extensions, European Journal of Operational Research, Vol. 99, pp. 221–235

[56] ELMAGHRABY, S.E., (1963), A Note on the 'Explosion' and 'Netting' Problems in the Planning of Materials Requirements, Operations Research, Vol. 11, pp. 530–535

[57] ELMAGHRABY, S.E., (1978), The Economic Lot Scheduling Problem (ELSP): Review and Extensions, Management Science, Vol. 24, pp. 587–598

[58] ELMAGHRABY, S.E., GINSBERG, A.S., (1964), A Dynamic Model for Optimal Loading of Linear Multi–Operation Shops, Management Technology, Vol. 4, pp. 47–58

[59] FEDERGRUEN, A., TZUR, M., (1991), A Simple Forward Algorithm to Solve General Dynamic Lot Sizing Models with n Periods in $O(n \log n)$ or $O(n)$ Time, Management Science, Vol. 37, pp. 909–925

[60] FLEISCHMANN, B., (1988), Operations–Research–Modelle und –Verfahren in der Produktionsplanung, Zeitschrift für betriebswirtschaftliche Forschung, Vol. 58, pp. 347–372

[61] GABBAY, H., (1979), Multi–Stage Production Planning, Management Science, Vol. 25, pp. 1138–1149

[62] GONÇALVES, J.F., LEACHMAN, R.C., GASCON, A., XIONG, Z.K., (1994), Heuristic Scheduling Policy for Multi–Item, Multi–Machine Production Systems with Time–Varying, Stochastic Demands, Management Science, Vol. 40, pp. 1455–1468

[63] GORENSTEIN, S., (1970), Planning Tire Production, Management Science, Vol. 17, pp. B72–B82

[64] GOYAL, S.K., GUNASEKARAN, A., (1990), Multi–Stage Production–Inventory Systems, European Journal of Operational Research, Vol. 46, pp. 1–20

[65] GRAVES, S.C., (1981), Multi–Stage Lot–Sizing: An Iterative Procedure, in: Schwarz, L.B., (ed.) Multi–Level Production/Inventory Control Systems: Theory and Practice, Studies in the Management Science, Vol. 16, Amsterdam, North–Holland, pp. 95–109

[66] GRAVES, S.C., MEAL, H.C., DASU, S.D., QUI, Y., (1986), Two–Stage Production Planning in a Dynamic Environment, in: Axsäter, S., Schneeweiss, C., Silver, E.,

(eds.), Multi–Stage Production Planning and Inventory Control, Berlin, Springer, pp. 9–43

[67] GUPTA, S.M., BRENNAN, L., (1992), Lot Sizing and Backordering in Multi–Level Product Structures, Production and Inventory Management Journal, Vol. 33, No. 1, pp. 27–35

[68] GUPTA, Y.P., KEUNG, Y.K., (1990), A Review of Multi–Stage Lot–Sizing Models, International Journal of Production Management, Vol. 10, pp. 57–73

[69] HAASE, K., (1994), Lotsizing and Scheduling for Production Planning, Lecture Notes in Economics and Mathematical Systems, Vol. 408, Berlin, Springer

[70] HAEHLING VON LANZENAUER, C., (1970), A Production Scheduling Model by Bivalent Linear Programming, Management Science, Vol. 17, pp. 105–111

[71] HAEHLING VON LANZENAUER, C., (1970), Production and Employment Scheduling in Multistage Production Systems, Naval Research Logistics, Vol. 17, pp. 193–198

[72] HARL, J.E., RITZMAN, L.P., (1985), An Algorithm to Smooth Near–Term Capacity Requirements Generated by MRP Systems, Journal of Operations Management, Vol. 5, pp. 309–326

[73] HARRIS, F.W., (1913), How Many Parts to Make at Once, Factory, The Magazine of Management, Vol. 10, pp. 135–136, 152, reprinted (1990), Operations Research, Vol. 38, pp. 947–950

[74] HECHTFISCHER, R., (1991), Kapazitätsorientierte Verfahren der Losgrössenplanung, Wiesbaden, Deutscher Universitäts–Verlag

[75] HEINRICH, C.E., (1987), Mehrstufige Losgrössenplanung in hierarchisch strukturierten Produktionsplanungssystemen, Berlin, Springer

[76] HEINRICH, C.E., SCHNEEWEISS, C., (1986), Multi–Stage Lot–Sizing for General Production Systems, in: Axsäter, S., Schneeweiss, C., Silver, E., (eds.), Multi–Stage Production Planning and Inventory Control, Berlin, Springer, pp. 150–181

[77] HELBER, S., (1994), Kapazitätsorientierte Losgrössenplanung in PPS–Systemen, Stuttgart, M & P

[78] HELBER, S., (1995), Lot Sizing in Capacitated Production Planning and Control Systems, OR Spektrum, Vol. 17, pp. 5–18

[79] HO, C., (1993), Evaluating Lot–Sizing Performance in Multi–Level MRP–Systems: A Comparative Analysis of Multiple Performance Measures, International Journal of Operations and Production Management, Vol. 13, pp. 52–79

[80] JACOBS, F.R., KHUMAWALA, B.M., (1980), Multi–Level Lot Sizing: An Experimental Analysis, Proceedings of the American Institute of Decision Sciences, Atlanta, GA, p. 288

[81] JACOBS, F.R., KHUMAWALA, B.M., (1982), Multi–Level Lot Sizing in Material Requirements Planning: An Empirical Investigation, Computers & Operations Research, Vol. 9, pp. 139–144

[82] JAHNKE, H., (1995), Produktion bei Unsicherheit — Elemente einer betriebswirtschaftlichen Produktionslehre bei Unsicherheit, Physica–Schriften zur Betriebswirtschaft, Vol. 50, Heidelberg, Physica

[83] JONEJA, D., (1991), Multi–Echelon Assembly Systems with Non–Stationary Demands: Heuristics and Worst Case Performance Bounds, Operations Research, Vol. 39, pp. 512–518

[84] JORDAN, C., (1996), Batching and Scheduling — Models and Methods for Several Problem Classes, Lecture Notes

in Economics and Mathematical Systems, Vol. 437, Berlin, Springer

[85] KALYMON, B.A., (1972), A Decomposition Algorithm for Arborescence Inventory Systems, Operations Research, Vol. 20, pp. 860–874

[86] KIMMS, A., (1996), Multi–Level, Single–Machine Lot Sizing and Scheduling (with Initial Inventory), European Journal of Operational Research, Vol. 89, pp. 86–99

[87] KIMMS, A., (1996), Competitive Methods for Multi–Level Lot Sizing and Scheduling: Tabu Search and Randomized Regrets, International Journal of Production Research, Vol. 34, pp. 2279–2298

[88] KIMMS, A., (1997), Demand Shuffle — A Method for Multi–Level Proportional Lot Sizing and Scheduling, Naval Research Logistics, Vol. 44, pp. 319–340

[89] KIMMS, A., (1997), Multi–Level Lot Sizing and Scheduling — Methods for Capacitated, Dynamic, and Deterministic Models, Heidelberg, Physica

[90] KRAJEWSKI, L.J., RITZMAN, L.P., WONG, D. S., (1980), The Relationship between Product Structure and Multistage Lot Sizes: A Preliminary Analysis, Proceedings of the American Institute of Decision Sciences, Atlanta, GA, pp. 6–8

[91] KUIK, R., SALOMON, M., (1990), Multi–Level Lot–Sizing Problem: Evaluation of a Simulated–Annealing Heuristic, European Journal of Operational Research, Vol. 45, pp. 25–37

[92] KUIK, R., SALOMON, M., VAN WASSENHOVE, L.N., MAES, J., (1993), Linear Programming, Simulated Annealing and Tabu Search Heuristics for Lotsizing in Bottleneck Assembly Systems, IIE Transactions, Vol. 25, No. 1, pp. 62–72

[93] Kuik, R., Salomon, M., van Wassenhove, L.N., (1994), Batching Decisions: Structure and Models, European Journal of Operational Research, Vol. 75, pp. 243–263

[94] LaForge, R.L., (1982), MRP and the Part Period Algorithm, Journal of Purchasing and Materials Management, Vol. 18, pp. 21–26

[95] LaForge, R.L., (1985), A Decision Rule for Creating Planned Orders in MRP, Production and Inventory Management Journal, Vol. 26, pp. 115–126

[96] Lambrecht, M.R., Vander Eecken, J., (1978), A Facilities in Series Capacity Constrained Dynamic Lot–Size Model, European Journal of Operational Research, Vol. 2, pp. 42–49

[97] Lambrecht, M.R., Vander Eecken, J., Vanderveken, H., (1981), Review of Optimal and Heuristic Methods for a Class of Facilities in Series Dynamic Lot–Size Problems, in: Schwarz, L.B., (ed.), Multi–Level Production / Inventory Control Systems: Theory and Practice, Studies in the Management Science, Vol. 16, Amsterdam, North–Holland, pp. 69–94

[98] Lambrecht, M.R., Vander Eecken, J., Vanderveken, H., (1983), A Comparative Study of Lot Sizing Procedures for Multi–Stage Assembly Systems, OR Spektrum, Vol. 5, pp. 33–43

[99] Lasserre, J.B., (1992), An Integrated Model for Job-Shop Planning and Scheduling, Management Science, Vol. 38, pp. 1201–1211

[100] Love, S.F., (1972), A Facilities in Series Inventory Model with Nested Schedules, Management Science, Vol. 18, pp. 327–338

[101] Maes, J., (1987), Capacitated Lotsizing Techniques in Manufacturing Resource Planning, Ph.D. dissertation, University of Leuven

[102] Maes, J., McClain, J.O., van Wassenhove, L.N., (1991), Multilevel Capacitated Lotsizing Complexity and LP–Based Heuristics, European Journal of Operational Research, Vol. 53, pp. 131–148

[103] Maes, J., van Wassenhove, L.N., (1991), Capacitated Dynamic Lot–Sizing Heuristics for Serial Systems, International Journal of Production Research, Vol. 29, pp. 1235–1249

[104] Mathes, H.D., (1993), Some Valid Constraints for the Capacitated Assembly Line Lotsizing Problem, in: Fandel, G., Gulledge, Th., Jones, A., Operations Research in Production Planning and Control, Berlin, Springer, pp. 444–458

[105] McClain, J.O., Maxwell, W.L., Muckstadt, J.A., Thomas, L.J., Weiss, E.N., (1982), On MRP Lot Sizing, Management Science, Vol. 28, pp. 582–584

[106] McKnew, M.A., Saydam, C., Coleman, B. J., (1991), An Efficient Zero–One Formulation of the Multilevel Lot–Sizing Problem, Decision Sciences, Vol. 22, pp. 280–295

[107] McLaren, B.J., (1976), A Study of Multiple Level Lotsizing Procedures for Material Requirements Planning Systems, Ph.D. dissertation, Purdue University

[108] New, C.C., (1974), Lot–Sizing in Multi–Level Requirements Planning Systems, Production and Inventory Management Journal, Vol. 15, No. 4, pp. 57–72

[109] Patterson, J.H., (1973), Alternate Methods of Project Scheduling with Limited Resources, Naval Research Logistics, Vol. 20, pp. 767–784

[110] PENG, K., (1985), Lot–Sizing Heuristics for Multi–Echelon Assembly Systems, Engineering Costs and Production Economics, Vol. 9, pp. 51–57

[111] POCHET, Y., WOLSEY, L.A., (1991), Solving Multi–Item Lot–Sizing Problems Using Strong Cutting Planes, Management Science, Vol. 37, pp. 53–67

[112] RAJAGOPALAN, S., (1992), A Note on "An Efficient Zero–One Formulation of the Multilevel Lot–Sizing Problem", Decision Sciences, Vol. 23, pp. 1023–1025

[113] RAMSEY, T.E., RARDIN, R.R., (1983), Heuristics for Multistage Production Planning Problems, Journal of the Operational Research Society, Vol. 34, pp. 61–70

[114] RAO, V.V., (1981), Optimal Lot Sizing for Acyclic Multi–Stage Production Systems, Ph.D. dissertation, Georgia Institute of Technology

[115] RATURI, A., S., HILL, A.V., (1988), An Experimental Analysis of Capacity–Sensitive Setup Parameters for MRP Lot Sizing, Decision Sciences, Vol. 19, pp. 782–800

[116] REHMANI, Q., STEINBERG, E., (1982), Simple Single Pass, Multiple Level Lot Sizing Heuristics, Proceedings of the American Institute of Decision Sciences, Atlanta, GA, pp. 124–126

[117] RICHTER, K., VÖRÖS, J., (1989), On the Stability Region for Multi–Level Inventory Problems, European Journal of Operational Research, Vol. 41, pp. 169–173

[118] ROLL, Y., KARNI, R., (1991), Multi–Item, Multi–Level Lot Sizing with an Aggregate Capacity Constraint, European Journal of Operational Research, Vol. 51, pp. 73–87

[119] ROSLING, K., (1986), Optimal Lot–Sizing for Dynamic Assembly Systems, in: Axsäter, S., Schneeweiss, C., Silver, E.A., (eds.), Multi–Stage Production Planning and Inventory Control, Berlin, Springer, pp. 119–131

[120] Roundy, R.O., (1986), A 98%–Effective Lot–Sizing Rule for a Multi–Product, Multi–Stage Production/Inventory System, Mathematics of Operations Research, Vol. 11, pp. 699–727

[121] Roundy, R.O., (1993), Efficient, Effective Lot Sizing for Multistage Production Systems, Operations Research, Vol. 41, pp. 371–385

[122] Salomon, M., (1991), Deterministic Lotsizing Models for Production Planning, Lecture Notes in Economics and Mathematical Systems, Vol. 355, Berlin, Springer

[123] Salomon, M., Kuik, R., van Wassenhove, L.N., (1993), Statistical Search Methods for Lotsizing Problems, Annals of Operations Research, Vol. 41, pp. 453–468

[124] Simpson, N.C., Erenguc, S.S., (1994), Multiple Stage Production Planning Research: History and Opportunities, Working Paper, State University of New York at Buffalo

[125] Simpson, N.C., Erenguc, S.S., (1994), Improved Heuristic Methods for Multiple Stage Production Planning, Working Paper, State University of New York at Buffalo

[126] Stadtler, H., (1996), Mixed Integer Programming Model Formulations for Dynamic Multi–Item Multi–Level Capacitated Lotsizing, European Journal of Operational Research, Vol. 94, pp. 558–579

[127] Stådtler, H., (1997), Reformulations of the Shortest Route Model for Dynamic Multi-Item Multi-Level Capacitated Lotsizing, OR Spektrum, Vol. 19, pp. 87–96

[128] Steinberg, E., Napier, H.A., (1980), Optimal Multi–Level Lot Sizing for Requirements Planning Systems, Management Science, Vol. 26, pp. 1258–1271

[129] Steinberg, E., Napier, H.A., (1982), On "A Note on MRP Lot Sizing", Management Science, Vol. 28, pp. 585–586

[130] SUM, C.C., HILL, A.V., (1993), A New Framework for Manufacturing Planning and Control Systems, Decision Sciences, Vol. 24, pp. 739–760

[131] SUM, C.C., PNG, D.O.S., YANG, K.K., (1993), Effects of Product Structure Complexity on Multi–Level Lot Sizing, Decision Sciences, Vol. 24, pp. 1135–1156

[132] TEMPELMEIER, H., (1995), Material–Logistik — Grundlagen der Bedarfs– und Losgrössenplanung in PPS–Systemen, Berlin, Springer, 3rd edition

[133] TEMPELMEIER, H., DERSTROFF, M., (1993), Mehrstufige Mehrprodukt–Losgrössenplanung bei beschränkten Ressourcen und genereller Erzeugnisstruktur, OR Spektrum, Vol. 15, pp. 63–73

[134] TEMPELMEIER, H., DERSTROFF, M., (1996), A Lagrangean–Based Heuristic for Dynamic Multi–Level Multi–Item Constrained Lotsizing with Setup Times, Management Science, Vol. 42, pp. 738–757

[135] TEMPELMEIER, H., HELBER, S., (1994), A Heuristic for Dynamic Multi–Item Multi–Level Capacitated Lotsizing for General Product Structures, European Journal of Operational Research, Vol. 75, pp. 296–311

[136] THOMPSON, G.L., (1965), On the Parts Requirement Problem, Operations Research, Vol. 13, pp. 453–461

[137] TOKLU, B., WILSON, J.M., (1992), A Heuristic for Multi–Level Lot–Sizing Problems with a Bottleneck, International Journal of Production Research, Vol. 30, pp. 787–798

[138] VAZSONYI, A., (1958), Scientific Programming in Business and Industry, New York, Wiley

[139] VEINOTT, A.F., (1969), Minimum Concave–Cost Solution of Leontief Substitution Models of Multi–Facility Inventory Systems, Operations Research, Vol. 17, pp. 262–291

[140] VERAL, E.A., LAFORGE, R.L., (1985), The Performance of a Simple Incremental Lot–Sizing Rule in a Multilevel Inventory Environment, Decision Sciences, Vol. 16, pp. 57–72

[141] VICKERY, S.K., MARKLAND, R.E., (1986), Multi–Stage Lot–Sizing in a Serial Production System, International Journal of Production Research, Vol. 24, pp. 517–534

[142] VÖRÖS, J., CHAND, S., (1992), Improved Lot Sizing Heuristics for Multi–Stage Inventory Models with Backlogging, International Journal of Production Economics, Vol. 28, pp. 283–288

[143] VÖRÖS, J., (1995), Setup Cost Stability Region for the Multi–Level Dynamic Lot Sizing Problem, European Journal of Operational Research, Vol. 87, pp. 132–141

[144] WAGELMANS, A., VAN HOESEL, S., KOLEN, A., (1992), Economic Lot Sizing: An $O(n \log n)$ Algorithm that Runs in Linear Time in the Wagner–Whitin Case, Operations Research, Vol. 40, pp. S145–S156

[145] WAGNER, H.M., WHITIN, T.M., (1958), Dynamic Version of the Economic Lot Size Model, Management Science, Vol. 5, pp. 89–96

[146] WEMMERLÖV, U., (1982), An Experimental Analysis of the Use of Echelon Holding Costs and Single–Stage Lot–Sizing Procedures in Multi–Stage Production/Inventory Systems, International Journal of Production Management, Vol. 2, pp. 42–54

[147] YANO, C.A., LEE, H.L., (1995), Lot Sizing with Random Yields: A Review, Operations Research, Vol. 43, pp. 311–333

[148] YELLE, L.E., (1976), Materials Requirements Lot Sizing: A Multi–Level Approach, International Journal of Production Research, Vol. 19, pp. 223–232

[149] ZÄPFEL, G., ATTMANN, J., (1980), Losgrössenplanung: Lösungsverfahren für den dynamischen Fall bei beschränkten Kapazitäten und mehrstufiger Fertigung, Das Wirtschaftsstudium, Vol. 9, pp. 122–126 (part I) and pp. 174–177 (part II)

[150] ZAHORIK, A., THOMAS, L.J., TRIGEIRO, W. W., (1984), Network Programming Models for Production Scheduling in Multi–Stage, Multi–Item Capacitated Systems, Management Science, Vol. 30, pp. 308–325

[151] ZANGWILL, W.I., (1966), A Deterministic Multiproduct, Multifacility Production and Inventory Model, Operations Research, Vol. 14, pp. 486–507

[152] ZANGWILL, W.I., (1969), A Backlogging Model and a Multi–Echelon Model of a Dynamic Economic Lot Size Production System — A Network Approach, Management Science, Vol. 15, pp. 506–527

[153] ZANGWILL, W.I., (1987), Eliminating Inventory in a Series Facility Production System, Management Science, Vol. 33, pp. 1150–1164

Chapter 4

Sequencing and Scheduling

A Branch and Bound Algorithm for the Job Shop Scheduling Problem

Jacek Błażewicz † *Erwin Pesch* ‡ *Małgorzata Sterna* †

† *Institute of Computing Science, Poznań University of Technology*

‡ *Institute of Economics and Business Administration, BWL 3, University of Bonn*

Abstract: The work is concerned with the deterministic case of the non-preemptive job shop scheduling problem. The presented approach is based on the problem representation in the form of a modified disjunctive graph extended by additional arcs. These modifications resulted from the analysis of time dependencies between feasible starting times of tasks. This technique makes it possible to enlarge a partial solution during the search for the optimal tasks' sequence and it is used as a part of a branch and bound algorithm solving the considered problem.

1 Introduction

1.1 The Job Shop Scheduling Problem

Scheduling [5] is defined as a process of assigning m machines from set $M=\{M_1, M_2, ..., M_m\}$ to n tasks from set $T=\{T_1, T_2, ..., T_{n-1}\}$ in order to complete all tasks under the given constraints. There are two general constraints in classical scheduling theory which require that each task can be processed by at most one machine at the same time and each machine can process at most one task at the same time.

The job shop scheduling problem [5] is a particular model of processing sets of tasks forming $\underline{n}$ ordered subsets called jobs on dedicated machines. Each task T_i, described by its processing time p_i, has to be executed without any interruption by a defined machine. The task sequences on machines are unknown and have to be determined in order to minimize a given optimality criterion such as the schedule length without violating any problem constraint.

The job shop model can be applied to manufacturing systems as well as to computer systems [5]. However, this scheduling problem is NP-hard in the strong sense [14,24]. There are only a few efficiently solvable cases of the job shop scheduling problem [15,16]. Others remain difficult even in the case of preemption [19]. To find the optimal solution of the job shop scheduling problem several methods based on a branch and bound algorithm have been developed [2,6,7,8,9,20] and because of the hardness of the problem, also many heuristics have been proposed [1,3,17,22,23,27].

1.2 The Disjunctive Graph

An instance of the job shop scheduling problem can be represented in the form of a disjunctive graph $G=(V, C \cup D)$ [25]. V denotes the set of vertices corresponding to tasks of jobs. This set contains two additional vertices: a source and a sink representing the start and the end of a schedule. The source is equivalent to a dummy task preceding all other tasks and the sink to a dummy task succeeding all other tasks. C is the set of conjunctive arcs which reflect the precedence constraints. Disjunctive edges belonging to set D connect mutually unordered tasks which require the same machine for their execution (a disjunctive edge can be modeled by two opposite directed arcs). Each arc is labeled by the positive weight equal to the processing time of the task where the arc has its beginning.

The job shop scheduling problem requires to find an order of all tasks on machines, i.e. to select one arc in each disjunction, such that the resulting graph is acyclic and the length of the longest path from the source to the sink determining the makespan is minimum.

1.3 An Example

Consider an example of the job shop consisting of a set of 3 machines $M = \{ M_1, M_2, M_3 \}$ and a set of 3 jobs $J = \{ J_1, J_2, J_3 \}$ described by 3 chains of tasks:

$$J_1 : T_1 \rightarrow T_2 \rightarrow T_3, \quad J_2 : T_4 \rightarrow T_5, \quad J_3 : T_6 \rightarrow T_7 \rightarrow T_8$$

For any task T_i the required machine $M(T_i)$ and processing time p_i are presented in Table 1.

task	T_1	T_2	T_3	T_4	T_5	T_6	T_7	T_8
$M(T_i)$	M_1	M_2	M_3	M_3	M_2	M_2	M_1	M_3
p_i	3	2	3	3	4	6	3	2

Table 1. Task requirements

The corresponding disjunctive graph is presented in Figure 1 (weights for disjunctive edges are omitted for simplicity)

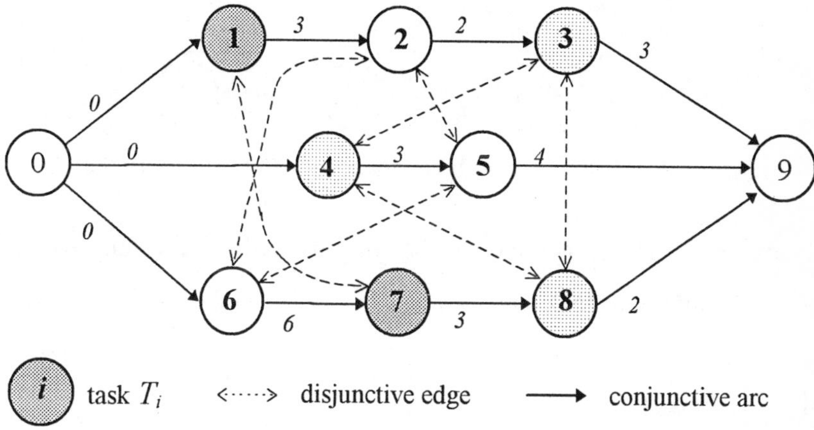

Figure 1. Disjunctive graph example.

For this instance an optimal schedule is presented on the Gantt chart in Figure 2 and has the length $C^*_{\max}$ equal to 13 time units.

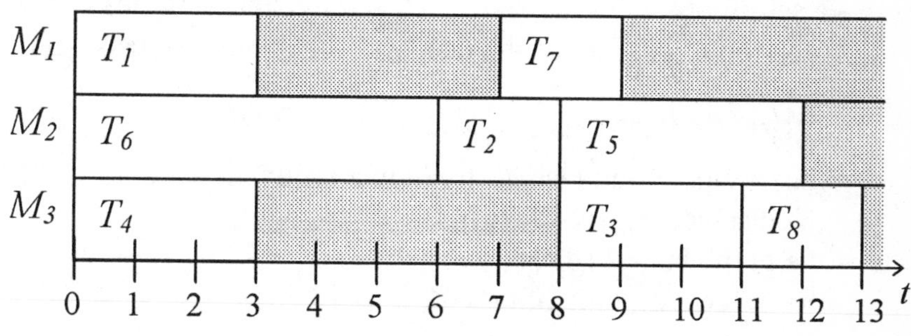

Figure 2. Gantt chart.

The disjunctive graph representing this solution is given in Figure 3. The direction of all disjunctive edges has been fixed causing no cycle in the graph.

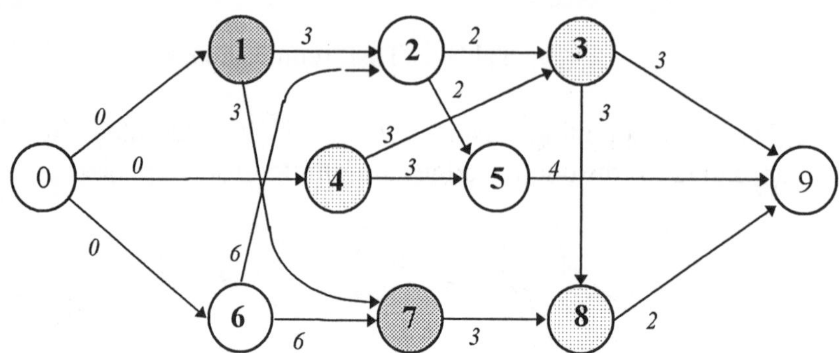

Figure 3. Disjunctive graph representing the complete schedule.

The makespan is determined by the longest path of the length equal to C^*_{max}, containing tasks $T_0 \rightarrow T_6 \rightarrow T_2 \rightarrow T_3 \rightarrow T_8 \rightarrow T_9$.

1.4 Notation

In the sequel we will use the following notation:

M - a set of machines $M = \{ M_1, \ldots, M_k, \ldots, M_m \}$, the number of machines $|M| = m$,

T_i - the i-th task,
T_0 / T_n - a dummy task preceding / succeeding all tasks,

J - a set of jobs $J = \{ J_1, \ldots, J_j, \ldots, J_{\underline{n}} \}$, the number of jobs $|J| = \underline{n}$, each job is a subset of the task set, the number of tasks equals $\sum_{j=1}^{\underline{n}} |J_j| = n-1$,

d_{ij} - the distance from task T_i to T_j in a disjunctive graph,

D - the distance matrix $[d_{ij}]_{(n+1)\mathrm{x}(n+1)}$,

G - the graph matrix $[g_{ij}]_{(n+1)\mathrm{x}(n+1)}$,

p_i - the processing time of task T_i,

st_i - the starting time of task T_i,

est_i - the earliest starting time of task T_i,

lst_i - the latest starting time of task T_i,

ct_i - the completion time of task T_i,
ect_i - the earliest completion time of task T_i,
lct_i - the latest completion time of task T_i,
$\rightarrow$ - a precedence relation between tasks, if $T_i \rightarrow T_j$ then T_i is executed before T_j,
$M(T_i)$ - the machine required by task T_i for its execution,
C_{max} - the schedule length (makespan) equal to the maximal completion time of all tasks, $C_{max} = \max_{T_i}\{ct_i\}$,

C^*_{max} - the length of the optimal schedule,
UB - an upper bound of the schedule length where $UB \geq C^*_{max}$,
$\uparrow,\downarrow$ - increasing / decreasing of variable's value (e.g. $d_{ij}\uparrow$).

Based on the job descriptions, the following sets for each task T_i are also defined:

$Successor(T_i) = \{T_k : T_i \rightarrow T_k\}$
$Predecessor(T_i) = \{T_k : T_k \rightarrow T_i\}$
$Unknown(T_i) = \{T_k : T_k \notin Successor(T_i) \wedge T_k \notin Predecessor(T_i)\}$

2 Distance Analysis

The distance analysis is a kind of pre-processing technique which turns implicit information hidden in time dependencies into explicit ones expressed in the form of precedence relations. In this way it can shorten the process of searching for a solution. The distance analysis is strictly related to the immediate selection approach proposed by Carlier and Pinson [7,8] and developed by Brucker et al. [6] as well as to an arc-path consistency analysis [10,21] applied to constraint satisfaction problems [23]. This approach is based on a representation of the problem in the form of a modified disjunctive graph. The graph contains an additional arc from dummy task T_n to dummy task T_0 the weight of which is $-UB+1$. The distance analysis makes it possible to determine the order of some tasks in a job shop based on distances between them in a graph. Distances between tasks reflect time dependencies between their feasible starting times and fixing the execution sequence of two tasks is equivalent to replacing the disjunctive undirected edge (an opposite directed arc pair) by a newly directed arc which represents new information in the job shop description. The analysis applies to all tasks, not only those which require the same machine. Consequently, a graph is also extended by arcs between tasks belonging to different jobs and requiring different machines for their execution. These additional arcs often allow to fix the order of tasks connected by disjunctive edges which are more important for the final schedule.

2.1 Distance Interpretation

The decreased upper bound, introduced into a disjunctive graph in the form of the reverse arc corresponds to a deadline common for all tasks. Because the deadline is smaller than the length of an initial or the best known solution of a considered problem, it is possible to obtain a solution of length shorter than a given upper bound. No path between dummy tasks T_0 and T_n in a feasible solution can be longer than this deadline.
Distances between all pairs of tasks in a graph build a matrix of distances $D = [d_{ij}]_{(n+1)\text{x}(n+1)}$. A single entry d_{ij} of the distance matrix is equal to the length of a longest path from vertex i to j that corresponds to the longest path from task T_i to T_j. The path contains only conjunctive arcs (i.e. the arcs derived from job descriptions and arcs selected during the solution process; disjunctions which are still undirected are not considered). Distances between T_0 and T_i or between T_i and T_n have a special interpretation. Distance d_{0i} is equal to the head of task T_i and distance d_{in} is equal to the tail of task T_i increased by its processing time p_i. Thus d_{0i} is equivalent to the earliest starting time of T_i denoted by est_i. Additionally the earliest completion time of T_i denoted by ect_i can be expressed as $d_{0i} + p_i$. On the other hand, all tasks have to be performed before their common deadline equal to $-d_{n0}$, so each task T_i has to be started at least d_{in} time units before the deadline. Thus the latest starting time of T_i denoted by lst_i can be expressed as $-d_{n0} - d_{in}$ and consequently its latest completion time lct_i is equal to the value $-d_{n0} - d_{in} + p_i$. Summing up, because all tasks T_i have to be executed before their common deadline, the following bounds of starting times st_i and completion times ct_i have to be kept:

$st_i \in [\, est_i, lst_i \,]$,where $est_i = d_{0i}$, $lst_i = -d_{n0} - d_{in}$
$ct_i \in [\, ect_i, lct_i \,]$,where $ect_i = d_{0i} + p_i$, $lct_i = -d_{n0} - d_{in} + p_i$
If the order between two tasks T_i and T_j is fixed and T_i precedes T_j then d_{ij} denotes the minimum time distance between the starting moment of T_i and the starting moment of T_j , i.e. $st_i + d_{ij} \leq st_j$. In this way d_{ij} is equal to the minimum feasible interval between the start of T_i and the start of T_j in a current partial or complete schedule.
If an order between two tasks T_i and T_j is unknown or T_i succeeds T_j then the interpretation of d_{ij} is changed. In this situation the path connecting tasks T_i and T_j has to include the reverse arc and distance d_{ij} is computed as follows: $d_{ij}=d_{in}+d_{n0}+d_{0j} = d_{0j} -(-d_{n0}-d_{in}) = est_j - lst_i$
Thus the distance d_{ij} denotes the interval between the earliest starting time of T_j and the latest starting time of T_i.

2.2 Distance Matrix Initialization

Based on the interpretation of distances d_{ij}, some rules of their quick initialization can be formulated. They allow to compute distance values according to the job shop description (task sequence in a job).

$$d_{0n} = \max_{J_j \in J} \left\{ \sum_{T_i \in J_j} p_i \right\} \tag{1}$$

The rule determines the duration of the longest job (d_{0n} corresponds to the length of the critical path and the job bound).

$$d_{n0} = -UB + 1 \tag{2}$$

The rule fixes the length of the reverse arc in a disjunctive graph.

$$d_{00} = d_{nn} = d_{0n} + d_{n0} \tag{3}$$

$$\underset{\substack{T_i \in J_j \\ J_j \in J}}{\forall} \; d_{0i} = 0 + \sum_{\substack{T_k \in J_j \\ T_k \to T_i}} p_k, \; d_{ni} = d_{n0} + d_{0i} \tag{4}$$

The rule computes heads of all tasks (and connected with them derivative values d_{ni}) as the sum of processing times of tasks preceding the considered one in a job.

$$\underset{\substack{T_i \in J_j \\ J_j \in J}}{\forall} \; d_{in} = p_i + \sum_{\substack{T_k \in J_j \\ T_i \to T_k}} p_k \; , \; d_{i0} = d_{in} + d_{n0} \tag{5}$$

The rule sets values of modified tasks' tails (and connected with them derivative values d_{i0}). Precisely, the sum of the task's tail and its processing time is computed.

$$\underset{J_j \in J}{\forall} \; \underset{T_k \in J_j}{\exists} \; \underset{\substack{T_l \in J_j \\ l \neq k}}{\forall} \left(T_k \to T_l\right): \underset{i \in [1,n-1]}{\forall} d_{ik} = d_{i0} \tag{6}$$

The rule computes distances to the first task (T_k) of each job, to which the distance is the same as to dummy task T_0. T_0 directly precedes the first tasks of all jobs and it has the zero processing time.

$$\underset{\substack{T_i \in J_j \\ J_j \in J}}{\forall} T_i : \underset{T_l \in J_j}{\exists} T_l \to T_i \underset{\substack{T_k \in J_r \\ J_r \in J}}{\forall} T_k : \left(j \neq r\right) \vee \left(i = k\right) \vee \left(j = r \wedge T_i \to T_k\right)$$

$$d_{ki} = d_{kn} + d_{n0} + d_{0i} \tag{7}$$

The rule fixes distances between tasks not connected by conjunctive arcs (i.e. d_{ki} for T_k not preceding T_i). In this case the path between tasks includes the additional reverse arc of the length d_{n0}.

$$\underset{\substack{T_i \in J_j \\ J_j \in J}}{\forall} T_i : \underset{T_l \in J_j}{\exists} T_l \to T_i \underset{T_k \in J_j}{\forall} T_k : T_k \to T_i \; d_{ki} = d_{0i} - d_{0k} \tag{8}$$

The rule computes distances between tasks for which a precedence constraint exists. It is equal to the sum of processing times of tasks building the path between T_k and T_i.

2.3 Distance Analysis Rules

The purpose of the distance analysis is to introduce new explicit arcs into a disjunctive graph and in this way to extend a current partial solution of the considered job shop problem. At the beginning of the search process, only conjunctive arcs are considered. Obviously, the graph contains also not yet chosen disjunctions but they do not deliver any information about the tasks order. New arcs are obtained directly from time dependencies among starting times of tasks (described by distances among them) and they can relate to conflicting tasks requiring the same machine as well as completely independent tasks belonging to different jobs and performed by different machines. The time order of non-conflict tasks sometimes influences the order of others in conflict. Therefore the distance analysis can be treated as a process of collecting all accessible information about the tasks' order based on the relations among admissible starting times of tasks.

2.3.1 Basic Relations

The following three lemmas present basic relations between distances and starting times of tasks.

Lemma 1
For any pair of tasks T_i and T_j in a disjunctive graph the inequality $st_i + d_{ij} \leq st_j$ is true.

Proof
Assume that T_i precedes T_j. In this case T_j cannot be started before T_i, actually before finishing of T_i and other tasks building the longest path between T_i and T_j. The time necessary to complete all of these tasks is equal to d_{ij} because d_{ij} denotes the length of the longest path between T_i and T_j. Hence, we have: $st_i + d_{ij} \leq st_j$.
Consider the remaining cases when T_i succeeds T_j or their mutual order is unknown. For each task T_k, its st_k is bounded by the earliest and the latest starting time $est_k \leq st_k \leq lst_k$ therefore for T_i and T_j one has: $(est_j \leq st_j) \Rightarrow (0 \leq st_j - est_j)$ and $(st_i \leq lst_i) \Rightarrow (st_i - lst_i \leq 0)$
From both inequalities one achieves:
$(st_i - lst_i \leq 0 \leq st_j - est_j) \Rightarrow (st_i - lst_i \leq st_j - est_j) \Rightarrow (st_i + (est_j - lst_i) \leq st_j)$
and finally taking into consideration the distance interpretation one obtains: $st_i + d_{ij} \leq st_j$.

Lemma 2
For any two tasks T_i and T_j in a disjunctive graph the inequality $d_{ij} \leq - d_{ji}$ is true.

Proof
Basing on Lemma 1 one gets:
$(st_i + d_{ij} \leq st_j) \wedge (st_j + d_{ji} \leq st_i) \Rightarrow (st_i + d_{ij} \leq st_j) \wedge (st_j \leq st_i - d_{ji})$
$\Rightarrow (st_i + d_{ij} \leq st_j \leq st_i - d_{ji}) \Rightarrow (st_i + d_{ij} \leq st_i - d_{ji}) \Rightarrow (d_{ij} \leq - d_{ji})$

Lemma 3

For any pair of tasks T_i and T_j such that T_i precedes T_j in a disjunctive graph the following implication is true: $(T_i \to T_j) \Leftrightarrow (st_i + p_i \leq st_j)$.

Proof

If T_i precedes T_j then task T_j cannot be started before finishing task T_i that is before time moment $st_i + p_i$. Analogously if T_j starts after completion of T_i i.e. $st_i + p_i \leq st_j$ then T_j follows T_i.

2.3.2 Task Pair Analysis Rules

From the practical point of view, especially these distances d_{ij} are interesting which relate to tasks of an unknown mutual order. The analysis of these values often makes it possible to introduce new explicit precedence relations which are implicitly hidden in time dependencies (distances) between tasks. In other words, the analysis of d_{ij}, d_{ji} values for mutually not ordered tasks T_i, T_j can deliver useful information about the sequence of their execution as well as about the whole schedule, the global order of tasks.

The dependencies between distance values and task orders are described by the following theorems.

Theorem 1

For any pair of tasks T_i and T_j in a disjunctive graph the following implication is true: $(d_{ij} > -p_j) \Rightarrow \neg(T_j \to T_i)$.

Proof

Using the logic transposition rule: $(\alpha \Rightarrow \beta) \Leftrightarrow (\neg\beta \Rightarrow \neg\alpha)$ it is enough to prove that: $\neg(\neg(T_j \to T_i)) \Rightarrow \neg(d_{ij} > -p_j)$ which can be formulated as follows: $(T_j \to T_i) \Rightarrow (d_{ij} \leq -p_j)$

From Lemma 3 one gets: $(T_j \to T_i) \Leftrightarrow (st_j + p_j \leq st_i)$ and from Lemma 1: $st_i + d_{ij} \leq st_j$.

From both equations and the assumption one has:

$$(T_j \to T_i) \wedge (st_i + d_{ij} \leq st_j) \Rightarrow (st_j + p_j \leq st_i) \wedge (st_i + d_{ij} \leq st_j)$$
$$\Rightarrow (st_i + d_{ij} \leq st_j) \wedge (st_j \leq st_i - p_j) \Rightarrow (st_i + d_{ij} \leq st_j \leq st_i - p_j) \Rightarrow$$
$$\Rightarrow (st_i + d_{ij} \leq st_i - p_j) \Rightarrow (d_{ij} \leq -p_j)$$

Moreover Theorem 1 denotes that:
$(d_{ij} > - p_j) \Rightarrow ((est_j - lst_i) > - p_j) \Rightarrow (est_j + p_j > lst_i) \Rightarrow (ect_j > lst_i)$
Because the earliest completion time of task T_j is greater than the latest starting time of task T_i so T_i has to be started before T_j can be finished. That means that T_j cannot precede T_i, it can only succeed T_j or these tasks overlap.

Remark

Introducing the additional assumption that both considered tasks require the same machine for their execution one achieves:
$(d_{ij} > - p_j) \wedge (M(T_i)=M(T_j)) \Rightarrow (T_i \rightarrow T_j)$

Theorem 2

For any pair of tasks T_i and T_j in a disjunctive graph the following implication is true: $(d_{ij} \geq p_i) \Rightarrow (T_i \rightarrow T_j)$.

Proof

Using the logic transposition rule: $(\alpha \Rightarrow \beta) \Leftrightarrow (\neg \beta \Rightarrow \neg \alpha)$ it is enough to prove that: $\neg(T_i \rightarrow T_j) \Rightarrow \neg(d_{ij} \geq p_i)$ which can be formulated as follows: $\neg(T_i \rightarrow T_j) \Rightarrow (d_{ij} < p_i)$
From Lemma 3 one gets: $\neg(T_i \rightarrow T_j) \Leftrightarrow \neg(st_i + p_i \leq st_j) \Leftrightarrow (st_i + p_i > st_j)$
and from Lemma 1: $st_i + d_{ij} \leq st_j$
From both equations and the assumption one has:
$(st_i + d_{ij} \leq st_j) \wedge \neg(T_i \rightarrow T_j) \Rightarrow (st_i + d_{ij} \leq st_j) \wedge (st_j < st_i + p_i)$
$\Rightarrow (st_i + d_{ij} \leq st_j < st_i + p_i) \Rightarrow (st_i + d_{ij} < st_i + p_i) \Rightarrow (d_{ij} < p_i)$
Analogously, Theorem 2 denotes that:
$(d_{ij} \geq p_i) \Rightarrow ((est_j - lst_i) \geq p_i) \Rightarrow (est_j \geq lst_i + p_i) \Rightarrow (est_j \geq lct_i)$
Because the earliest starting time of task T_j is greater or equal to the latest completion time of task T_i we see that T_i is completed before T_j is started. That means that T_i precedes T_j.

Theorems 1 and 2 show that a very small negative value d_{ij}, such that $d_{ij} \leq - p_j$, does not deliver any useful information about a schedule. A value $d_{ij} \leq - p_j$ implies only that $est_j < lst_i$ which does not allow to fix the task order. Additionally, one has to remember that distances d_{ij} and d_{ji} for any two tasks T_i and T_j are mutually correlated but do not always provide the same information although this information is never in contradiction. For example d_{ij} makes it possible to extend a current partial solution while d_{ij} is too small to fix the tasks order.

2.3.3 Task Clique Analysis Rules

A clique [7] is a subset of the task set that contains at least two tasks. From the practical point of view especially those cliques are interesting which include only tasks with an unknown execution order and performed by the same machine. The considered cliques are built by tasks (vertices) mutually connected by a pair of disjunctive arcs (see Figure 4).

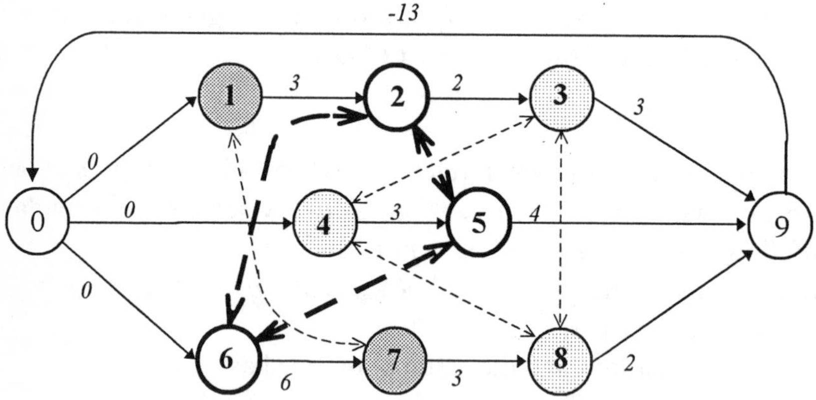

Figure 4. Task clique in a disjunctive graph.

The task belonging to a clique which is executed as the first one is called the input of a clique and the task executed as the last one is called the output of a clique.

For the subsets of tasks described above the following lemmas and theorems can be formulated [7].

Lemma 4

If T_i is a task that belongs to clique C and

$$d_{0i} + \sum_{T_k \in C} p_k + \min_{\substack{T_k \in C \\ k \neq i}} \{d_{kn} - p_k\} > -d_{n0}$$

then T_i cannot be the input of clique C.

Lemma 5

If T_i is a task that belongs to clique C and

$$\min_{\substack{T_k \in C \\ k \neq i}} \{d_{0i}\} + \sum_{T_k \in C} p_k + (d_{in} - p_i) > -d_{n0}$$

then T_i cannot be the output of clique C.

Lemma 6

If T_i is a task that belongs to clique C and

$$\min_{\substack{T_k \in C \\ k \neq i}} \{d_{0i}\} + \sum_{T_k \in C} p_k + \min_{\substack{T_k \in C \\ k \neq i}} \{d_{kn} - p_k\} > -d_{n0}$$

then T_i has to be performed before or after all the other tasks from clique C.

Theorem 3

If T_i is a task that belongs to clique C and Lemma 4 and 6 are satisfied then T_i has to be the output of clique C that means $\forall_{\substack{T_k \in C \\ k \neq i}} T_k \rightarrow T_i$.

Theorem 4

If T_i is a task that belongs to clique C and Lemma 5 and 6 are satisfied then T_i has to be the input of clique C that means $\forall_{\substack{T_k \in C \\ k \neq i}} T_i \rightarrow T_k$.

2.3.4 Remarks

Resuming the presented theorems, the distance analysis is based on the following propositions which are applied for every pair or clique of unordered tasks.

Proposition 1

$$(d_{ij} \geq p_i) \Rightarrow (T_i \rightarrow T_j)$$

Proposition 2

$$(d_{ij} > - p_j) \wedge (M(T_i)=M(T_j)) \Rightarrow (T_i \rightarrow T_j))$$

Proposition 3 (clique output determination rule)

$$\{\ d_{0i} + \sum_{T_k \in C} p_k + \min_{\substack{T_k \in C \\ k \neq i}}\{d_{kn} - p_k\} > -d_{n0}\ \} \ \wedge$$

$$\{\ \min_{\substack{T_k \in C \\ k \neq i}}\{d_{0i}\} + \sum_{T_k \in C} p_k + \min_{\substack{T_k \in C \\ k \neq i}}\{d_{kn} - p_k\} > -d_{n0}\ \} \Rightarrow \forall_{\substack{T_k \in C \\ k \neq i}} T_k \rightarrow T_i$$

Proposition 4 (clique input determination rule)

$$\{\ \min_{\substack{T_k \in C \\ k \neq i}}\{d_{0i}\} + \sum_{T_k \in C} p_k + (d_{in} - p_i) > -d_{n0}\ \} \ \wedge$$

$$\{\ \min_{\substack{T_k \in C \\ k \neq i}}\{d_{0i}\} + \sum_{T_k \in C} p_k + \min_{\substack{T_k \in C \\ k \neq i}}\{d_{kn} - p_k\} > -d_{n0}\ \} \Rightarrow \forall_{\substack{T_k \in C \\ k \neq i}} T_i \rightarrow T_k$$

A new arc (T_i, T_j) between task T_i and T_j can be added if such an order is determined by these propositions or if there exists a task T_k such that ($T_i \rightarrow T_k$) and ($T_k \rightarrow T_j$) or if there exist two tasks T_k, T_l such that ($T_i \rightarrow T_k$), ($T_k \rightarrow T_l$) and ($T_l \rightarrow T_j$). It is important that tasks T_i, T_j, T_k and T_l are completely arbitrary, particularly they do not require the same machine.

2.4 Distance Matrix Updating

The introduction of a new conjunctive arc into a disjunctive graph representing a partial solution of the job shop scheduling problem can change, precisely increase, distances d_{ij} between tasks.

If T_i precedes T_j before adding a new arc then this order is kept in a new extended partial solution. In both cases, before and after changing a disjunctive graph, d_{ij} is equal to the length of the longest path from T_i to T_j and adding a new arc into a graph can only increase the path length. That means, heads and tails of tasks increase or do not change during extension of a partial solution. On the other hand, changing of heads and tails can cause only narrowing starting time bounds. Thus, we have

$d_{0i}\uparrow \Rightarrow est_i \uparrow$

$d_{in}\uparrow \Rightarrow (- d_{n0} - d_{in})\downarrow \Rightarrow lst_i \downarrow$

and these bounds influence distances d_{ij} as well. If T_i succeeds T_j or their mutual order is unknown then $d_{ij} = d_{in} + d_{n0} + d_{0j}$. It is obvious that d_{ij} is growing with an increase of T_i's tail or T_j's head.
In summary, extending a partial solution by fixing the tasks' order (i.e. choosing one conjunctive arc from a disjunctive arc pair) can limit domains of feasible starting times of tasks and increase distances between them. For any two tasks T_i, T_j the following relations are kept (primed variables correspond to the considered values after solution extension):

$$est_i \leq est_i', \qquad lst_i \geq lst_i', \qquad d_{ij} \leq d_{ij}'$$

The domains of tasks' starting times can only become tighter after the distance analysis and an obtained order of tasks' executions sometimes can be equivalent to a complete schedule. But in most cases, the distance analysis is not sufficient to find the solution of a problem and for this reason it is only a part of a branch and bound algorithm which arbitrarily fixes the sequence of some tasks.
Because an introduction of new arcs changes a disjunctive graph it is necessary to calculate new distances between all task pairs which can be achieved with respect to a modification of Floyd's algorithm.

2.4.1 Modified Floyd's Algorithm

A distance matrix updating procedure based on the Floyd's algorithm consists of two phases. In the first phase, the reverse arc is not taken into account, it is temporarily removed from a graph. In this way, the graph contains only arcs of the positive weights and it becomes acyclic. Thus, the modified Floyd's algorithm computing longest (originally the shortest) distances between all pairs of vertices can be applied [13]. It determines distances d_{ij} between tasks T_i and T_j such that $T_i \rightarrow T_j$. Remaining distances can be computed only if the reverse arc, closing the cycle between the last and the first task, is enabled. They are determined during the second phase according to the presented rules:

$$d_{n0} = -UB + 1 \tag{1}$$

$$d_{00} = d_{nn} = d_{0n} + d_{n0} \tag{2}$$

The rules determine the weight of the enabled reverse arc and the loop lengths d_{ii} for both dummy tasks T_0, T_n.

$$\underset{\substack{T_i \in J_j \\ J_j \in J}}{\forall} d_{ii} = d_{in} + d_{n0} + d_{0i}$$

$$d_{ni} = d_{n0} + d_{0i} \tag{3}$$

$$d_{i0} = d_{in} + d_{n0}$$

The rules determine the distances between each task and both dummy tasks T_0, T_n and the length of its loop.

$$\underset{\substack{T_i, T_j \in J \\ \neg(T_i \to T_j)}}{\forall} d_{ij} = d_{in} + d_{n0} + d_{0j} \tag{4}$$

The rules compute distances between tasks connected by the reverse arc.

The presented procedure, based on the Floyd's algorithm is rather time-consuming. It has the complexity of the shortest path procedure of the order $O(n^3)$ [13] and moreover it calculates distances between all pairs of tasks although some of them have not changed after introduction of a new arc into a disjunctive graph.

2.4.2 Distance Matrix Updating Rules

Taking into account the whole knowledge about a given instance of the job shop scheduling problem, it is possible to formulate some specialized distance matrix updating rules. Based on the information about a current partial schedule, they calculate only these distances which really have to be updated after introduction of a new arc (T_i, T_j) into the graph.

Distance Matrix Updating Procedure

step 1: $d_{ij} = p_i$;

step 2: $d_{in}' = d_{ij} + d_{jn}$;
if $(d_{in}' > d_{in})$ *then*
begin
    $Tail = Tail \cup \{T_i\}$;
    $d_{in} = d_{in}'$;
    $d_{i0} = d_{in} + d_{n0}$;
    for each $T_k \in Predecessor(T_i)$ *do*
    begin
        $d_{kn}' = d_{ki} + d_{in}$;

```
            if ( d_kn' > d_kn) then
            begin
                Tail = Tail ∪ { T_k };
                d_kn = d_kn';
                d_k0 = d_kn + d_n0 ;
            end
        end
    end

step 3:  d_0j' = d_0i + d_ij ;
    if ( d_0j' > d_0j) then
    begin
        Head = Head ∪ { T_j };
        d_0j = d_0j';
        d_nj = d_n0 + d_0j ;
        for each T_k ∈ Successor ( T_j ) do
        begin
            d_0k' = d_0j + d_jk ;
            if ( d_0k' > d_0k) then
            begin
                Head = Head ∪ { T_k };
                d_0k = d_0k';
                d_nk = d_n0 + d_0k ;
            end
        end
    end

step 4:  LP = max over T_i, 0<i<n { d_0i + d_in } ;
    if ( LP > d_0n ) then
    begin
        d_0n = LP;
        d_00 = LP + d_n0 ;
        d_nn = LP + d_n0 ;
    end

step 5:  for each T_k ∈ Tail do
    begin
        d_kk = d_kn + d_n0 + d_0k ;
        for each T_l ∈ { Predecessor ( T_k ) ∪ Unknown ( T_k ) } do
            d_kl = max { d_kn + d_n0 + d_0l , d_kl } ;
    end
```

Step 4 in notation: $LP = \max_{\substack{T_i \\ 0<i<n}} \{d_{0i} + d_{in}\}$;

step 6: *for each* $T_k \in$ *Head do*
begin
$d_{kk} = d_{kn} + d_{n0} + d_{0k}$;
for each $T_l \in$ { *Successor* (T_k) $\cup$ *Unknown* (T_k) } *do*
$d_{lk} = max \{ d_{ln} + d_{n0} + d_{0k} , d_{lk} \}$;
end

step 7: *for each* $T_k \in$ *Successor* (T_j) *do*
$d_{ik} = max \{ d_{ij} + d_{jk} , d_{ik} \}$;

step 8: *for each* $T_k \in$ *Predecessor* (T_i) *do*
$d_{kj} = max \{ d_{ki} + d_{ij} , d_{kj} \}$;

step 9: *for each* $T_k \in$ *Predecessor* (T_i) *do*
for each $T_l \in$ *Successor* (T_j) *do*
$d_{kl} = max \{ d_{ki} + d_{ij} + d_{jl}, d_{kl} \}$;

In the first step of the procedure the distance between newly connected tasks T_i, T_j is determined. It is equal to the processing time of task T_i where the new conjunctive arc has its origin.
Connecting task T_i to task T_j may increase T_i's tail (d_{in}) by changing the path between it and dummy task T_n. In this case tails of T_i's predecessors may also increase. These modifications are made in step 2. All tasks of which tails have been updated are added into set *Tail* for the further analysis. Analogously, a new arc (T_i, T_j) may cause an increase of T_j's head (d_{0j}) and, as a result, heads of T_j's successors. All necessary updates are made in step 3 and tasks with modified heads are kept in set *Head.* In step 4 the length of the longest path in a newly extended graph is determined. If this value is greater than the current d_{0n} then three entries of the distance matrix have to be changed: d_{0n}, d_{00}, d_{nn}. Actually calculating of the longest path's length can be performed in steps 2 and 3 by checking sum ($d_{0i}+d_{in}$) for those tasks of which heads or tails increased. Simultaneously, there is a possibility to detect contradictions. If the new longest path (i.e. any value $d_{0i} + d_{in}$) exceeds the value ($-d_{n0}$), that means that a current (usually partial) schedule violates the deadline and has to be suppressed. Increasing of a task's tail implies that distances between this task and all other tasks which do not succeed it have to be updated because these paths are partly built by the analyzed tail. The mentioned modifications are made in step 5 for

all tasks belonging to set *Tail*, that is for all tasks of which tails increased after introduction of a new arc. Similar updates have to be done in step 6 for tasks from set *Head* of which heads increased. In this case, distances between all non-succeeding tasks and the considered one have to be calculated. Steps 5 and 6 determine distances between tasks connected by the reverse arc. The next steps update distances for paths consisting only of conjunctive arcs. An introduction of arc (T_i, T_j) implies that there appear new connections between task T_i and T_j's successors which contain task T_j and these distances may need an update (step 7). Analogously, paths between task T_j and T_i's predecessors containing task T_i have to be analyzed (step 8) and, finally, connections between T_i's predecessors and T_j's successors which contain both tasks T_i, T_j are considered (step 9).

The presented procedure analyses only these distances which might change when a new arc has been introduced into a disjunctive graph. It does not update all n^2 entries of a distance matrix. It considerably limits the number of analyzed task pairs although it still has a complexity of the order $O(n^2)$.

3 Disjunctive Graph Machine Representation

The disjunctive graph contains all information in order to describe a partial or complete solution of the job shop scheduling problem, so its proper machine representation significantly influences the efficiency of the whole algorithm. The graph matrix is a specialized machine representation for a disjunctive graph which combines three classical graph representations [11]: a neighborhood matrix, predecessor and successor lists.

3.1 Graph Matrix Description

A graph matrix $G = [g_{ij}]_{(n+1) \times (n+1)}$ represents a disjunctive graph according to the following assumptions:

$(-n+1) \leq g_{ij} < 0 \quad \Leftrightarrow T_j \in Unknown(T_i)$

$0 \leq g_{ij} \leq (n-1) \quad \Leftrightarrow T_j \in Predecessor(T_i)$

$(n-1) < g_{ij} \leq 2*(n-1) \Leftrightarrow T_j \in Successor(T_i)$

The value of entry g_{ij} provides information on the order of any two tasks in a job shop, i.e. any two vertices in a graph.

Because of the general assumption that T_0 precedes all other tasks and T_n succeeds all other tasks in a job shop, entries g_{i0}, g_{0i}, g_{in}, g_{ni} and g_{ii} can be used to store additional information.

Entry g_{i0} (g_{0i},) contains the index of the first (last) element of a predecessors list for task T_i and entry g_{in} (g_{ni},) contains the index of the first (last) element of a successors list for task T_i. In this way, it is possible to browse both lists and add new elements to their ends (solving the job shop scheduling problem, predecessors and successors lists can only be enlarged with new tasks as a result of the solution's extension). Entry g_{ii} contains the index of the first element of a list containing these tasks which belong neither to T_i's predecessors list nor to T_i's successors list. The information about the last task of this list is not important because tasks are only removed from it. During the search for a solution tasks can be only moved from the *Unknown* list to *Successor* or *Predecessor* lists.

Other entries g_{ij} of a graph matrix have a double meaning. As it has been already mentioned, they describe the mutual relation between tasks T_i, T_j and in addition they are elements of one out of three lists: *Unknown*(T_i), *Successor*(T_i) or *Predecessor*(T_i).

For each task T_i the predecessors list is organized according to the following rules:

$(g_{i0} = 0) \wedge (g_{0i} = 0) \Leftrightarrow Predecessor(T_i) = \varnothing$

$(g_{i0} = f) \Leftrightarrow T_f$ is the first element of the list *Predecessor*(T_i)

$(g_{0i} = l) \Leftrightarrow T_l$ is the last element of the list *Predecessor*(T_i) and has to be $(g_{il} = l)$

$(g_{ij} = k) \wedge (j \neq l) \Leftrightarrow T_k$ is the next element of the list *Predecessor*(T_i) following T_j

The successors list is defined in a similar way:

$(g_{in} = n-1) \wedge (g_{ni} = n-1) \Leftrightarrow Successor(T_i) = \varnothing$

$(g_{in} = f) \Leftrightarrow T_f$ is the first element of the list *Successor*(T_i)

$(g_{ni} = l) \Leftrightarrow T_l$ is the last element of the list *Successor*(T_i) and has to be $(g_{il} = n-1+l)$

$(g_{ij} = n - 1 + k) \wedge (j \neq l) \Leftrightarrow T_k$ is the next element of the list $Successor(T_i)$ following T_j

The remaining tasks belong to the list $Unknown(T_i)$ which is organized as follows:

$(g_{ii} = -i) \Leftrightarrow Unknown(T_i) = \varnothing$

$(g_{ii} = -f) \Leftrightarrow T_f$ is the first element of the list $Unknown(T_i)$

$(g_{ij} = -j) \Leftrightarrow T_j$ is the last element of the list $Unknown(T_i)$

$(g_{ij} = -k) \wedge (k \neq j) \Leftrightarrow T_k$ is the next element of the list $Unknown(T_i)$ following T_j

For a complete matrix determination, the values of four entries have to be defined: $g_{00} = 0$, $g_{nn} = -n$, $g_{n0} = 0$, $g_{0n} = 2n$.

3.2 Graph Matrix Updating Rules

The introduction of a new arc (T_i, T_j) into a disjunctive graph requires its update according to the following recursive procedure:

Graph Matrix Updating Procedure

step 1: $Successor(T_i) = Successor(T_i) \cup \{ T_j \}$;
$Unknown(T_i) = Unknown(T_i) \setminus \{ T_j \}$;

step 2: $Predecessor(T_j) = Predecessor(T_j) \cup \{ T_i \}$;
$Unknown(T_j) = Unknown(T_j) \setminus \{ T_i \}$;

step 3: *for each* $T_k \in Predecessor(T_i)$ *such that* $k \neq j$ *do*
steps 1 and 2 ;

step 4: *for each* $T_l \in Successor(T_j)$ *such that* $l \neq i$ *do*
steps 1 and 2 ;

step 5: *for each* $T_k \in Predecessor(T_i)$ *such that* $k \neq j$ *do*
for each $T_l \in Successor(T_j)$ *such that* $l \neq i$ *do*
steps 1 and 2 ;

The mentioned sets (lists) *Unknown*, *Successor*, *Predecessor* can be easily implemented by using a graph matrix.

4 The Branch and Bound Algorithm

The branch and bound algorithm [5] belongs to a class of general enumerative algorithms which can be adjusted for solving different, usually hard, combinatorial problems.

The presented method is based on a depth first search strategy and uses the distance analysis for extending a partial solution on each node of the search tree. Searching for the most efficient implementation, a few versions of the algorithm have been proposed differing in their branching scheme but the main outline of the method is standard [5].

4.1 Upper Bound

The upper bound of the schedule length is computed by the priority dispatching rules [9] which construct a schedule chronologically selecting tasks one after the other and performing them as soon as possible.

The algorithm is repeated 10 times with a different selection rule according to which the next task for execution is chosen. The implemented rules are as follows:

- *FIFO* - First In First Out rule;
- *EST* - Earliest Starting Time rule;
- *LST* - Latest Starting Time rule;
- *EFT* - Earliest Finishing Time rule;
- *LFT* - Latest Finishing Time rule;
- *SPT* - Shortest Processing Time rule;
- *LPT* - Longest Processing Time rule;
- *MTR* - Most Task Remaining rule;
- *MWR* - Most Work Remaining rule;
- *RAND* - Random rule.

Because the results obtained by using these rules are usually different the upper bound is equal to the best (the lowest) computed bound among 10 values.

4.2 Lower Bounds

The job and machine bounds roughly estimate the lower bound value. The job bound is equal to the length of the longest job, i.e. to the maximal sum of processing times of tasks belonging to the same job:

$$JB = \max_{J_j \in J} \left\{ \sum_{T_i \in J_j} p_i \right\}$$

The machine bound corresponds to the maximal sum of processing times of tasks requiring the same machine for their execution:

$$MB = \max_{M_k \in M} \left\{ \sum_{T_i : M(T_i) = M_k} p_i \right\}$$

The right lower bound of the schedule length, used by the branch and bound algorithm, is computed as follows:

$$LB = \max \{LB_1, LB_2\}$$

LB_1 is equal to the maximum value among the optimal makespans of the preemptive one-machine scheduling problem associated with machines [7]:

$$LB_1 = \max_{M_k \in M} LB_k^1$$

$$LB_k^1 = \min_{T_i : M(T_i) = M_k} \{d_{0i}\} + \sum_{T_i : M(T_i) = M_k} p_i + \min_{T_i : M(T_i) = M_k} \{d_{in} - p_i\}$$

The second estimation LB_2 is equal to the length of the longest path in a disjunctive graph:

$$LB_2 = \max_{\substack{T_i \in J_j \\ J_j \in J}} \{d_{0i} + d_{in}\}$$

The lower bound can be computed based on the distance matrix.

4.3 Current Solution Extension

To reduce the number of steps necessary to find the optimal solution of the job shop scheduling problem, the branch and bound algorithm tries to extend a current partial schedule through the distance analysis. It is processed for each node of the search tree (obviously if

its lower bound is not greater than the decreased upper bound). The purpose of this computations is to make explicit as many arcs in the graph as possible, enlarge a schedule and reduce the number of branches (arbitrarily chosen disjunctions) necessary to find the optimal solution.
The distance analysis concerns mostly conflict tasks requiring the same machine for their execution but it can also take into account independent tasks. It runs in two stages, task pairs analysis and task clique analysis. In the first stage, distances d_{ij}, d_{ji} for all unordered pairs of conflict tasks are analyzed using *Propositions 1* and *2* (see Sections 2.3.2 and 2.3.4). Then cliques of conflict tasks on each machine are considered using *Propositions 3* and *4* (see Sections 2.3.3 and 2.3.4) to determine inputs and outputs of groups of tasks. Each precedence relation disclosed by the propositions is introduced into the graph in the form of a newly directed arc. The process is continued until no new information can be obtained, a partial schedule becomes complete or contradictions appear (they are caused by an infeasible upper bound).

4.4 The Branching Scheme

The distance analysis can deliver a complete schedule but usually the order of tasks is not well determined. In that case, branching of the search tree is necessary.

4.4.1 The Binary-Branching Scheme - BBS

Computations of a branch and bound algorithm using a binary-branching scheme are represented by a binary search tree. In order to create the next level, the method chooses one disjunction and creates two new nodes of the search tree corresponding to two possible orders of considered tasks (connected by the chosen disjunction). Only those pairs of tasks are considered which require the critical machine for their execution. The critical machine is the one with the best lower bound and still not completely scheduled. The branching decision chooses a disjunction connecting a task pair where both tasks cannot be the input of a clique or the output of a clique. Fixing the sequence of such tasks should cause the biggest changes in a current partial schedule. If such a pair of tasks cannot be specified then the Bertier & Roy's rule is applied [4]. It chooses the disjunction between unordered conflict tasks T_i and T_j with maximum

$$v_{ij} = | q_{ij} - q_{ji} |$$

or in case of ties, the disjunction with maximum

$$a_{ij} = max\{ q_{ij}, q_{ji} \}$$

where

$$q_{ij} = max \{ 0, d_{0i} + p_i + d_{jn} - LB \}$$

$$q_{ji} = max \{ 0, d_{0j} + p_j + d_{in} - LB \}$$

and LB denotes the lower bound of the schedule length on the machine required by the considered tasks. Such a disjunction contains those tasks T_i and T_j, for which the difference between the lengths of the two possible schedules, where $T_i \rightarrow T_j$ or $T_i \rightarrow T_j$, is the biggest one.

If a disjunction for branching is still not chosen, tasks which both cannot be executed in the middle of their clique or finally two not ordered conflict tasks, are taken into consideration.

4.4.2 The Multi-Branching Scheme - MBS

A branch and bound algorithm using a multi-branching scheme represents a different approach to solve the job shop scheduling problem. In each step of the algorithm, the branching procedure chooses the task from a set of available tasks which has the minimum head d_{0i} (the minimum feasible starting time). The set of available tasks is determined for the machine which is required by the maximum number of not yet scheduled tasks and has maximum lower bound.

The number of branches, of newly created nodes of the search tree, is equal to the number of available tasks on the given machine. A task is available if it is unscheduled and all its predecessors have been already assigned to their machines.

At each level of the search tree one task is scheduled on its machine, so after n-1 times of branching one can obtain a solution of the problem that may change the initial upper bound. On the other hand, the multi-branching scheme requires slightly different data structures than the binary version and complicates the structure of the search tree.

4.5 Contradiction Detection

Some partial schedules obtained during the search process can be infeasible because they exceed the upper bound. The upper bound is equal to the schedule length of the best solution found. Contradictions are detected when the common deadline represented by the reverse arc in a disjunctive graph is violated by tasks constituting the longest path in the graph.

The contradiction detection does not require any special algorithm. If a new arc closes a cycle then for at least one pair of tasks the contradiction occurs ($T_i \rightarrow T_j$ and $T_j \rightarrow T_i$). Thus a cycle says that a newly introduced arc tries to turn the previous tasks order into the reverse one. This fact is easily detected during updating the graph matrix G (see Section 3.2). The distance matrix D also reflects the feasibility (and the quality) of a current schedule. If for any task entry d_{ii} exceeds zero that means that the deadline is violated, the length of the longest path exceeds the upper bound. The crucial entries can be analyzed during a distance matrix updating (see Section 2.4.2) without any additional computational effort.

5 Computational Results

Computational experiments were performed for different instances of the job shop scheduling problem obtained from the *OR-Library* of Imperial College Management School at the University of London (*http://mscmga.ms.ic.ac.uk*). It contains 82 instances of the job shop scheduling problem (see Table 2) and some out of them were used for our computational experiments. We dropped those problems, where we could not find on optimal solution within a reasonable time bound. Computations were performed on a supercomputer SGI Power Challenge XL at the Supercomputing and Networking Center of Poznań.

instance name	reference
abz5 - abz9	J.Adams, E.Balas, D.Zawack; 1988; [1]
ft06, ft10, ft20	H.Fisher, G.L.Thompson; 1963; [12]
la01 - la40	S. Lawrence; 1984; [18]
orb01 - orb10	D.Applegate, W.Cook; 1991; [2]
swv01- swv20	R.H.Storer, S.D.Wu, R.Vaccari; 1992; [26]
yn1 - yn4	T. Yamada, R. Nakano; 1992; [28]

Table 2. Job shop scheduling problem instances in the *OR-Library*.

5.1 Priority Dispatching Rules

The quality of the upper bound strongly influences the efficiency of the whole algorithm. To find the best priority dispatching rule 30 arbitrarily selected instances of the job shop scheduling problem were analyzed.

Particular priority dispatching rules apparently differ in the quality of produced schedules which can be measured as the average distance (difference) between the upper bound and the optimal value. Some useful information about the rule efficiency delivers also the number of instances for which the generated schedule is actually the optimal one and for which the obtained makespan determines the final upper bound for a branch and bound algorithm (utilization factor). These numbers of instances are given in [%], with respect to the whole number of analyzed ones. The mentioned quality factors are presented in Table 3.

in [%]	priority dispatching rule									
	FIFO	EST	LST	EFT	LFT	SPT	LPT	MTR	MWR	RAND
distance from optimum	18,1	15,4	27,3	12,7	33,2	18,1	28,7	15,5	10,1	21,3
number of optimums	10,0	3,3	0,0	13,3	0,0	0,0	0,0	23,3	20,0	0,0
utilization	16,7	6,7	0,0	23,3	0,0	13,3	0,0	30,0	50,0	0,0

Table 3. Result quality for different priority dispatching rules.

Taking into account the distance from the optimum, the best results are obtained by application of *MWR* rule. On average, it finds schedules of length only 10% above the optimal solution. The worst solutions are obtained by the *LFT* rule. Their lengths were on average more than 33% from the optimum. Actually the *LFT* as well as the *LPT* and the *LST* rules create schedules with makespans worse than those from the trivial *Random* rule.

Only 5 among 10 tested priority dispatching rules found at least one optimal schedule: *FIFO*, *EST*, *EFT*, *MTR* and *MWR*. According to this criterion the *MTR* rule is the best one. It found optimal solutions for more than 23% of analyzed instances. The *MWR* rule finds better (i.e. shorter) schedules but the *MTR* rule more often reaches an optimum. Actually, the *MTR* rule is less stable, because it considers only the number of tasks which have to be executed without consideration of their processing times.

In general, these rules have a non-zero utilization factor which guarantees a better quality of the obtained solutions than by the trivial *Random* rule.

However, the number of optimums generated should not be used as a main criterion of the validity estimation of priority dispatching rules. The utilization factor seems to be more adequate. Thus, *MWR* is apparently the best rule. It determines the upper bound of the schedule length for 50 % of considered cases. *LST*, *LFT*, *LPT* and *RAND* rules never decided about the upper bound value. Solutions obtained by them were always worse than those of the remaining rules.

Hence, priority dispatching rules can be divided into two main groups. The first one contains *MWR*, *MTR*, *EST*, *EFT*, *SPT* and *FIFO* rules which characterize the high quality of obtained schedules. The remaining rules are rather useless for the upper bound determination process.

5.2 Distance Analysis

The best criterion to estimate the influence of the distance analysis on the search process is to compare the number of arcs introduced into a disjunctive graph by our propositions used during the distance analysis to the number of disjunctions fixed by the branching procedure. This relation slightly differs for particular branching schemes applied inside the branch and bound algorithm (see Tables 4 and 5).

The utilization of particular propositions depends also on the order in which they are applied because some time dependencies can be disclosed by different theorems. In such cases only the order of applications decides about the number of arcs introduced by particular propositions.

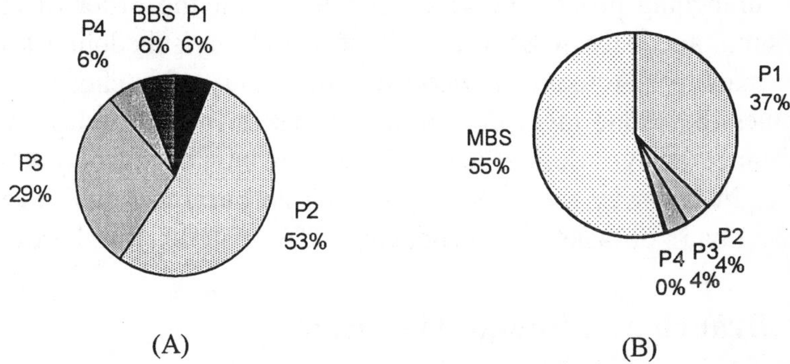

Figure 5. The number of arcs introduced by particular propositions and branching schemes BBS (A), MBS (B) in [%].

The binary-branching scheme ensures much higher efficiency of the distance analysis than the multi-branching version. Using this approach, 94% of all arcs introduced into a disjunctive graph during the search process were provided by our propositions (see Figure 5-A). Only 6% of arcs were determined by the branching scheme.

The most important rule seems to be *Proposition 2* which decides about the order of a task pair (see Section 2.3.4). It allowed to introduce 53% of all arcs (see Figure 5-A). The second task pair analysis rule, *Proposition 1,* was applied relatively rarely because it requires a very strong relation between tasks which can be disclosed only for tight upper bounds (6% of all arcs). *Proposition 3* was responsible for 29% of all introduced arcs. The superiority of the clique output detection over the clique input detection rule (6%) arises from the order of their application during the analysis procedure.

Summing up, both parts of the distance analysis (task pair analysis and task clique analysis) influence on the amount of information about a schedule which is disclosed by this pre-processing technique. But this observation is not longer true for the multi-branching scheme.

An application of the multi-branching scheme significantly decreases the number of arcs introduced by the distance analysis to 46% (see Figure 5-B). Almost 55% of the precedence relations are determined by the branching procedure which decides about the order of many tasks simultaneously. MBS creates a new node by scheduling a task on its machine. Actually, it chooses the input of a clique which resembles the clique input detection and makes this second phase of the distance analysis less efficient. Indeed, it introduced only 4% of all arcs. In this way the most important rule became *Proposition 2* analyzing pairs of tasks and introducing 37% of all disclosed arcs.

5.3 Branch and Bound Algorithm

The first implementation of the branch and bound method used the classical disjunctive graph representation and Floyds's algorithm as a procedure for distance matrix updating (see Section 2.4.1.). Moreover two different method configurations were tested with respect to the binary-branching scheme (see Section 4.4.1) and the multi-branching scheme (see Section 4.4.2).
Although the results obtained are not competitive in many cases however they revealed useful insight into the intrinsic problem structure.

The branch and bound method with the binary-branching scheme found the optimal solutions for some classical job shop instances (see Table 4). A few instances of Lawrence (entries la05, la06, la09-14) were solved only by the distance analysis procedure without running the main search process. The pre-processing technique significantly shortens the search process by disclosing a lot of precedence relations hidden in a disjunctive graph (see Section 5.2). Any branching scheme strongly influences on the efficiency of the distance analysis.

instance		node	number of arcs introduced by					time
name	size	number	P1	P2	P3	P4	BBS	[sec.]
ft06	6x 6	119	117	666	172	62	59	4,41
la01	10x 5	179	58	575	1434	75	92	10,23
la02	10x 5	33808	5679	111286	53148	4949	16984	2008,25
la03	10x 5	35410	22887	239592	55157	19149	17796	2831,45
la04	10x 5	277579	134702	1305601	519924	88279	138884	20167,92
la05	15x 5	1	0	0	0	0	0	0,04
la06	15x 5	1	0	0	0	0	0	0,04
la09	15x 5	1	0	0	0	0	0	0,04
la10	15x 5	1	0	0	0	0	0	0,04
la11	20x 5	1	0	0	0	0	0	0,04
la12	20x 5	1	0	0	0	0	0	0,04
la13	20x 5	1	0	0	0	0	0	0,05
la14	20x 5	1	0	0	0	0	0	0,05
la17	10x10	11132	10621	39849	12803	6205	5799	4026,36
la19	10x10	3461681	2036630	19331120	10535679	3974343	1730990	1439925,00
la20	10x10	857018	634552	6139990	2654177	713128	428593	453705,19

Table 4. Results of the branch and bound algorithm with the BBS scheme.

The BBS introduces only one arc into a disjunctive graph during each branching of the search tree and thus it determines the execution order of only one pair of tasks. The MBS represents a different approach. It decides about the order of a task and all other unscheduled tasks on the same machine. Intuitively, this second approach should shorten a process of searching for the optimal solution. However, computational experiments show that the binary-branching scheme is more efficient than the multi-branching version (see Table 5).

Using this second approach fewer instances from a test set were solved because in the most cases the run time as well as the number of analyzed search tree nodes increased. Even more interesting, the multi-branching scheme made the distance analysis less efficient.

Instance		node	number of arcs introduced by					time
name	size	number	P1	P2	P3	P4	MBS1	[sec.]
ft06	6x6	264	1175	140	43	13	842	4,85
la01	10x5	38	249	43	16	0	178	1,19
la02	10x5	68204	179941	1252	5	10	455159	2576,29
la05	15x5	1	0	0	0	0	0	0,04
la06	15x5	1	0	0	0	0	0	0,05
la09	15x5	1	0	0	0	0	0	0,06
la10	15x5	1	0	0	0	0	0	0,06
la11	20x5	1	0	0	0	0	0	0,07
la12	20x5	1	0	0	0	0	0	0,09
la13	20x5	1	0	0	0	0	0	0,09
la14	20x5	1	0	0	0	0	0	0,09
la17	10x10	205648	265935	18055	160296	178	1196211	36210,95

Table 5. Results of the branch and bound algorithm with the MBS scheme.

Summing up, the binary-branching scheme seems to be more adequate for the branch and bound algorithm based on the disjunctive graph representation of the job shop scheduling problem. The multi-branching scheme is probably better for a schedule oriented approach determining starting times of tasks.

6 Conclusions

New approaches to solve the job shop scheduling problem have been proposed. They are based on a classical disjunctive graph representation which has been modified by introduction of a additional reverse arc closing cycles among vertices in the graph and corresponding to the upper bound of the schedule length. Such a form of the deadline representation made it possible to describe a job shop by means of a distance matrix. The specialized rules for initialization and updating this data structure have been proposed and moreover the new data structure for a disjunctive graph representation has been mentioned. Furthermore, the new auxiliary

technique i.e. the distance analysis, which supports methods of solving the job shop scheduling problem, has been developed. The concept of task cliques introduced by Carlier and Pinson [7] has been adjusted to the new approach which enlarges a partial solution of the problem by disclosing hidden information about task orders in the two phases task pair and clique analysis.
Computational results have shown the superiority of the binary-branching scheme which determines the direction of a single disjunction. It seems to be more suitable for the approach based on the disjunctive graph model which mainly reflects binary relations between pairs of tasks. The branch and bound algorithm with the multi-branching scheme determining the clique input on the critical machine requires additional run time and is less efficient with respect to the distance analysis.

References

[1] ADAMS J., BALAS E., ZAWACK D. (1988), The shifting bottleneck procedure for job shop scheduling, *Management Science*, Vol. 34, pp. 391-401

[2] APPLEGATE D., COOK W. (1991), A computational study of the job shop scheduling problem, *ORSA Journal of Computing*, Vol. 3, pp. 149-156

[3] BALAS E., VAZACOPOULOS A. (1994), Guided Local Search with Shifting Bottleneck for Job Shop Scheduling, *Management Science Research Report #MSRR-609*, Graduate School of Industrial Administration, Carnegie Melon University, Pittsburgh PA.

[4] BERTIER P., ROY B., (1965), Trois exemples numeriques d'application de la procedure SEP, Note de travaile n°32 de la Direction Scientifique de la SEMA

[5] BŁAŻEWICZ J.,ECKER K.,PESCH E.,SCHMIDT G.,WĘGLARZ J. (1996), *Scheduling Computer and Manufacturing Processes*, Heidelberg, Springer

[6] BRUCKER P., JURISCH B., SIEVERS B. (1994), A branch and bound algorithm for the job shop scheduling problem, *Discrete Applied Mathematics*, Vol. 49, pp. 107-127

[7] CARLIER J., PINSON E. (1989), An algorithm for solving the job-shop problem, *Management Science*, Vol. 35, pp. 164-176

[8] CARLIER J., PINSON E. (1994), Adjustment of heads and tails for the job-shop problem, *European Journal of Operational Research*, Vol. 78, pp. 146-161

[9] CASEAU Y., LABURTHE F. (1995), Disjunctive Scheduling with Task Intervals, Laboratoire d'Informatique de l'Ecole Normale Supérieure, Liens-95-25, Paris

[10] CHING-CHIH HAN, CHIA-HOANG LEE (1988), Comments on Mohr and Henderson's path consistency algorithm, *Artificial Intelligence*, Vol. 36, pp. 125-130

[11] DEO N. (1974), *Graph Theory with Applications to Engineering and Computer Science*, Englewood Cliffs, Prentice Hall Inc.

[12] FISHER H., THOMPSON G.L. (1963), Probabilistic learning combinations of local job-shop scheduling rules, In: (Muth J.F., Thompson G. eds.) *Industrial Scheduling*, Englewood Cliffs, Prentice Hall, pp. 225-251

[13] FLOYD R.W. (1962), Algorithm 97: shortest path, *Communications ACM*, Vol. 5, pp. 345

[14] GAREY M.R., JOHNSON D.S. (1979), *Computers and Intractability*, San Francisco, Freeman and Company

[15] HEFETZ H., AIDIRI I. (1982), An efficient optimal algorithm for two-machine unit-time job shop schedule length problem, *Mathematics of Operations Research*, Vol. 7, pp.354-360

[16] JACKSON J. R. (1965), An extension of Johnson's results on job shop scheduling, *Naval Research Logistic Quarterly*, Vol. 3, pp. 201-203

[17] LAARHOVEN P.J.M., AARTS E.H.L., LENSTRA J.K. (1992), Job shop scheduling by simulated annealing, *Operations Research*, Vol. 40, pp. 113-125

[18] LAWRENCE S. (1984), Resource constrained project scheduling: an experimental investigation of heuristic scheduling techniques (Supplement), Graduate School of Industrial Administration, Carnegie-Mellon University, Pittsburgh, Pennsylvania

[19] LENSTRA J.K.,RINNOOY KAN A.H.G. (1979), Computational complexity of discrete optimization problems, *Annals of Discrete Mathematics*, Vol. 4, pp. 121-140

[20] MCMAHON G., FLORIAN M. (1975), On scheduling with ready times and due dates to minimize maximum lateness, *Operations Research*, Vol. 23, pp. 475-482

[21] MOHR R., HENDERSON T.C. (1986), Arc and path consistency revisited, *Artificial Intelligence*, Vol. 28, pp. 225-233

[22] NOWICKI E., SMUTNICKI C. (1996), A fast taboo search algorithm for the job-shop problem, *Management Science*, Vol. 46, pp. 797-813

[23] PESCH E., (1994), *Learning in Automated Manufacturing. A Local Search Approach*, Heidelberg, Physica

[24] RINNOOY KAN A.H.G. (1976), *Machine Scheduling Problems: Classification, Complexity and Computations*, The Hague, Nijhoff

[25] ROY B., SUSSMANN B. (1964), Les problémes d'ordonnancement avec constraintes disjoncties, SEMA, Note D.S., No 9, Paris

[26] STORER R.H., WU S.D, VACCARI R. (1992), New search spaces for sequencing instances with application to job shop scheduling, *Management Science*, Vol. 38, pp. 1495-1509

[27] WIDMER M. (1989), Job-shop scheduling with tooling constraints: a tabu search approach, OR working paper 89/22, Départment of Mathématiques, Ecole Polytechnique Fédérale de Lausanne

[28] YAMADA T., NAKANO R. (1992), A genetic algorithm applicable to large-scale job-shop problems, In: (Männer R., Manderick B. eds.) *Parallel Problem Solving from Nature 2*, Amsterdam, Elsevier Publisher, pp. 281-290

A Prioritized Re-Insertion Approach to Production Rescheduling

Can Akkan
Koç University

Abstract: Given a pre-schedule (or a reservation schedule), we examine the problem of rescheduling operations on a single workcenter due to a disruption. We propose an approach where a set of operations around the disruption are un-scheduled and re-inserted one at a time and the goal is to minimize both total tardiness and the ripple (or domino) effect. For determining the order of re-insertion ten priority rules are proposed and tested. Two local adjustment heuristics are developed to improve the schedule after re-insertion. The computational study showed that the approach of using re-insertion heuristics combined with local adjustment heuristics performs very well.

1 Introduction

MRP-based systems have been the traditional basic paradigm of production planning and control for several decades. In this paradigm production planning and control is a batch process which treats time as a sequence of discrete periods (called time buckets). It is essentially a hierarchical approach which first 'loads' some work to a certain time period and then tries to assign the elements of this work to certain manufacturing resources. The same process is repeated every period.

Due to the divide-and-conquer nature of this approach, planning is done independent of more detailed low-level decisions such as lot sizing and scheduling. This creates a difference between the planned and the actual performance. In fact, quite often the planned workload turns out to be infeasible. In addition, this inaccurate hierarchical approach leads to poor work-order flow-time performance. Moreover, these systems do not have explicit ways of incorporating feedback from the shop-floor back to the planning process (see Kanet [17] for a more detailed discussion on flaws of MRP-based systems).

In response to these shortcomings, Finite Capacity Schedulers (FCS) have been developed [16, 18]. Some FCS are designed to be attached to the back end of an MRP system. In other words, these schedulers get the output of MRP, a load profile for a series of time buckets, and then try to fit the assigned work load to each time bucket using a Gantt chart. We believe that any benefits that might be gained by using such a system are limited because those systems still rely on the infinite loading scheme of MRP with its inflated lead times. They still can neither guarantee a feasible load profile nor provide considerable reduction in flow times of the work-orders.

Using detailed finite-capacity scheduling for both short-term and medium-term planning purposes is one alternative to these approaches [10, 11, 20]. In this approach arriving work-orders are inserted in real-time into a detailed schedule to reserve capacity on work-centers. While the short-term schedule is used for execution, medium-term schedule is used for planning. A schedule created for planning purposes is called a predictive, advance, or reservation schedule or a pre-schedule.

There are several benefits of creating reservation schedules. From the schedulers'/planners' point of view it helps determining the potential performance of the system, giving insight into possible improvements. ¿From the customers' point of view, a reservation schedule gives them information on potential earliness/tardiness of their order, which typically is their most important concern.

However, just preparing a reservation schedule is not enough. There are many events that require the revision of the reservation schedule. The ability to revise a schedule properly is so important that lack of it makes scheduling systems impractical (e.g. see Morton and Pentico [22, page 557]). Graves [14] states that 'a frequent comment heard in many scheduling shops is that there is no scheduling problem but rather a rescheduling problem.' Based on research on the practice of scheduling, McKay et al. [21] state that 'a shop is seldom stable for more than half an hour [and they] have documented about 16 types of variability that can affect a scheduler's world.'

Now, with the rapid development in scheduling software, production rescheduling has become practically possible. Especially for using detailed scheduling for planning production, the ability to do efficient and less-disruptive rescheduling is crucial. In this research, we have looked into a group of rescheduling algorithms, which we call re-insertion-based algorithms. In the next section we will go over the related work that has appeared in the literature. Then in Section 3 we will present a detailed description of the problem studied, followed by a discussion on two different mixed-integer programming formulations in Section 4. In Section 5 we will describe re-insertion based rescheduling approach, followed by local adjustment heuristics used to improve a given schedule in Section 6. We will conclude the paper with the presentation of the experimental results in Section 7 and our conclusions in Section 8.

2 Literature Review

Production rescheduling has received relatively less attention from the research community compared to production scheduling (i.e. generation of schedules from scratch). The most widely studied and applied method of dynamically rescheduling production is based on the use of dispatch priorities which has been the classical approach to dealing with uncertanties in production scheduling. Priority dispatching rules are mostly simple myopic decision rules which are used to determine which operation is to be done next when a work-center becomes available [24]. In general there are two types of such priorities: dynamic (i.e. time-dependent) and static. Adam and Surkis [1] looked into the effects of the frequency of updating dynamic dispatch priorities. They analyzed four types of events that trigger the update of the dispatch priorities, which are: a machine becoming available, an arrival of a job, ends of prespecified time intervals, and a combination of the above.

The second way of dealing with production uncertanties is to create a detailed schedule for execution and planning purposes (a reservation schedule or pre-schedule) and update it (i.e. do rescheduling) at certain points in time. There are quite a number of systems that have been developed recently (both in academia and in the industry) that employ some form of this approach. Some of these use Operations Research-based algorithms and others use constraint propagation methods based on Artificial Intelligence [3].

Once one takes the scheduling/rescheduling approach, when to do rescheduling rises as a critical question. Some researchers looked into periodic rescheduling, where there is a fixed time-interval between schedule updates [7, 23]. This is similar to the rolling schedules used in production planning [19].

Farn and Muhlemann [7] studied the effects of frequency of rescheduling on a single machine scheduling problem with sequence dependent setup times. Their findings are interesting since they determine that the best heuristic for the static case is not necessarly so in the dynamic situation. When they compared the rescheduling heuristics based on the total number of tool changes and total cpu time spent on rescheduling, they found out that a fast heuristic that performs worse than a slower heuristic on a static problem, can be used more frequently than the latter one in the dynamic case, yielding about the same total cpu time over a scheduling horizon with better tool change performance. Muhlemann et al. [23] studied a similar problem where a shop is rescheduled periodically using dispatching priorities.

Works of Adam and Surkis [1], Farn and Muhlemann [7] and Muhlemann et al. [23] all show that increasing rescheduling frequency improves the schedule quality (such as average tardiness). However, none of these take schedule stability (or minimization of ripple effect) into account.

Instead of periodic rescheduling, triggering of rescheduling can be based on certain events, such as arrival of a high-priority work-order, machine breakdown, etc. Yamamoto and Nof [27], propose rescheduling after the actual schedule deviates from the advance schedule by more than a prescribed amount (how this amount is determined is not stated explicitly). Wu et al. [26] address the problem of rescheduling work on a single work-center after a disruption to minimize the weighted sum of makespan and deviation from the pre-schedule (which is a surrogate measure for the ripple effect). One way deviation from the pre-schedule is measured is the sum of absolute values of the difference between the start-time of each operation in the pre-schedule and after rescheduling. In its method of dealing with schedule stability this work is very similar to ours.

Bean et al. [5] develop the conditions for optimality of following a pre-schedule until a disruption and then reconstructing a part of it until it matches up with the pre-schedule.

Church and Uzsoy [8] compare two rescheduling policies (event-driven and periodic rescheduling) in the face of dynamic work-order arrivals and determine under what conditions they perform well. In their work, their presumption is that frequency of rescheduling is a good measure of schedule stability. However, as argued in Wu et al.'s [26] research and will be demostrated here, if rescheduling algorithms take into account the goal of schedule stability, frequent rescheduling may not be very disruptive.

3 Problem Definition

This research addresses a problem where a disruption has occured on a single work-center and a rescheduling action is necessary to restore feasibility. The purpose of the rescheduling action is to find a new schedule that is not only feasible but also has minimum total tardiness and deviation from the pre-disruption schedule. The disruption could be a machine break-down or a high-priority work-order/operation. Each operation on the work-center has a release-time and due-time. In addition to the obvious definition of these terms for a single work-center, in a more general job-shop setting, if an operation has no predecessors, its release-time would be the work-order's release-time, otherwise it could be the maximum of the operation's predecessors' completion-times. If an operation has no successors (root of the product tree) its due-time is that of the work-order, otherwise it is the operation's successor's start-time. We assume that release-time is a hard-constraint, whereas due-time is a soft-constraint. In other words, we do not allow a left-ward ripple effect on the Gantt chart.

As stated above our objective function is the sum of total tardiness for the rescheduled operations and the deviation from the pre-disruption schedule. Schedule deviation is used as a measure of schedule stability or the extent of possible ripple effect caused by rescheduling. We measure schedule deviation as the difference (absolute value) between the start-time of operations before disruption and after rescheduling (this measure was first proposed by Wu et al. [26]). We assume equal weights for the tardiness and deviation measures, but our approach could be generalized for the case of unequal weights.

4 Mixed Integer Programming (MIP) Formulations

There are two MIP formulations that we have experimented with. For Formulation I, we need to define the following variables and parameters:

Variables:

$$X_{i,j} = \begin{cases} 1, & \text{if operation } i \text{ precedes operation } j \\ 0, & \text{otherwise} \end{cases}$$

δ_i : change in the start-time of i

$$a_i = \begin{cases} 1, & \text{if } \delta_i > 0 \\ 0, & \text{otherwise} \end{cases}$$

τ_i : tardiness of operation i

Y_i : $\delta_i a_i$

Parameters:

s_i : the start-time of operation i before disruption

p_i : the total-time of operation i

r_i : the release-time of operation i

d_i : the due-time of operation i

M : a large enough number for the constraint it is used in

Formulation I:

$$z = \min \sum_i \tau_i + 2Y_i \quad - \quad \delta_i$$

s.t.

$$
\begin{aligned}
X_{i,j} + X_{j,i} &= 1 \; \forall i,j \; i \neq j && (1) \\
s_i + \delta_i + p_i &\leq s_j + \delta_j + M(1 - X_{i,j}) && \\
& \quad \forall i,j \; i \neq j && (2) \\
r_i &\leq s_i + \delta_i \; \forall i && (3) \\
\tau_i &\geq s_i + \delta_i + p_i - d_i \; \forall i && (4) \\
\tau_i &\geq 0 \; \forall i && (5) \\
Y_i - Ma_i &\leq 0 \; \forall i && (6) \\
Y_i - \delta_i &\leq 0 \; \forall i && (7) \\
\delta_i - Y_i + Ma_i &\leq M \; \forall i && (8) \\
X_{i,j} &= 0 \; or \; 1 && \\
a_i &= 0 \; or \; 1 && \\
Y_i \; \text{and} \; \delta_i & \; \text{unrestricted in sign} &&
\end{aligned}
$$

In this model, the objective function is equivalent to $\sum_i \tau_i + |\delta_i|$. Thus any deviation from the pre-disruption start-time and tardiness is penalized with equal weight. Constraint (1) simply requires that of any two operations one should precede the other. If an operation i precedes an operation j, Constraint (2) disallows overlapping of the two. Constraint (3) implies that no operation can start before its release-time. On the other hand, constraints (4) and (5) are equivalent to $\tau_i = \max\{s_i + \delta_i + p_i - d_i; 0\}$. Constraints (6), (7), and (8) are used to linearize the nonlinear term $Y_i = \delta_i a_i$ (e.g. see Williams [25, page 178]).

We can arrive at an alternative formulation, if we see the rescheduling problem as one of finding optimal length idle times between operations. As a first step of arriving at this formulation, we can develop a linear programming model for determining the optimal insertion of idle time for a given sequence. In addition to the previously defined parameters and variables, this formulation uses the following variables:

γ_i : inserted idle time (gap) before operation i
ϵ_i : earliness of operation i
δ_i^+ : positive change in start-time of operation i
δ_i^- : negative change in start-time of operation i
σ_i : start-time of operation i

With this notation, we can write the linear programming formulation as follows:

$$z = \min \sum_i \delta_i^+ + \delta_i^- + \tau_i$$

s.t.

$$\sigma_i \geq r_i \ \forall i \tag{9}$$

$$s_i + \delta_i^+ - \delta_i^- = \sigma_i \ \forall i \tag{10}$$

$$\sigma_i + p_i + \gamma_{i+1} = \sigma_{i+1} \ \forall i \tag{11}$$

$$\sigma_i + p_i + \epsilon_i - \tau_i = d_i \ \forall i \tag{12}$$

$$\delta_i^+, \delta_i^-, \epsilon_i, \tau_i, \sigma_i, \gamma_i \geq 0$$

As in the previous formulation, the objective here is to minimize the sum of changes in the start-time and tardiness of operations. Constraint (9) simply requires each operation's start-time after rescheduling not to be earlier than its release-time. Constraint (10) models the relationship between the start-time of operations before and after rescheduling. For each operation at most one of δ_i^+ and δ_i^- can be positive. Constraint (11) is the precedence cosntraint. In this formulation, γ_i variables can be done away with by changing constraint (11) to $\sigma_i + p_i \leq \sigma_{i+1}$. However, γ_i are used here because this model is also used in discussing idle-time insertion algorithms in Section 6 and the Appendix where γ_i are the decision variables. Finally, constraint (12) models earlines/tardiness of operations.

As stated above, this linear programming formulation determines the optimal start-times of operations for a given sequence. So, as the second and final step towards our second formulation, Formulation II, the L.P. model can be extended into a mixed-integer programming formulation which determines both the optimal sequence and the timing of operations. This can be done by replacing equation (11) with equation (13) and adding equations (14) and (15).

$$\sigma_i + p_i \leq \sigma_j + M(1 - P_{i,j}) \; \forall i, j \tag{13}$$

$$\sum_{j \neq i} P_{i,j} = 1 \; \forall i \tag{14}$$

$$\sum_{i \neq j} P_{i,j} = 1 \; \forall j \tag{15}$$

where,

$$P_{i,j} = \begin{cases} 1, & \text{if } i^{th} \text{ operation is the immediate predecessor of } j^{th} \\ 0, & \text{otherwise} \end{cases}$$

It should be noted that in order to include constraints (13), (14) and (15) in the model, one has to define a dummy start activity, say 0, and a dummy end activity, say $n+1$, such that $\sum_{j<n+1} P_{n+1,j} = 0$ and $\sum_{i>0} P_{i,0} = 0$ hold.

5 Re-Insertion Approach

The term insertion of a work-order or an operation is used to refer to scheduling of such an element into gap(s) (idle and available time segments) of an existing schedule without creating any infeasibilities. An early example of this approach is Crabill's job-at-a-time scheduling (see Conway et al. [9]). Some more recent work motivated by the developments in Finite Capacity Schedulers/Leitstands are Akkan [2] and Hastings and Yeh [15].

For the rescheduling problem discussed here, we looked into a set of re-insertion heuristics. These heuristics unschedule a set of operations and re-insert them, one-at-a-time while preserving feasibility.

There are three elementary decisions made regarding such heuristics:

1. Choice of operations to unschedule

2. Order of re-insertion

3. Method of re-insertion

5.1 Choice of Operations to Unschedule

When a set of operations are unscheduled, the rest are 'locked' (i.e. they can not be rescheduled). So the choice of operations to be unscheduled is critical in finding good feasible solutions.

One way of doing this is unscheduling operations within an interval (on a single work-center) of the Gantt chart, which encompasses the disruption. An obvious criterion in determining the size of this interval is the amount of idle time within the interval. The total idle time within the interval should be enough to provide a feasible schedule given the disruption. Theoretical motivation for this approach is given by the work done by Bean et al. [5] where they showed that assuming enough idle time is present in the pre-disruption schedule one can always 'match-up' to this schedule. In our experimentations we chose the interval using the following procedure:

Let l and r be the sequence indices of the operations at the left-most and right-most end of the rescheduling interval, respectively. Let, d be the sequence index of the disruption and p_d its duration. Finally, let γ_i be the idle time (gap) between the two operations in the ith and $(i-1)^{th}$ position in the sequence. We determine l and r as follows:

$$l = \max\{i | \textstyle\sum_{j=(i+1)}^{d} \gamma_j \geq 2p_d\}$$
$$r = \min\{i | \textstyle\sum_{j=(d+1)}^{i} \gamma_j \geq 2p_d\}$$

so that the total idle-time within the interval is at least four times the duration of the disruption.

It should be noted that unscheduling an operation does not necessarily mean that it will be re-scheduled at a different time. Due to the method of re-insertion discussed below, many unscheduled operations could be scheduled at their pre-disruption times. Besides this favorable property, unscheduling provides an empty workspace for the rescheduling algorithm(s) with a well-defined upper limit on the amount of rescheduling done on the work-center.

5.2 Order of Re-insertion

To determine the order with which the unscheduled operations are re-inserted into the schedule we used certain well-known dispatch heuristics after modifying them to account for the objective of keeping the schedule as little changed as possible.

Specifically, we experimented with SPT (Shortest Processing Time), EDD (Earliest Due Date) and MSL (Minimum Slack). These dispatching priorities are reported to be used for scheduling with release-dates and due-dates, and perform relatively well (e.g. Bean et al. [5]). Note that, unlike classical dispatching, in our approach these priorities are used to determine the order of re-insertion rather than sequencing.

We modified the three dispatching priorities mentioned above in two different ways. Therefore, we indexed the dispatch heuristics to refer to the type (or lack) of modification, as SPT_t, EDD_t and MSL_t.

For t equal to 0 the definitions are as follows:

$$SPT_0 = 1/p_i$$
$$EDD_0 = s_{[1]}/d_i$$
$$MSL_0 = d_i - r_i - p_i - u\_c$$

where, $s_{[1]}$ is the start-time of the left-most operation unscheduled in the pre-schedule and u_c is the portion of the disruption that overlaps with the interval $[r_i, d_i]$.

In all of the re-insertion heuristics, we sorted the operations in the order of decreasing priorities.

The modification of the basic priorities were done by multiplying each by a factor. We experimented with two such factors, F_1 and F_2. For $t = 1$ and $t = 2$, the expressions given above for $t = 0$ are multiplied by F_1 and F_2, respectively.

For F_1, we used a measure of the 'distance from the disruption'. There may be several possible ways of defining the 'distance from the disruption'. We experimented with a couple and the following performed well:

$$dd(i) = \max\{\max\{s_d - c_i; 0\}; \max\{s_i - c_d; 0\}\}$$

where, s_i and c_i are start-time and completion-time of operation i before disruption occured, respectively, and sub-script d refers to the disruption. Then,

$$F_1(i) \;=\; e^{-d_i \div \overline{p}}$$

where $\overline{p}$ is the average processing time of operations on the work-center.

The exponentially decreasing nature of $F_1(i)$ is intuitively appealing since if two operations are quite far from the disruption, it should not matter which one is re-inserted first. Also due to the definition of $dd(i)$, any operation that has an overlap with the disruption has zero distance from it, irrespective of the size and location of the overlap.

The second type of modification done on the classical dispatch priorities is based on the start-times of the operations in the pre-disruption schedule. Other things being constant this tries to re-insert the operations in the order of their original sequence. If we let $s_{[i]}$ to be the start-time of the ith operation from left to right in the pre-disruption schedule, we calculate 'relative start-time' (the distance from the start of the rescheduling range) as:

$$rs(i) \;=\; s_{[i]} - s_{[1]}$$

hence, we define $F_2(i)$ as,

$$F_2(i) \;=\; e^{-rs(i) \div (s_{[n]} - s_{[1]})}$$

assuming there are n operations unscheduled within the rescheduling range.

In addition to these nine re-insertion priorities we also experimented with 'distance from disruption' (DiDi) as a priority. DiDi is simply equal to $1/dd(i)$ where $dd(i)$ is as defined above. So, according to this rule, the closer an operation is to the disruption the sooner it is re-inserted.

5.3 Method of Re-insertion

When we re-insert an operation i, we would like to do it so that it's start-time is as close to its initial start-time s_i (using the notation in Section 4) as possible. When operation i cannot be re-inserted to start at s_i it could either be scheduled to start after or before that. We propose the choice between the two to be based on the total penalty associated with inserting i, which is defined as the sum of:

total slide : $|s_i - \sigma_i|$
tardiness : $\max\{0; (\sigma_i + p_i) - d_i\}$
overlap : overlap of candidate schedule of operation i with the pre-disruption schedule of the yet un-inserted operations.

'Overlap' is included in the calculations as a myopic measure of the ripple effect that might be caused and showed to be beneficial in preliminary computational experiments.

6 Local Adjustment Algorithms

To supplement the re-insertion heuristics (or any heuristic that creates a sequence), we developed two local adjustment algorithms. These algorithms start with a given sequence and try to improve it by local changes. The first method is a pairwise interchange method which we called *swap*. The second one is called *jump* and it moves an operation from one side of the disruption to the other.

After each change in the sequence (either due to a single pairwise interchange or a *jump*) the optimal start-times of operations for that sequence are determined.

Determining the optimal start-times for a given sequence is equivalent to determining the optimal idle time to be inserted between operations. This can be done by solving the Linear Programming model discussed in Section 4. However, instead of using the simplex method, one can use a single-pass algorithm to solve this problem. Such an algorithm was developed by Fry et al. [13] for a very similar problem. We started with Fry et al.'s algorithm and made some changes to accomodate certain features of our problem (pseudo-code of the algorithm is given in the Appendix).

The pairwise interchange algorithm *swap* starts with the left-most operation in the schedule and swaps two adjacent operations (i and j) if:

1. j follows i with $\sigma_i + p_i = \sigma_j$.

2. Neither i nor j is the disruption (which is inserted into the schedule just like an operation except the fact that it is 'locked' so that it can not be rescheduled).

3. $\sigma_i \geq r_j$ so that as a result of the interchange j could start at σ_i

4. total cost (sum of tardiness and schedule deviation) of operations i and j after swap is less than total cost before.

Due to conditions 1 and 2 above, the above procedure does not move an operation on one side of the disruption to the other side. However, under certain conditions it might be beneficial to do so. Therefore, the following procedure, called *jump*, is implemented:

procedure *jump*
{
Find the right-most operation b, before disruption d,
such that $\sigma_d + p_d + p_b \leq d_b$
Find the left-most operation a, after disruption d,
such that $\sigma_d - p_a \geq r_a$
if $(\sum_i \delta_i^- > 1.10(\sum_i(\delta_i^+ + \tau_i))$ AND operation b exists)
    resequence b to immediately follow d
else
    if $(\sum_i \delta_i^- < 0.90(\sum_i(\delta_i^+ + \tau_i))$ AND operation a exists)
        resequence a to be the immediate predecessor of d
}

As the "if ... else" statements above show, the decision on whether an operation should be *jump*ed from left of the disruption to its right, or vice versa, is based on the costs of the starting schedule. If the total tardiness and late-start costs are more than early-start costs by a certain margin, an attempt is made to move an operation that is currently scheduled after the disruption to a position immediately preceding the disruption. A symmetric method is followed for *jumps* in the other direction.

After *swap* is finished, *jump* is called repeatedly until there is no cost improvement using the following procedure:

```
procedure call_jump
let last_cost be the cost of the current schedule
cont = TRUE
while (cont = TRUE) {
        call jump
        if (jump results in a sequence change) {
                insert optimum idle time
                if (sequence is feasible)
                        let sch_cost be the cost of the new schedule
                else
                        let sch_cost be a very large number
        }
        if (last_cost ≤ sch_cost)
                cont = FALSE
        else
                last_cost = sch_cost
}
```

7 Experimental Results

To test the performance of the heuristics developed we randomly generated 50 test problems.

Total processing times (total-time) of operations were generated from a normal distribution with mean 4 hours and standard deviation 1 hour.

Arrival times of operations were generated according to a Poisson process. To have a high utilization factor (0.90), which would help generating complicated rescheduling problems, the mean arrival rate was set to be 1 order per 4.444 hours. For each operation, it's release-time was set to be its arrival time.

Due-time allowance of each operation was set to be proportional to its total-time. That is, we let $d_i = r_i + \nu * p_i$, where ν has a normal distribution with mean 10 and standard deviation of 2. Although this may not be the best due-date quoting method (e.g. see [6]), it could be used to model due-date requests given by the customers (and to a certain extent negotiated by the manufacturer).

We assumed that disruptions occur according to a Poisson process with a mean occurrence rate of 1 per 24 hours. The length of disruptions follow normal distribution with mean 10 hrs and std dev 2 hours.

Test problems were generated in the following manner: A disruption was generated to occur at time 550 + *idt* where *idt* comes from an exponential distribution with mean 24. Then starting from time 0 new operations were generated to arrive at the shop, and inserted backward from their due-time. Backward insertion of the operations continued until the total slack (or sum of length of gaps) after the disruption reached two times the length of the disruption. Then starting from the right-most inserted operation, a backward search continued until the total slack between the right-most operation and the left-most operation was equal to four times the length of the disruption. Backward insertion (scheduling) of operations was preferred over forward insertion (scheduling), because while forward insertion creates schedules with high utilization, backward insertion creates schedules with gaps (segments of idle time) ([4, 15]). Naturally, the existence of gaps makes the schedule more robust in case of disruptions.

The codes for the random problem generation and the heuristics were written in C programming language and were run on a SUN ULTRA SPARC 140 (with 143 Mhz processor, 64 Mb RAM and SUNOS 5.5 operating system). The run-times were all less than a second, making them appropriate for real-time use.

Table 1 shows the results of the simulation experiments (the average of 50 runs). The 'initial' column shows the results ob-

Table 1: Comparison of Heuristics

re-insertion	initial	swap	jump
SPT_0	100.50	88.34	84.70
SPT_1	67.06	61.34	58.84
SPT_2	100.10	88.16	83.54
EDD_0	81.08	75.26	58.60
EDD_1	69.06	64.58	61.50
EDD_2	81.30	74.46	57.90
MSL_0	91.66	77.94	75.78
MSL_1	68.04	63.32	61.98
MSL_2	80.82	74.12	58.32
$DiDi$	70.64	65.40	61.28
$None$	108.32	101.20	69.02

tained after re-insertion (except for the row $None$ where the pre-disruption sequence is kept) before any local adjustment is done. The 'swap' column shows the results obtained after applying the *swap* heuristic to the schedules corresponding to the 'initial' column. Similarly, the 'jump' column gives the results obtained by the *jump* heuristic applied to the schedules that correspond to the 'swap' column.

We can see that modifying the classical dispatch priorities by the multiplier F_1 clearly improved the solution quality in all cases. For example, the best three performers in the 'initial' and 'swap' columns are SPT_1, MSL_1 and EDD_1. It is also interesting to note that the performance (67.06, 68.04, and 69.06, respectively) of these three re-insertion heuristics is statistically equivalent to the average performance (69.02) of the local adjustment heuristics applied to the pre-disruption schedule.

The results in Table 1 also show that when the *jump* heuristic is taken into account, EDD_2, MSL_2 became the best performers. So, to see the benefits of doing re-insertion, we compared the best-performing re-insertion algorithms with the 'no re-insertion' case. For this comparison we used the last column of Table 1. By using a paired t-test we compared each one of EDD_2, MSL_2, SPT_1, and $DiDi$ with $None$, which lead to p-values of 0.013,

0.013, 0.083, and 0.133, respectively. Hence, we can say that instead of applying some simple local adjustment heuristics to an existing schedule, changing their sequence by re-insertion and then applying local adjustment heuristics is a better method.

Besides the above comparison of the heuristics, we tried to solve the mixed integer programming formulations of the test problems using CPLEX [12]. After some preliminary experiments we found out that Formulation I performed better than Formulation II (see Section 4). Also as a result of these preliminary experiments we decided to use branching strategy 1 and variable selection strategy 3 among the given branch-and-bound strategies in CPLEX. Before we ran CPLEX for each problem we entered the best objective function value obtained by heuristics as an upper bound. This reduced the size of the branch-and-bound tree considerably. In addition to these we limited the run-time of CPLEX to 1 hour. Given these, 20 of the 50 test problems were solved to optimality by CPLEX. For these 20 problems at least one of the heuristics found the optimal solution in all but one case, and in that one, the best heuristic solution was within 1.6% of the optimum. For 28 problems CPLEX could not find any solution better than the upper bound provided by heuristics within the time-limit of 1 hour. For the remaining 2 problems the best objective function values obtained by the heuristics were 30.2% and 11.5% larger than those found by CPLEX (in both cases CPLEX's solution was not the guaranteed optimum). So, to summarize, of the 50 randomly generated problems in only 3 of them the branch-and-bound procedure of CPLEX found a better solution than the best performing heuristic.

8 Conclusions

Given a pre-schedule (or a reservation schedule), that is used for planning purposes, we examined the problem of rescheduling operations on a single work-center due to a disruption.

Since ripple-effect or schedule instability due to rescheduling actions is believed to be one of the most important drawbacks of schedule-based planning, developing algorithms that explicitly account for schedule stability are of utmost importance.

Therefore, in this research, we looked into combining schedule stability and tardiness goals in a re-insertion-based rescheduling approach in which a set of operations around the disruption are unscheduled and re-inserted one at a time. For further improvement of the schedules, we implemented two local adjustment heuristics. The proposed algortihms required running times of less than a second, making them suitable for real-time use and the results showed that these algorithms could produce satisfactory schedules.

We believe with further research on these kinds of rescheduling heuristics finite-capacity schedule-based planning could become a successful production planning method.

Acknowledgements

We would like to thank the referee for recommendations which helped us improve the presentation of this paper.

References

[1] ADAM, N. R., and SURKIS, J., (1980) "Priority Update Intervals and Anomalies in Dynamic Ratio Type Job Shop Scheduling Rules", *Management Science*, Vol. 26, pp. 1227-1237.

[2] AKKAN, C., (1996) "Overtime Scheduling: An Application in Finite-Capacity Real-Time Scheduling", *Journal of the Operational Research Society*, Vol. 47, pp. 1137-1149.

[3] AKKAN, C., (1996) "Production Rescheduling: A Review", Working Paper No. 1996/13, Koç University, Istanbul, Turkey.

[4] AKKAN, C., (1997) "Finite-Capacity Scheduling-Based Planning for Revenue-Based Capacity Management", *European Journal of Operational Research*, Vol. 100, pp.170-179.

[5] BEAN, J. C., BIRGE, J. R., MITTENTHAL, J., and NOON, C. E., (1991) "Matchup Scheduling with Multiple Resources, Release Dates and Disruptions", *Operations Research*, Vol. 39, pp. 470-483.

[6] BERTRAND, J. W. M., (1983) "The Effect of Workload Dependent Due-Dates on Job Shop Performance", *Management Science*, Vol. 29, pp. 799-816.

[7] FARN, C.-K., and MUHLEMANN, A. P., (1979) "The Dynamic Aspects of a Production Scheduling Problem", *International Journal of Production Research*, Vol. 17, pp. 15-21.

[8] CHURCH, L. K., and UZSOY, R., (1992) "Analysis of Periodic and Event-Driven Rescheduling Policies in Dynamic Shops", *International Journal of Computer Integrated Manufacturing*, Vol. 5, pp. 153-163.

[9] CONWAY, R. W., MAXWELL, W. L., and MILLER, L. W., (1967) *Theory of Scheduling.* Addison Wesley, Reading, Massachusetts.

[10] CONWAY, R. W., and MAXWELL, W. L., (1987), "Low-level Interactive Scheduling", *Proc. of National Bureau of Standards on Real Time Optimization for Automated Manufacturing Facilities*, pp. 99-107.

[11] CONWAY, R. W., and MAXWELL, W. L., (1993) *PRS: Production Reservation System: User's Guide.* C-Way Associates, Ithaca, New York.

[12] CPLEX Optimization Inc., (1995) *Using the CPLEX Callable Library, Ver. 4.0.* Nevada, U.S.A.

[13] FRY, T. D., ARMSTRONG, R. D., and BLACKSTONE, J. H., (1987) "Minimizing Weighted Absolute Deviation in

Single Machine Scheduling", *IIE Transactions*, Vol. 19, pp. 445-450

[14] GRAVES, S. C., (1981) "A Review of Production Scheduling", *Operations Research*, Vol. 29, pp. 646-675.

[15] HASTINGS, N. A. J., and YEH, C.-H, (1990) "Job Oriented Production Scheduling", *European Journal of Operational Research*, Vol. 47, pp. 35-48.

[16] JAIN, S., BARBER, K., and OSTERFELD, D., (1990) "Expert Simulation for On-Line Scheduling", *Communications of the ACM*, Vol. 33, pp. 55-60.

[17] KANET, J. J., (1988) "MRP 96: Time to Rethink Manufacturing Logistics", *Production and Inventory Management Journal*, 2nd Quarter, pp. 57-61.

[18] KANET, J. J., and SRIDHARAN, V., (1990) "The Electronic Leitstand: A New Tool for Shop Scheduling", *Manufacturing Review*, Vol. 3, pp. 161-170.

[19] KIMMS, A., (1996) "Stability Measures for Rolling Schedules with Applications to Capacity Expansion Planning, Master Production Scheduling, and Lot Sizing", Working Paper No. 418, Lehrstuhl für Produktion und Logistik, Institut für Betriebwirtschaftslehre, Christian-Albrechts- Universität zu Kiel, Germany.

[20] McCARTHY, S. W., and BARBER, K. D., (1990) "Medium to Short Term Finite Capacity Scheduling: A Planning Methodology for Capacity Constrained Workshops", *Engineering Costs and Production Economics*, Vol. 19, pp. 189-199.

[21] McKAY, K. N., SAFAYENI, F. R., and BUZACOTT, J. A., (1988) "Scheduling Theory: What's Relevant", *Interfaces*, Vol. 18, pp. 84-90.

[22] MORTON, T. E., and PENTICO, D. W., (1993) *Heuristic Scheduling Systems: With Applications to Production Systems and Project Management*, Wiley Series in Engineering and Technology Management, New York, New York.

[23] MUHLEMANN, A. P., LOCKETT, A. G., and FARN, C. K., (1982) "Job Shop Scheduling Heuristics and Frequencies of Scheduling", *International Journal of Production Research*, Vol. 20, pp. 227-241.

[24] PANWALKER, S. S., and ISKANDER, W., (1977) "A Survey of Scheduling Rules", *Operations Research*, Vol. 25, pp. 45-61.

[25] WILLIAMS, H. P., (1993) *Model Building in Mathematical Programming*, Wiley-Interscience, Chichester, England.

[26] WU, S. D., STORER, R. H., and CHANG, P.-C., (1993) "One-Machine Rescheduling Heuristics with Efficiency and Stability as Criteria", *Computers and Operations Research*, Vol. 20, pp. 1-14.

[27] YAMAMOTO, M., and NOF, S. Y., (1985) "Scheduling/rescheduling in the Manufacturing Operating System Environment", *International Journal of Production Research*, Vol. 23, pp. 705-722.

A Appendix

Optimum Idle Time Insertion Algorithm
In this algorithm the disruption is scheduled like an operation and it is locked (it's schedule cannot be changed). Besides, its due-time is assumed to be equal to its completion-time.

The following notation is used:

N : number of operations (jobs)
P : position of the operation of the current iteration
B_k : the set of operations between gaps k and $k+1$
I_k : the length of gap k

$S(p)$	:	equals j if operation j is in the pth position in the sequence
G_j	:	gap in front of operation j in the earliest start-time schedule for the inital sequence
φ_j	:	finish time of operation j
σ_j	:	start time of operation j
σ_j^o	:	start time of operation j before disruption
d_j	:	due-time of job j
B_k^t	:	set of operations j with $\varphi(j) \geq d_j$
B_k^e	:	set of operations j with $\sigma_j < \sigma_j^o$
B_k^l	:	set of operations j with $\sigma_j \geq \sigma_j^o$
D_k	:	$\sum_{S(j)\in B_k^t} 1 + \sum_{S(j)\in B_k^l} 1 - \sum_{S(j)\in B_k^e} 1$
Δ	:	maximum amount of idle time to be inserted

(1)
Compute φ_j and G_j based on the initial earliest time schedule.
Let P^* be the position of the disruption in the sequence.
Let, $P = N; k = 1; D_k = 0; I_0 = Big\_M$ and
$B_k = B_k^t = B_k^e = B_k^l = \emptyset$.
(2)
$B_k = B_k \cup \{S(P)\}$
Let, $I_k = G_{S(P)}$;
if $(P == P^*)$
then $D_k = Big\_M$;
$B_k^t \cup \{S(P)\}; B_k^l \cup \{S(P)\}$ GOTO (5)
if $(\phi_{S(i)} \geq d_{S(P)})$
then $D_k = D_k + 2$
$B_k^t \cup \{S(P)\}; B_k^l \cup \{S(P)\}$ GOTO (5)
(3)
if $(\sigma_{S(P)} < \sigma_{S(P)}^o)$
then $D_k = D_k - 1$;
$B_k^e = B_k^e \cup \{S(P)\}$
else $D_k = D_k + 1$;
$B_k^l = B_k^l \cup \{S(P)\}$
if $(D_k \geq 0)$
then GOTO (5)
(4)

Let $\Delta = \min\{\min_{l \notin B_k^t}(d_l - \varphi_l); I_{k-1}; \min_{l \in B_k^e}(\sigma_l^o - \sigma_l)\}$
$\varphi_j = \varphi_j + \Delta$ and $\sigma_j = \sigma_j + \Delta$ for $j \in B_k$
Case 1: $\{\Delta == d_l - \varphi_l); l \notin B_k^t; \Delta \neq \sigma_l^o - \sigma_l\}$
$D_k = D_k + 1;\ I_k = I_k + \Delta;\ I_{k-1} = I_{k-1} - \Delta;$
$B_k^t = B_k^t \cup \{l\}$
if $(D_k \geq 0)$
then GOTO (5)
else GOTO (4)
Case 2: $\{\Delta == d_l - \varphi_l); l \notin B_k^t; \Delta == \sigma_l^o - \sigma_l\}$
$D_k = D_k + 2; I_k = I_k + \Delta; I_{k-1} = I_{k-1} - \Delta;$
$B_k^t = B_k^t \cup \{l\}\ B_k^e = B_k^e \setminus \{l\}\ B_k^l = B_k^l \cup \{l\}$
if $(D_k \geq 0)$
then GOTO (5)
else GOTO (4)
Case 3: $\{\Delta \neq d_l - \varphi_l); l \in B_k^e; \Delta == \sigma_l^o - \sigma_l\}$
$D_k = D_k + 2;\ I_k = I_k + \Delta;\ I_{k-1} = I_{k-1} - \Delta;$
$B_k^e = B_k^e \setminus \{l\}\ B_k^l = B_k^l \cup \{l\}$
if $(D_k \geq 0)$
then GOTO (5)
else GOTO (4)
Case 4: $\{\Delta = I_{k-1}\}$
$I_{k-1} = I_k + I_{k-1};\ D_{k-1} = D_k + D_{k-1};$
$B_{k-1} = B_{k-1} \cup B_k;\ B_{k-1}^e = B_{k-1}^e \cup B_k^e;\ B_{k-1}^l = B_{k-1}^l \cup B_k^l;$
$B_{k-1}^t = B_{k-1}^t \cup B_k^t;\ k = k - 1;$
if $(D_k \geq 0)$
then GOTO (5)
else GOTO (4)
(5)
if $(I_k > 0)$
then /* there is gap (initial or new)*/
$k = k + 1;\ D_k = 0;$
$B_k = B_k^e = B_k^t = B_k^l = \emptyset$
(6)
$P = P - 1;$
if $(P > 1)$
then GOTO (2)
else STOP.

Maintaining Robust Schedules by Fuzzy Reasoning

Jürgen Dorn — TU Wien
Roger Kerr — Univ. of New South Wales
Gabi Thalhammer — IBM Austria

Abstract: Practical scheduling usually has to react to many unpredictable events and uncertainties in the production environment. Although often possible in theory, it is undesirable to reschedule from scratch in such cases. Since the supplier of raw materials and clients will be prepared for the predicted schedule it is important to change only those features of the schedule that are necessary.

We show how on one side fuzzy logic can be used to support the construction of schedules that are robust with respect to changes due to certain types of events. On the other side we show how a reaction can be restricted to a small environment by means of fuzzy constraints and a repair-based problem-solving strategy.

We demonstrate the proposed representation and problem-solving method by introducing a scheduling application in a steelmaking plant. We construct a preliminary schedule by taking into account only the most likely duration of operations. This schedule is iteratively "repaired" until some threshold evaluation is found. A repair is found with a local search procedure based on tabu search. Finally, we show which events can lead to reactive scheduling and how this is supported by the repair strategy.

1 Introduction

Scheduling is a hard problem both in theory and practice. Theoretical scheduling problems, which are concerned with searching for optimal schedules subject to a limited number of constraints, suffer from excessive combinatorial complexity and are mostly NP-hard. The optimization aspects of the scheduling problem have been a subject of investigation by the Operations Research community for many years [18]. However, there are relatively few reports of the implementations of such systems in manufacturing practice. Practical problems although more highly constrained, are complex due to the number and variety of the constraints themselves, many of which are “soft”, e.g. potentially relaxable human preference constraints, rather than “hard” physical constraints. In addition, an effective schedule often needs to be evaluated against a number of potentially conflicting goals which themselves may not be precisely defined, and many of the scheduling parameters such as processing times, material arrival times, resource availability etc. are also subject to uncertainty. The optimization-oriented approaches of Operations Research have rather limited ability to express goals and constraints of this nature, and lead to problem representations that are often inadequate in terms of their correspondence with shop floor reality, and with the ways in which human schedulers can relate to the schedule construction process. The importance of adequate constraint representation has lead to the increasing use in scheduling of knowledge representation techniques developed by the Artificial Intelligence (AI) community [4, 7, 15, 17, 22, 27]. Such techniques allow for the explicit representation of imprecision, uncertainty, and relative importance in goals and constraints. Recent work by the AI community on the solution of constraint satisfaction problems is also of direct relevance to scheduling if the latter is regarded as the incremental construction (in terms of progressive assignment of start and finish times to operations) of a solution that satisfies the constraints in a problem space in which each additional assignment imposes a new set of constraints on the remainder of the solution.

In the dynamic environment of the shop floor, a variety of unexpected events are continually occurring and any schedule must in practice be subject to frequent revision to ensure it is in line with changing shop floor status. Scheduling is thus an ongoing and continuous process. The problem of updating schedules in the most effective way when the constraints or assumptions on which they are based are changed or invalidated is one that is receiving increasing attention amongst both researchers and practitioners, and is generally termed “reactive scheduling”.

The main alternatives to the revision of a schedule in the presence of real time shop floor feedback are either to incorporate the new information by completely regenerating the original schedule from scratch, or by "repairing" the previous schedule in some way. The first approach might in principle be better capable of maintaining optimal solutions, but as pointed out above, such solutions are rarely achievable in practice, and computation times are likely to be prohibitive. Furthermore, frequent schedule regeneration can result in instability and lack of continuity in detailed shop floor plans, resulting in increased costs attributable to what has been termed "shop floor nervousness". Thus most approaches to reactive scheduling are based on infrequent regeneration of a basic predictive schedule which then maintains continuity by serving as a nominal reference for the identification and specification of schedule changes as it is progressively modified [3, 20, 25, 30]. Predictive and reactive scheduling may thus be seen as complementary activities [29].

An important issue in which this complementary relationship between predictive and reactive scheduling is highlighted is that of *schedule robustness*. A robust predictive schedule is one which is likely to remain valid under a wide variety of different types of disturbances. Robustness is clearly a desirable attribute of a predictive schedule as it will reduce the number of subsequent reactive scheduling decisions required as the schedule is executed.

Robustness can generally be increased by the avoidance of unnecessary computational reaction to disturbances in a predictive schedule that has been specified to unrealistic degrees of precision. Although least commitment strategies are potentially useful in this respect, an alternative is to generate schedules in which the presence of uncertainty is recognized and explicitly represented. Temporal propagation of the uncertainty allows the possibility of the effects of unexpected events being localized to that part of the schedule where it is most certain they will have a significant impact rather than being propagated to far regions of the planning horizon where in reality they would be swamped by cumulative uncertainty arriving from other sources. Furthermore, evaluation of schedules incorporating uncertain processing times can provide a framework in which situations where critical events are tightly scheduled with little slack will be given a lower robustness score than those where slack exists. Although some work has been reported on the explicit representation of uncertain events in scheduling [2, 14, 23], the problem of reactive scheduling and schedule robustness has not been explicitly addressed in this context.

In spite of the fact that there exist some approaches in AI for combining and propagating uncertainty, there are many types of knowledge where a probabilistic interpretation of uncertainty is inappropriate. In particular, the assumption of independence of the contributing conditions for the firing of a rule (which is required for a probabilistic interpretation in which uncertainty factors are multiplied) does not hold in many circumstances. The incorporation of conditional or joint probability distributions to describe probabilities of occurrence of dependent events is usually impossible because the distributions are not known. Probabilistic interpretations are generally associated with randomness in data, in which facts or events are precisely defined, but the degree of certainty of their occurrence is not. However, a great deal of knowledge is imprecise, not because of randomness in the occurrence of precisely defined facts or events, but through vagueness or imprecision in the definitions of the facts or events themselves. Much of human knowledge processing in manufacturing companies is based on concepts, ideas and associations which are neither crisply defined nor describable in terms of probability distributions. For example, consider the following piece of knowledge:

If a job is *late* and of *priority* then classify job as *urgent*.

Here neither the concept of lateness, priority, or urgency are precisely defined in numerical terms. Yet to handle this rule computationally we should need to make the artificially precise division of jobs into the clear-cut classes of late or non-late, priority or non-priority jobs.

The imprecision of statements like this can be represented in a systematic manner using the concept of fuzzy sets, first formulated by Zadeh [32]. *Fuzzy sets* are classes of objects for which the transition from membership to non membership is gradual rather than being sharply defined. If we consider the set J of all jobs j_1, j_2, j_3, etc. on the shop floor, then fuzzy subset L of all late jobs can be defined by the membership function $f_L(j)$, which associates with each member j of the set J of jobs a degree of membership in the range [0, 1] of the fuzzy subset of late jobs. Every job on the shop floor thereby has some degree of lateness associated with it, in contrast to normal set theory where a job would have a degree of membership of either 0 or 1 of the subset of late jobs. The degree of membership of the job of the fuzzy subset of late jobs provides a measure of the degree of satisfaction of the conditional part of the rule referring to lateness. A good introduction to the use of fuzzy sets in decision making is given in [33].

In this paper we examine an application of fuzzy temporal reasoning to represent and propagate schedule uncertainty within the context of scheduling the operations of a steelmaking plant. This will be seen to assist in the generation of robust schedules which result in considerable reduction in the amount of reactive scheduling required. The examples are taken from an application at the Böhler Kapfenberg works, one of the most important European producers of high-grade steel. An expert system that supports the technical staff in the steel making plant in generating schedules for steel heats for one week has been previously developed [10]. However, the rule based nature of this system only allowed shallow modeling of the expert's knowledge, and several enhancements have been proposed to generalize the problem-solving method and to use deep knowledge instead of shallow rules [6, 7] and iterative improvement methods for problem solving [8, 11]. These enhancements include explicit representation of uncertainty and vagueness in scheduling parameters using fuzzy sets. Recently, a reusable (object-oriented) framework for the construction of intelligent scheduling systems was realized with these techniques [13].

2 The Steelmaking Process

The steel plant under consideration produces high-grade steel from crude steel using an alloying process for subsequent delivery to other plants which are concerned with forging and rolling. The basic processes to be scheduled are shown in block form in Figure 1. Crude steel is first melted with the relevant alloying metals such as manganese, tungsten, chromium, or others in an electric arc furnace (EAF). The melted steel is then processed in the secondary metallurgy (SM) first by pouring into ladles that are transported by a crane to a ladle furnace where fine alloying takes place, then on to a special treatment in a vacuum oxygen decarburation unit. The next step is the processing of the steel in a continuous caster (CC) to form slabs or the casting of it into moulds to form ingots. For jobs using the CC, if two sequential jobs using the same format are processed, the second job must be delivered in time to the caster to allow it to cast continuously otherwise a set-up time must be considered.

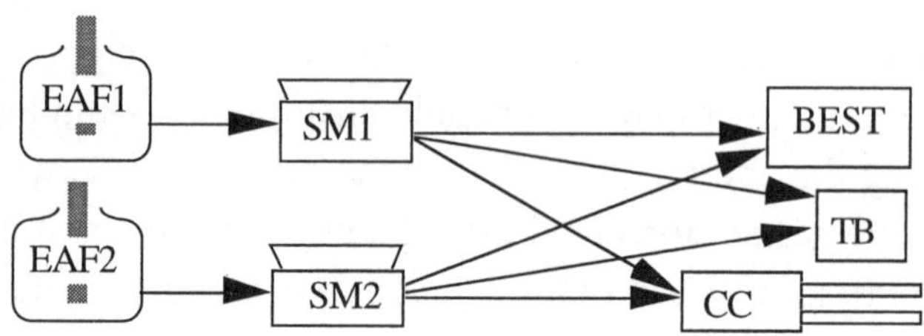

Figure 1: Simplified production lines of the plant

For solidification of ingots, the moulds require space in a teeming bay (TB). For some large ingots, the Böhler-Electro-Slag-Topping (BEST)-technology (a special casting technology for ingots) is used. One space in the TB is dedicated to such ingots. The steel produced is then delivered to other plants including different rolling mills and forges which order slabs or ingots of a certain quality. Sometimes products are stored for several days in intermediate stockyards, because the plant that has ordered the product cannot process the jobs in the same sequence as the steel making plant. The differing sequencing criteria of jobs cause considerable costs for the company since a large amount of capital is invested in these semi-finished goods. Moreover, since the steel cools down it must be reheated in the next plant. To reduce costs and to improve the quality, some orders have due dates at the subsequent plant which must be met as closely as possible.

As indicated in Figure 1, two production lines share the CC and the TBs. For every steel quality there is a process plan that prescribes on which line the job will be produced and which processes it will require. Once a week, engineers of the different plants meet to discuss orders for the next week. Compatible orders that may have different destinations may be combined to form jobs. For each production line of the steelmaking plant a list of jobs for one week is worked out manually. The task of the scheduler is to find a possible sequence for all jobs which violate as few compatibility constraints as possible and to allocate resources over time without violating temporal and capacitive constraints. The result of this scheduling process may be that some orders are rejected and shifted to the next week.

2.1 Constraints in Scheduling

During the construction of a schedule several constraints have to be considered. These constraints are often vague and conflicting. The engineer has no pretension to generate an optimal schedule because he knows that the uncertainty in the execution of his plan would break this optimality, but he does know that some schedules are better than others. Three kinds of constraints may be distinguished: compatibility, temporal, and capacitive constraints. Compatibility constraints between the permissible relative chemical compositions of sequentially processed jobs are the most important constraints on the schedule. Residuals of one heat in the EAF may pollute the next heat and so the engineers use as a rule of thumb that 3 % of a chemical element in a heat remain in the wall of the EAF and 3 % of the difference of the elements in two consecutive heats will be assimilated by the second heat. The 3 % is in fact always on the safe side and in some cases the expert relaxes this factor. However, since these constraints are not central to the issue of reactive scheduling, we do not consider them further here and refer to [7] for more details.

Since some steel qualities require the cast steel to be hot for subsequent treatments such as forging, due dates exist that should preferably be met to within a tolerance of ± 2 hours. The average number of jobs with such due dates is about 10 %. These due dates are soft rather than hard constraints since they may be relaxed through negotiations with the subsequent plant.

Waiting time between operations means either that the steel cools down and must be reheated or it must be heated continuously during waiting. Both result in costs and for some steel grades, unfavorable effects on the quality may result. The scheduler must therefore guarantee that slack times between the successive operations of a job may not exceed a certain limit. An objective for production is thus to have slack intervals as small as possible, resulting in minimization of the flow time and reduced production costs. Also, jobs on the CC should be scheduled either continuously or with sufficiently large breaks to allow set-up and maintenance. Since the CC is a resource which is shared between the two production lines, it is important to schedule it effectively and to ensure that jobs arrive at the correct time.

BEST-ingots introduce some problems. Typically there are groups of such jobs which tend to be low alloyed, forging grade ingots with the same chemical quality requirements. From the compatibility view point they should be produced in sequence. However, since only one place exists for them in the TB, and the solidification time tends to be considerably longer than the other process times, they cannot be scheduled immediately one after the other. This can present a problem especially for big ingots which have a very long solidification time. Typically they need to be scheduled with sufficient gap between them to allow for solidification of one job in the TB to be complete before the next job arrives.

2.2 Uncertain Process Times and Reactive Scheduling

Individual processing times of jobs depend on the type and quantity of steel being processed. The duration of the SM and the CC processes are to some extent controllable so that slack time can be reduced. This controllability will be ignored in the discussion that follows which centers on the uncontrollable and unpredictable components of the process time only. All processes have an inescapable element of variability, which may be captured in the form of possibility distributions. Assuming jobs of average size, the diagrams in Figure 2 represent estimates by experienced schedulers of the possibility distributions of the times required for different processes. Each diagram represents the degree of membership of the natural numbers within the indicated range, of the fuzzy set of time durations for that process. Time durations are thus measured by trapezoidal fuzzy numbers. However, for the purpose of improving the understandability of temporal propagation in subsequent discussion we also show in Table 1 three characteristic durations, i.e. the minimal, the most likely, and the maximal duration. The most likely duration can be computed from the fuzzy sets by determination of the center of gravity.

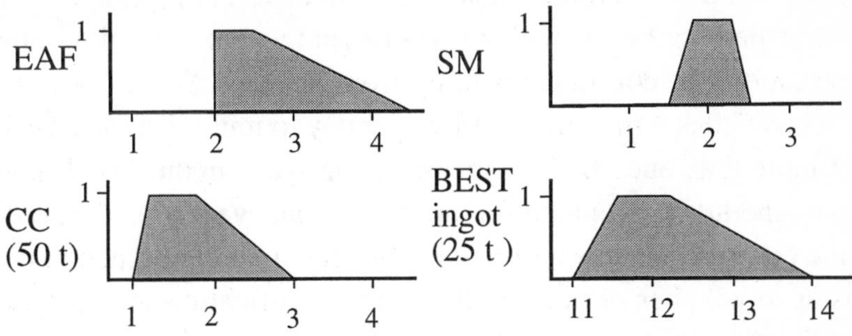

Figure 2: Process durations modeled by fuzzy numbers

duration \ aggregate	EAF	SM	CC	BEST
minimal	2	1.5	1	11
most likely	2.5	2	1.5	12
maximal	4.5	2.5	3	14

Table 1: Durations of operations

From the extent of the variability shown by these process times, it is clear that potential problems can occur as a result of jobs taking longer than their expected process times. A schedule that has been constructed to satisfy the relevant constraints can be invalidated due to constraint violations when jobs take other than the process times assumed for the purposes of constructing the predictive schedule. Since the bottleneck resources tend to be the EAFs, a single job taking longer than normal to process on the EAF can result in all subsequent jobs being later than their scheduled times, thus causing due dates to be missed. This can also have serious implications for jobs processed on the CC. For example if a job arrives from one production line for processing on the CC but this is still processing a late job from the other production line, the first job may be destroyed if it has to wait too long. The same type of problem may occur with jobs using the BEST-technology where if the dedicated TB is still occupied by the previous "BEST"-job which is late, the job requiring the TB must wait and will block other jobs in the SM.

In practice, the engineers who schedule the plant attempt to construct schedules that are reasonably robust with respect to the variability in the process time. Thus jobs from different production lines are scheduled on the CC with gaps between them to try and minimize the possibility of conflicts resulting from lateness. "BEST"-jobs are also scheduled at a maximum of two per day to minimize the possibility of more than one "BEST"-job requiring space in the TB. The predictive schedule is prepared for a period of one week ahead, and during the course of its execution, the engineers make judgmental decisions as to whether or not it will require modification in the light of unfolding events. For example, a job that takes longer than usual to process on the EAF may, in a crisp evaluation of the schedule, result in a subsequent job being predicted to just miss its due date. However, rather than rescheduling this job, the engineer knows that the degree of uncertainty in the events still to occur before that job is processed will mean that this minor constraint violation is not yet significant at this stage and will not perform any rescheduling.

An expert scheduling system was constructed for use in the plant [10] which used crisply defined average process times for the construction and evaluation of the schedule. This was based on the use of simple heuristics similar to those used by the human scheduler to construct an initial schedule which was then evaluated against the applicable constraints, and subject to a "repair" strategy to attempt to improve it by reducing the number of constraint violations. This technique is clearly directly applicable also to the reactive scheduling problem in which unfolding events generate constraint violations which in principle could be subject to repair using the same strategy as the predictive scheduler. However, the assumption of crisp process times made in the original scheduler did not allow the possibility of grading reactively generated constraint violations according to the degree of uncertainty surrounding them, and of using this as a basis for selectively repairing only those violations that were significant. Furthermore, use of crisp process times also inhibited the evaluation of schedules according to their relative degrees of robustness against contingencies, and the establishment of a trade-off between robustness and cost (which the human schedulers performed implicitly). In the next section we describe, in the form of a simple example, a modified form of the predictive scheduler which uses fuzzy rather than crisp process times for the evaluation of schedules in order to partially overcome the problems outlined above.

3 Schedule Generation

The strategy used by the system to generate a schedule is to use constructive heuristics based on those used by human schedulers to produce an initial schedule assuming that all processing times take on their most likely values. Jobs are scheduled in order of their criticality (defined later) with no backtracking. Some constraints may hence be violated. However, when the schedule is evaluated for such constraint violations, the crisp processing times are replaced by their fuzzified possibility distributions, and temporal events are predictively propagated using rules for the addition of fuzzy numbers. The result of this is that temporally distant events whose timing has been computed by the addition of several fuzzy processing times will have broader possibility distributions than temporally close events whose timing is dependent on the execution of only one or two processes.

Temporal constraint violations are thus not expressed in a binary satisfied/unsatisfied form but as a matter of degree, with each temporal constraint having a degree of satisfaction in the range [0,1] based on the amount of overlap of the temporal estimates of events whose relationships are constrained. Thus the extent to which a due date constraint is met is measured by the degree of overlap between the due date (which may itself be fuzzily defined) and the fuzzy estimate of the finishing time of the job.

Threshold values, which may be zero, can be defined for acceptable and unacceptable degrees of constraint satisfaction (which can vary according to the type of constraint). If the degree of satisfaction of a constraint is unacceptable, then the schedule must be repaired by rescheduling in such a way that the threshold value for the constraint is met. The evaluation of the total schedule is performed by combining the degree of satisfaction of the individual constraints (which may themselves be weighted) in some appropriate way. An iterative improvement method based on tabu search is used to attempt to improve the schedule score by progressive interchange of jobs.

This also allows the incorporation of robustness into schedule evaluation. For example, it is critical that jobs scheduled on the CC which originate from different production lines do not overlap. However, the closer they are scheduled together on the CC, the higher will be the degree of overlap of the fuzzy estimates of the time period over which they will require the CC, and the lower will be the robustness of the schedule. This fact can be incorporated into the schedule evaluation function with a weighting factor that gives the appropriate trade-off between robustness and cost. If a job takes longer than its normal processing time but (for example) this does not result in any due date constraint satisfaction dropping below its critical threshold, no rescheduling action need necessarily be taken other than the routine application of the iterative improvement method.

By means of a small demonstration we now show how a schedule is constructed, evaluated, and repaired when presented with unfolding events. We assume that within the planning horizon a preliminary schedule can be built from the given jobs for each production line. We start with one production line and later extend this example to the second line to highlight the concept of schedule robustness.

3.1 Generating a Preliminary Schedule with Domain Heuristics

For simplicity, we restrict ourselves to a planning horizon of one day and assume that we can schedule eight jobs. The preliminary schedule is constructed with simple heuristics which only implicitly consider the different constraints. The duration of the operations are represented crisply with a relatively large granulation. We will assume two and a half hour melting time in the EAF, two hours in the SM, one and a half hour on the CC and five and a half hour for normal ingots, and twelve hours for the BEST-ingots.

Jobs to be scheduled are displayed below in Table 2. The "job no" identifies the job (e.g. j_3^1 means first production line, third job). The column "type" reveals which product, a slab or an ingot with or without BEST-technology, should be processed. The "due date" imposes an additional time constraint on some jobs. The due date should be met by the finishing time of the last operation of the job. The "importance" of jobs is given a priori and is related to customer/market considerations. Linguistic variables have been used to describe job importance. From the importance of jobs and their specific characteristics a criticality value is computed that is used to decide in which sequence the jobs will be scheduled. An example for such a fuzzy rule is as follows:

IF the proportion of BEST-jobs is *greater* than 0.3 **AND** a job J is a BEST-job **AND** the *importance* of a job J is *high*
THEN the *criticality* of job J is *very high*

job no	type	due date	importance	criticality
j_1^1	BEST		*medium*	*very high*
j_2^1	BEST		*medium*	*very high*
j_3^1	slabs	6 am	*very high*	*high*
j_4^1	ingots	3 pm	*high*	*medium*
j_5^1	slabs		*medium*	*low*
j_6^1	ingots		*medium*	*very low*
j_7^1	ingots		*low*	*very low*
j_8^1	ingots		*very low*	*very low*

Table 2: List of jobs to be scheduled on the first production line

As there is only one place for BEST-ingots which need a long solidification time compared to the preceding processes, the jobs that produce BEST-ingots may not be scheduled immediately after another. An interval should be left between these jobs that allows the first ingot to have solidified by the time the second is cast. In this case the heuristic assumes an interval of twelve hours should be left between "BEST"-jobs (which can of course be later relaxed to some degree according to prevailing circumstances). The jobs processed on the CC should be scheduled in sequence so that the two heats can be cast continuously. Jobs with a due date will be scheduled next. As there is always enough room in the TB for normal ingots no capacity constraint is imposed there. The system now generates a schedule by scheduling the important or most difficult jobs as described below.

First we schedule the two "BEST"-jobs (j^1_1, j^1_2) as far apart as possible. The first starts at 2 pm which will be the start of our planning horizon and the second is scheduled at the end. Since we assume that each job takes two and a half hour on the EAF and we have eight jobs, the second "BEST"-job is scheduled at 7.30 am.

Next we schedule the caster job with a due date (j^1_3). We back schedule the operations of this job from the time due, assuming crisp processing times, and obtain a scheduled start time of 12.00 pm for this job. The second job with a due date is j^1_4. Again, scheduling backwards from the time due, and assuming the solidification time is five and a half hour, a starting time of 5 am is assigned to this job. Now the second job to be processed on the CC can be scheduled in sequence with the other "caster"-job. Finally, we also assign start times to the three remaining jobs in order of their criticality value and arrive at the schedule in Figure 3.

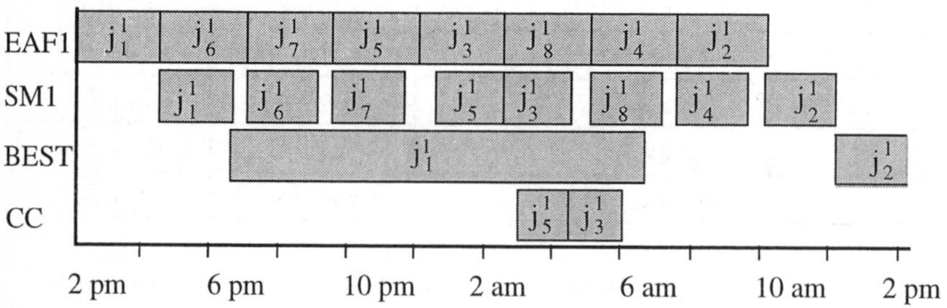

Figure 3: Preliminary schedule for first production line

3.2 Schedule Evaluation

A schedule is evaluated in terms of the degree of satisfaction of the applicable constraints. Schedules with high satisfaction scores are good schedules. A schedule for which the degree of satisfaction of any constraint is below the critical threshold for that constraint is not feasible. Since it is not always possible to find a feasible schedule for all jobs, we also evaluate the schedule in terms of the number of jobs successfully scheduled, weighted by their average importance.

We assume the importance of each job J_i is given by a fuzzy value denoted by the function *importance(J_i)*. Further, we assume that n is the number of existing jobs. We then compute the mean importance of all jobs that are scheduled. In our example we achieve a mean importance that is *medium*. The degree of satisfaction of a constraint is given by the function $satisfaction(C_j)$, where C_j is a constraint of type j from which m different exist. Since not all constraints have the same importance for the schedule evaluation, the satisfaction is weighted by a type-specific factor. Although a constraint satisfaction which is below the critical threshold (which can also represent violation of a hard constraint) should lead to a refusal of a schedule, in the evaluation function a strongly violated constraint which is nevertheless above its threshold may be neutralized by other unviolated constraints. The evaluation of a schedule S is then:

$$f(S) = \frac{1}{n}\sum_{i=1}^{n} (J_i \ isScheduledIn \ S \wedge importance(J_i)) +$$

$$\sum_{j=1}^{m} [Cj \ existIn \ S \wedge satisfaction(C_j) * weight\ (type(Cj))]$$

As mentioned previously, a very important set of constraints deals with chemical compatibility between adjacent jobs. However, to give better visibility to the temporal aspects of the reactive scheduling problem, we shall not consider them explicitly here. We thus have three explicit constraints in our problem. The jobs j^1_3 and j^1_4 have due dates and the constraint that j^1_1 and j^1_2 must be separated to the extent that the possibility they will conflict in their demand for space in the TB is acceptably low.

To model the evaluation of the due date constraint, we define a fuzzy set A which is the set of all jobs that approximately meet their due date dd within the interval of [dd-2, dd+2] hours. The membership function will evaluate to $m_A(x) = 1$ if the due date x is met within the interval of [dd - 1, dd + 1] hours which we define as an *in time* value for this constraint. It will evaluate to $m_A(x) = 0.6$ for a due date violation of ± 2 hours which we define as *early / late*. Finally, it will evaluate to $m_A(x) = 0.3$ for a violation of ± 3 hours which we define as *very early / very late*. Outside the fuzzy set A it will evaluate to $m_A(x) = 0$ for a violation larger then ± 4 hours. The membership function values in between can be linguistically described in terms of for example *almost very good* for 0.7 and so on.

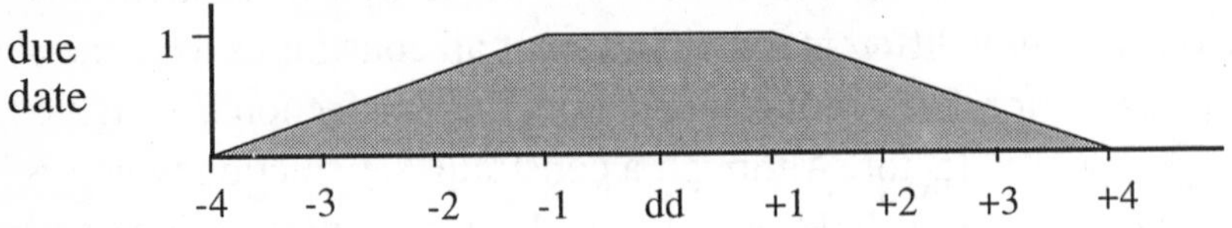

Figure 4: The fuzzy function for the evaluation of the due date constraint

For our preliminary schedule, events must now be propagated forward from the crisply defined start of the planning horizon to obtain a fuzzy estimate of the predicted completion times of the jobs with due dates. This is obtained by addition of the fuzzy process times defined in Figure 2 for all jobs that must be processed on the EAF prior to the jobs with due dates (which determine the fuzzy start times of these two jobs) and the fuzzy process times of the jobs, themselves. Addition of fuzzy process times is performed by rules defined in [21]. The boundaries on the resulting fuzzy finishing times for j_3^1 are shown in Figure 5 which displays the propagation of earliest start and latest finish times of jobs up to and including j_3^1. The resulting fuzzy finishing time of j_3^1 is shown in Figure 6 in relation to the fuzzy due date.

As four jobs precede j_3^1 and each adds about 2 hours possible delay on the EAF the starting of j_3^1 could be delayed at most 8 hours at the most (see Table 1). The fuzzy duration of j_3^1 shows that it can take +2 hours on the EAF, +0.5 hours on the SM and +1.5 hours on the CC. So the total latest finish time possibly predictable will be the scheduled finish time which is 6 am + 8 hours + 4 hours = 6 pm (which is almost impossible as expressed by the low membership degree in Figure 6). Even though the operation of the preceding jobs on the SM can also vary, it will not change the overall completion time as the worst duration on the EAF will dominate.

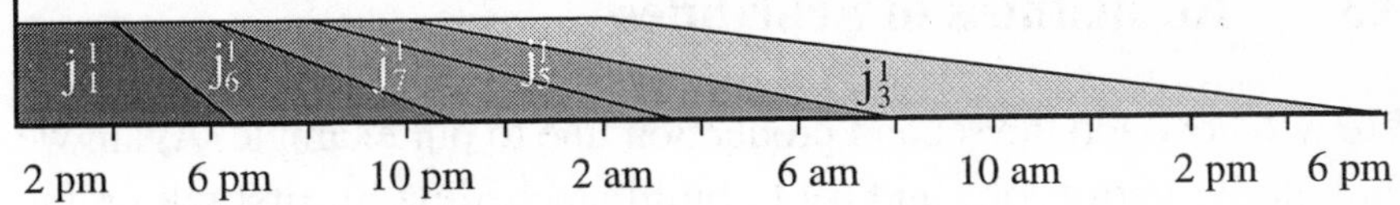

Figure 5: Fuzzy starting times of j_3^1 obtained by fuzzy temporal propagation

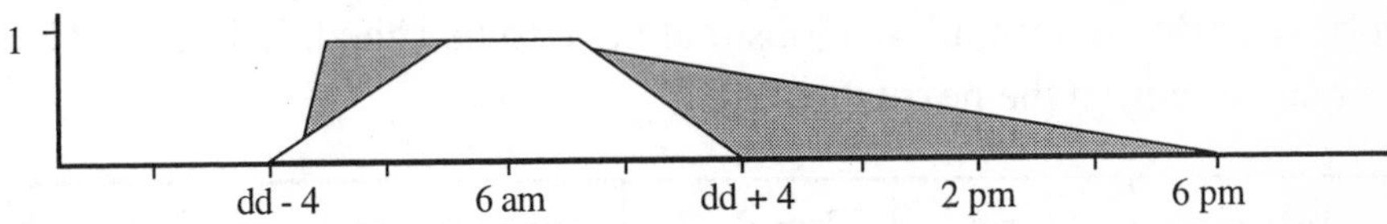

Figure 6: Degree of satisfying the due date constraint for j_3^1

To determine the degree of satisfaction of a due date, we apply a set of rules that map the topological features of the overlap between the fuzzy finishing time and the fuzzy due date to the interval [0,1]. These values can in turn be mapped to linguistic variables. We assume that our rules give a degree of satisfaction of 0.8 for the job j_3^1. If we assume that the critical satisfaction threshold for the due date constraint is set at 0.4 a further set of rules could map this degree of constraint satisfaction to the linguistic variable *high*. As seen in Figure 6 the overlap within the range of the most likely duration is very large.

Fuzzy temporal propagation is also used to determine whether there is enough time for the first BEST-ingot to solidify before the second one is cast. The overlaps of the fuzzy estimates of the finish time of the first "BEST"-job and the start time of the second are evaluated by event propagation. The rules which map the topological features of this overlap to the [0,1] interval generate a degree of satisfaction of 0.9 for this constraint, which is above the critical threshold of (say) 0.5 and is in turn again mapped to the linguistic variable *high* (i.e. resulting in a high satisfaction of this constraint). The overall evaluation of the preliminary schedule in terms of constraint satisfaction can then be determined using rules which combine the degrees of satisfaction of the individual (weighted) constraints. We assume that these rules result in a score of 0.9 for the overall schedule, which may also be mapped to the linguistic variable *high*.

3.3 Robustness of Schedules

We will now add the second production line to our example. Again we look at the list of jobs and try to build a schedule by first taking the important and difficult jobs into consideration. We will not explain how the normal ingots are scheduled and concentrate on the critical jobs which use the CC and the TB for BEST-ingots. We have two jobs with due dates and two jobs that have to be scheduled on the CC as can be seen in the next table.

job no	type	due date	importance	criticality
j_1^2	ingots	7 am	*high*	*high*
j_2^2	ingots	1 pm	*medium*	*high*
j_3^2	slabs		*low*	*very high*
j_4^2	slabs		*low*	*very high*
j_5^2	ingots		*medium*	*low*
j_6^2	ingots		*low*	*very low*
j_7^2	ingots		*very low*	*very low*
j_8^2	ingots		*very low*	*very low*

Table 3: List of jobs to be scheduled on the second production line

Since we have also two CC-jobs on the other production line these jobs are critical in terms of attempting to avoid conflicts for the CC. They should be scheduled in sequence. However, the due dates for the other two jobs must also be met if possible. We also have to guarantee that even if the jobs of the second production line scheduled on the CC are late they will not conflict with the jobs of the first production line. So the further we schedule those jobs apart the more robust our schedule will be. We assume we have constructed the schedule depicted in Figure 5 with our heuristics:

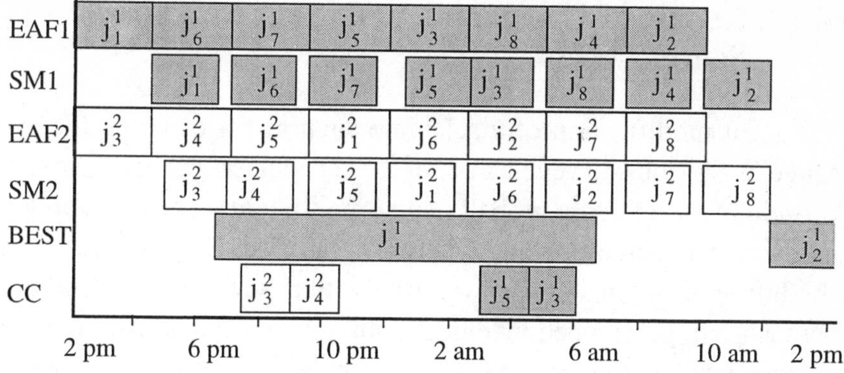

Figure 7: Preliminary schedule for both production lines

The predicted fuzzy finishing time of j_4^2 on the CC is shown in Figure 8 in relation to the predicted fuzzy start time on the CC of j_5^1. It will be seen that the presence of process time uncertainties leads to an overlap between these two events which implies the possibility of conflict. A set of rules is then invoked which reflects the human schedulers attitude to this degree of overlap in terms of schedule robustness, and we shall assume that it maps to a value of 0.3 in terms of the degree of robustness of the schedule with respect to this particular CC conflict constraint. If we further assume the mapping rules were developed on the basis of the acceptability criterion for degree of satisfaction of this particular robustness constraint being 0.5, then it is apparent that this schedule is unacceptable with respect to robustness, and must be repaired.

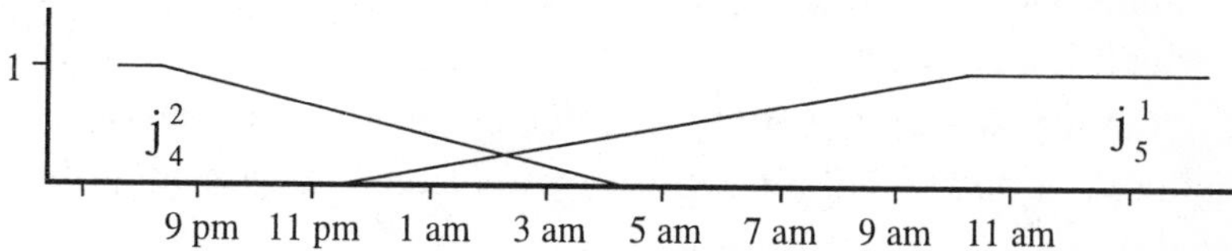

Figure 8: Potential conflict between jobs from different production lines that use the CC

3.4 Repair of Constraint Violations by Tabu Search

By a local modification of a schedule such as for example the exchange of two jobs, the evaluation of the schedule might be improved. To optimize a schedule globally such modifications can be applied iteratively until some stopping criterion is satisfied. However, for most scheduling problems it is very likely that such a hill-climbing approach will get trapped in a local optimum. To avoid this, iterative improvement techniques such as tabu search [1], simulated annealing [31], and genetic algorithms [16] have been applied to scheduling and related optimization problems. With these methods no optimal solutions can be guaranteed but usually the results achieved are very good. The latter two approaches apply a random choice of modification operators. For tabu search used here, we generate a sample of neighbors and then select the best of these schedules. How this sample is constructed will of course have considerable influence on the effectiveness of the search. We will sketch our approach here, but will not go into details since this is out of the scope of this paper. For details the reader is referred to [8] and [11].

Motivated by the min-conflicts strategy of Minton *et al.* [26] and the repair-based strategy of Zweben *et al.* [31] we try to repair the greatest constraint violation. Our repair-based strategy tries to improve the overall schedule evaluation in terms of fuzzy values. It is an iterative process consisting of single repair steps. The procedure first determines a significant constraint violation (i.e. a constraint satisfaction which is below its critical threshold) and tries to repair it. By applying the repair step to satisfy a violated constraint of our schedule, we achieve a new schedule whose evaluation function must be improved.

The repair process will iterate as long as there are any degrees of constraint satisfaction below their thresholds. Afterwards the repair is used to optimize the schedule further by attempting to increase the degree of satisfaction of remaining constraints that are above the threshold level. As a stopping criterion for the optimization we use the ratio between achieved improvement of a repair step and the number of applied repair steps. If this ratio becomes smaller than a predefined value we stop the search.

In its simplest form, a repair step is an exchange of two jobs that improves the schedule. For example, in the case of a due date constraint one job is involved. To repair an unacceptable constraint satisfaction value, the job is exchanged with any other job and the new schedule is evaluated. If the constraint satisfaction degree has increased to above its threshold, and no other constraint satisfaction has dropped below the corresponding threshold then the repair step is finished. Otherwise, other exchanges have to be tried. Sometimes we have more jobs than we can schedule. In this case an exchange with a job in the list of unscheduled jobs is also allowed.

In the example discussed, the only unacceptably satisfied constraint relates to the robustness issue which is a conflict that may occur on the CC. The repair-based algorithm tries to reschedule either j_4^2 or j_5^1. The exchange of the jobs j_5^1 and j_8^1 results in the best evaluation of the overall schedule. In this case, no other constraint will be violated and the violation of the robustness constraint is repaired.

In general, a repair step can consist of several exchange operations in order to escape local optima. Tabu search always selects the best neighbor, but to escape local optima it is also possible to select an operator that leads to a worse schedule. To avoid reversals and cycles, attributes of past solutions are stored in a tabu list. New generated solutions that have this attributes are "tabu" unless some aspiration criterion overrides this tabu status. Furthermore, after t modifications the tabu attribute is removed, because the list has a restricted length of n items.

We store a job and its old position in the schedule in the tabu list. Thus modifications that move a job back to its old position are tabu. After seven modifications the attribute is deleted and the job may return again to its old position. Moreover, our aspiration criterion allows the overriding of a tabu status if the new schedule achieves a new best evaluation. Initial experiments have shown that we always find a good feasible schedule within a quite small number of repair steps.

4 Reactive Scheduling

Repairing a schedule during its first construction is a similar process to the reaction to unexpected events. Such an event may cause a constraint to be violated. For example, if one operation takes longer than expected, a due date constraint may be violated. We can now evaluate this violation and if necessary, apply the repair strategy to solve it. We assume that through the occurrence of event *E* the evaluation of our schedule deteriorates. The optimal reaction would be that we find a set of repair steps so that the new evaluation would be as good or better than the old evaluation (before *E*).

However, to avoid frequent changes of the schedule due to very small deviations between predicted and actual times, we should define thresholds as to what events are significant. The first threshold is concerned with the event itself and is set so as to prevent re-evaluating the schedule as a result of events that are obviously trivial. Re-evaluation should only be performed if the event *E* is regarded as significant as measured for example by occurring at a time that deviates by an amount greater than some absolute threshold (e.g. 10 mins) from its predicted time. The second threshold relates to the degree of change in the schedule evaluation score, or in degrees of satisfaction of individual constraints, that can be tolerated before repair is considered necessary.

As a simple example, consider that the melting of job j_1^1 is finished in two hours rather than the expected 2.5 hours. If we do a new evaluation that consists of a new temporal propagation, the likelihood that the subsequent jobs will finish earlier than expected is greater, since they all start earlier. The fuzzy finish time of job j_3^1 can be shown in relation to its fuzzy due date. In Figure 9 job j_3^1 has a slightly changed latest finish time (compared with Figures 5 and 6). For the latest finish time of j_3^1 we assume all prior jobs to take 2 hours longer, but now we can reduce this uncertainty as job j_1^1 has definitely taken 2 hours instead of possibly four and a half hour. So we are gaining two and a half hour.

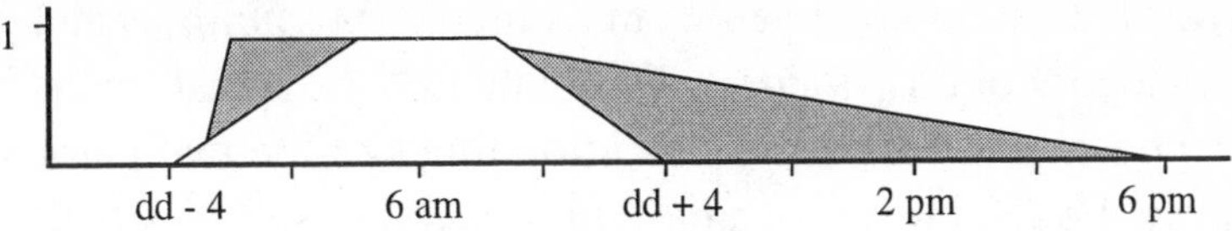

Figure 9: Degree of satisfying the due date constraint for j_3^1

The fuzzy finishing time does not extend so far down the time horizon since there are fewer uncertain process times remaining on which the timing of this event depends. However, our rules for mapping of the topology of the overlap with the fuzzy due dates on to [0,1] do not change the degree of constraint satisfaction, so no repair is necessary. This would contrast to the crisp case in which use of average times would lead to the prediction the job would be too early and would have to be rescheduled. As we progressively execute the schedule, the width of the fuzzy estimates of the finishing time of this job progressively narrows, to the point where it is sufficiently accurate as to tell us whether or not a significant problem exists and rescheduling should occur.

As a further example, consider the case in which melting of both jobs j_3^2 and j_6^1 are finished at 6 pm. In this case, the critical constraints are the degree of satisfaction of the robustness constraint relating to the CC, and the degree of satisfaction of the due date constraint for j_1^2, which is now more likely to be too late. The relevant fuzzy time estimates are again obtained by fuzzy temporal propagation. Figure 10 shows the evaluation of the robustness constraint in relation to the robustness of the repaired schedule.

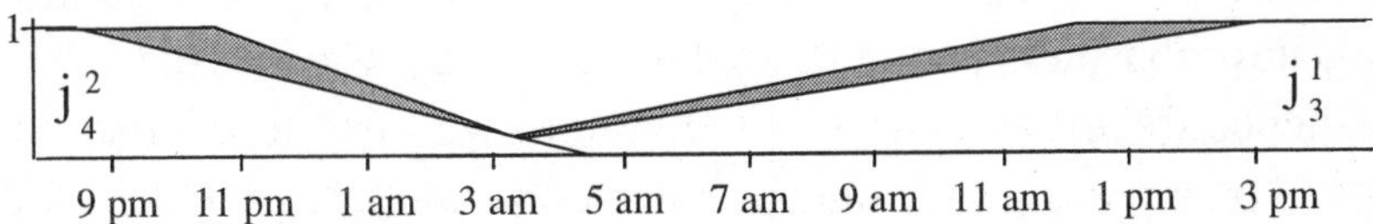

Figure 10: Potential conflict between jobs from different production lines that use the CC

Examination of the new overlap between the fuzzy finishing time of j_4^2 on the CC and the fuzzy start time of j_3^1, shows that this overlap has been slightly reduced due to the fact that some of the uncertainty in the schedule has been removed by the execution of three jobs (their process times being removed from a range of possibilities and placed in the category of crisp values). Thus although the actual events appear to have pushed the expected execution times of the two pairs of jobs requiring the CC closer together, the schedule itself is actually more robust than before because the event timings are known more precisely.

We find however that late finish of job j_3^2 results in a problem with j_1^2 now being too late (causing the due date constraint satisfaction function to drop below its threshold). Thus although no rescheduling is required from a robustness point of view, we do require to reschedule from the point of view of the due date constraint of j_1^2. As before, tabu search is applied to repair this constraint violation. The best choice is to exchange j_1^2 and j_5^2.

5 Conclusions

This paper has demonstrated by means of a simple example, how fuzzy temporal representation of schedules, coupled with a repair-based scheduling strategy, can assist in the explicit representation of the robustness of a predictive schedule, and through the setting of constraint satisfaction thresholds, in the evaluation of whether and the extent to which a predictive schedule needs to be modified during the progress of its execution. First prototype implementations are described in [5, 19]. At present, the degree of robustness of the schedule is measured simply in terms of degrees of overlap of certain critical events. Other measures of robustness are obviously possible and work is in progress on the development of a new system that evaluates robustness in a more sophisticated way. Clearly, the behavior and hence the effectiveness of the system will also to a large extent depend on the threshold values of each constraint. These values therefore require a considerable degree of fine tuning in an environment in which some measure of effectiveness of schedule reactivity can be systematically related to individual threshold values. It is anticipated that simulation will be used to assist in this task. First experiments on tuning weights are reported in [24] and [28]. The scheduling system is also in the process of extension to represent additional temporal constraints introduced by shifts, the working hours of subsequent plants, and production set-ups and maintenance operations. The issue of controllability of process times as discussed in [14] and the implications of this controllability for reactive scheduling and robustness are also being investigated.

Tabu search as well as the application of other iterative improvement techniques are still under investigation. Usually in iterative improvement techniques like tabu search [1], simulated annealing [31], or Min Conflicts strategy [26] only local modifications are tried. In this case we search for repair steps in the entire schedule. This can be very time consuming. Looking at the last example we can also see that more sophisticated repair steps are possible. Consider the case where the long duration of j_3^2 in the EAF had also violated the due date of j_2^2. The iterative improvement method would solve both constraints separately in two repair steps. However, an exchange of the jobs j_1^2, j_6^2, and j_2^2 with job j_5^2 would be more appropriate. Experiments with case-based reasoning have shown here benefits [9].

References

[1] Barnes, J.W. and Laguna, M. A. Tabu Search Experience in Production Scheduling. *Annals of Operations Research* **41**, pp. 141–156, 1993.

[2] Bel, G., Bensana, E., Dubois, D., Erschler, J. and Esquirol, P. A Knowledge Based Approach to Industrial Job Shop Scheduling, in A. Kusiak (ed.) *Knowledge Based Systems in Manufacturing*, Taylor and Francis, pp. 207–246, 1989.

[3] Burke, P. and Prosser, P. A Distributed Asynchronous System for Predictive and Reactive Scheduling, *Artificial Intelligence in Engineering* **6**, pp. 106–124, 1991.

[4] Collinot, A., LePape, C., and Pinoteau, G. SONIA: A Knowledge-based Scheduling System, *Artificial Intelligence in Engineering* **2** (2) pp. 86-94, 1988.

[5] Dagostar, A. S. A Decentralized Reactive Fuzzy Scheduling system for Cellular Manufacturing systems, Ph. D. Thesis Univ. of New South Wales, Australia, 1996.

[6] Dorn, J. Supporting Scheduling with Temporal Logic, *Proceedings of the IJCAI'93 Workshop on Production Planning, Scheduling and Control*, Chambéry, France, 1993.

[7] Dorn, J. and Slany, W. A Flow Shop with Compatibility Constraints in a Steel making Plant, in M. Zweben and M. Fox (eds.) *Intelligent Scheduling*, Morgan Kaufmann, pp. 629–654, 1994.

[8] Dorn, J. Iterative Improvement Methods for Knowledge-based Scheduling, *AICOM* Journal, pp. 20–34, March, 1995.

[9] Dorn, J. Case-based reactive scheduling in Roger Kerr and Elisabeth Szelke (Eds.) *Artificial Intelligence in Reactive Scheduling*, London: Chapman & Hall, pp. 32-50, 1995.

[10] Dorn, J. and Shams, R. Scheduling High-grade Steel Making, *IEEE Expert*, February, pp. 28-35, 1996.

[11] Dorn, J., Girsch, M., Skele, G. and Slany, W. Comparison of Iterative Improvement Techniques for Schedule Optimization, *European Journal on Operational Research* **94**, pp. 349–361, 1996.

[12] Dorn, J., Girsch, M. and Vidakis, N. DÉJÀ VU – A Reusable Framework for the Construction of Intelligent Interactive Schedulers, *Advances in Production Management Systems - Perspectives and Future Challenges* -, Okino et al. (eds.) Chapman & Hall, 1998.

[13] Drummond, M., Swanson, K. and Bresina, J. Robust Scheduling and Execution for Automatic Telescopes in *Intelligent Scheduling* Zweben and Fox (eds.) Morgan Kaufmann, pp. 629–654, 1994.

[14] Dubois, D. Fargier, H. and Prade H. The use of fuzzy constraints in job-shop scheduling. *Proceedings of the IJCAI'93 Workshop on Knowledge-based Production Planning, Scheduling, and Control*. 1993.

[15] Fox, M. S. *Constraint-Directed Search: A Case Study of Job-Shop Scheduling*, London: Pitman, 1987.

[16] Fox, B.R. and McMahon, M.B. Genetic Operators for Sequencing Problem, in G.J.E. Rawlings (ed.) *Foundations of Genetic Algorithms*, pp.284–300, 1991.

[17] Fox, M. S. ISIS: Retrospective, in Zweben and Fox (eds) *Intelligent Scheduling*, Morgan Kaufmann, pp. 3–28, 1994.

[18] French, S. *Sequencing and Scheduling – An Introduction to the Mathematics of the Job-Shop*. Chichester: Ellis Horwood, 1982.

[19] Glaser, J. Tabu-Suche für Reaktives Scheduling anhand eines Beispiels aus der Stahlindustrie, Diplomarbeit Technische Universität Wien, 1996.

[20] Hadavi, K. C. ReDS: A Real Time Production Scheduling System from Conception to Practice, in Zweben and Fox (eds) *Intelligent Scheduling*, Morgan Kaufmann, pp. 581–604, 1994.

[21] Kaufmann, A. and Gupta, M.M. *Intoduction to Fuzzy Arithmetic: Theory and Applications*. New York: van Nostrand Reinhold, 1985.

[22] Keng, N. P., and Yun, D. Y. Y. A Planning / Scheduling Methodology for the Constrained Resource Problem, *Proceedings of the 11th International Joint Conference on Artificial Intelligence*, pp. 998 - 1003, AAAI Press, 1989.

[23] Kerr, R.M. and Walker, R. N. A Job Shop Scheduling System Based on Fuzzy Arith metic. *Proceedings of the 2nd International Conference on Expert Systems and Leading Edge in Production and Operations Management,* Hilton Head Island, S.C., pp. 433–450, 1989.

[24] Klauč, R. Simulation für Reaktives Scheduling anhand eines Beispiels aus der Stahlindustrie, Diplomarbeit Technische Universität Wien, 1996.

[25] Le Pape, C. Experiments with a distributed architecture for predictive scheduling and execution monitoring, in *Artificial Intelligence in Reactive Scheduling*, Kerr and Szelke (Eds.), Chapman & Hall, pp. 129-145, 1995.

[26] Minton, S., Philipps, A., Johnston, M. and Laird, P. Solving Large Scale CSP and Scheduling Problems with a Heuristic Repair Method. *Proceedings of the 8th National Conference on Artificial Intelligence* (AAAI'90), pp. 17–24, 1990.

[27] Sadeh, N. *Look-ahead Techniques for Micro-opportunistic Job Shop Scheduling*, Ph. D. Thesis School of Computer Science, Carnegie Mellon University, Pittsburgh, 1991.

[28] Slany, W. Scheduling as a Fuzzy Multiple Criteria Optimization Problem, Ph. D. Thesis Technische Universität Wien, 1994.

[29] Smith S. F. OPIS: A Methodology and Architecture for Reactive Scheduling, in M. Zweben and M. Fox (eds) *Intelligent Scheduling*, Morgan Kaufmann, pp. 29–66, 1994.

[30] Tam, M. et al. A Predictive and Reactive Scheduling Tool Kit for Repetitive Manufacturing, in *Knowledge-based Reactive Scheduling*, E. Szelke and R.M. Kerr (Eds.), Elsevier Science, pp. 147-162, 1994.

[31] Zweben, M., Davis, E., Daun, B. and Deale, M.J. Scheduling and Rescheduling with Iterative Repair, *IEEE Transactions on Systems, Man, and Cybernetics* **23** *(6)*, pp. 1588–1596, 1992.

[32] Zadeh, L. Fuzzy Sets, *Information and Control* **8** p 338, 1965.

[33] Zimmermann, H.-J., Zadeh, L. A.and Gaines, B. R. *Fuzzy sets and decision analysis*, Amsterdam: North Holland, 1984.

Objectives for Order–Sequencing in Automobile Production

Christoph Engel and Jürgen Zimmermann
University of Karlsruhe

Alfons Steinhoff
IBM Informationssysteme GmbH

Abstract: The problem of sequencing units on a mixed–model assembly line can be viewed with several objectives in mind. This paper presents different optimization criteria and objectives for the order–sequencing problem. Former research has focused mainly on leveling procedures for model–sequencing and has emphasized material supply. In contrast, we provide a polynomial heuristic for order–sequencing by leveling the workload. In the context of automobile production we investigate different sequencing policies and introduce an extended heuristic for the case of color–batch–sequencing. For different types of objectives the performance of the heuristics presented is analyzed, taking known heuristics into consideration.

1 Introduction

Global competition forces enterprises, particularly in automobile industry, to increasing customer orientation and product diversification. With respect to the production system, this results in build–to–order production, out–sourcing of capacities, and the integration of pre–manufactured sub–systems.

Due to that development, there arise new requirements to material supply systems and sequencing procedures. In a *build–to–order production* the configuration of each product is determined by an individual selection of options corresponding to a customer order. In a *build–to–plan production* only a few model types are produced repeatedly. The variation in workload per order increases, if we consider build–to–order production of individual products instead of a build–to–plan production of a few model types.

In order to achieve a smooth workload distribution at the assembly stations, the extension of procedures for model–sequencing to the case of order–sequencing is necessary. A *model–sequence* is a production sequence where each unit represents a model type. In an *order–sequence* each unit of the sequence corresponds to a customer order and, therefore, is individual in its configuration. With respect to automobile production a *batch–sequence* may denote, e.g., a sequence where orders of a uniform color are combined to several color batches. For the basic concepts of assembly line sequencing we refer to [1, 12, 13, 23].

Increasing integration of sub–systems and build–to–order production result in an order–based component fabrication. In connection with Just–in–Time (JIT) production systems, this requires a sufficient look ahead of the order–sequence. In the case of order–sequencing, the underlying model demands, in general, are equal to one. However, many algorithms (approximately) solving the sequencing problem (cf. [10, 17, 24]) consider production rates of the underlying models by determining a model–sequence and, therefore, cannot be applied to the case of order–sequencing. Algorithms that consider production rates, which are not based on model types, can easily be adapted to the case of order–sequencing. In general, we can distinguish between algorithms for model– and order–sequencing or, with respect to the objective, between workload– and component–based approaches.

One of the first approaches considering the workload of models was presented by Thomopoulos [28], who treated the sequencing of assembly lines in combination with the balancing problem. Open and closed stations are described and four kinds of inefficiencies termed *idleness*, *deficiency*, *congestion*, and *utility work* are introduced. The unit to be scheduled next is determined by comparing penalty cost that are incurred by these inefficiencies. A drawback of this proceeding is the accumulation of models with high workload at the end of the sequence.

[28] motivates further research on workload based sequencing by Görke & Lentes [9] and Macaskill [16]. Macaskill alternately schedules a model with lowest penalty cost and a model with highest workload, which does not incur any utility work.

Dar–El [4] and Dar–El & Cother [5] studied the minimization of the line length by building model–sequences with minimal operator displacements from the left station border. These approaches are only useful in combination with the design and the balancing of assembly lines, since a change in the model–mix implies a change in the line length.

Okamura & Yamashina [21] proposed an improvement method for the minimization of the maximal operator displacement from the left station border, which is considered to be equivalent to the risk of stopping the conveyor. Tsai [29] introduced an optimal algorithm for minimizing the maximal operator displacement and the total utility work for the single station case. Finally, Sumichrast et al. [26, 27] transformed the algorithm of Monden [20] to an algorithm for workload leveling instead of leveling the usage of components.

Leveling the variation in component usage is the objective of a second category of algorithms. This idea has been introduced by Monden [20], who describes two scheduling algorithms used by Toyota. The first alternative, known as *Goal Chasing I*, consecutively schedules the model that incurs the minimal mean squared deviation between the expected accumulated component usage and the actual accumulated component usage. A simplified approach, termed *Goal Chasing II*, only takes the few critical parts into consideration and schedules the model that would, if not scheduled, incur the maximal deviation from the expected component usage. Miltenburg [17] adopted the idea of leveling the usage of components and suggested three improved scheduling heuristics. Miltenburg & Sinnamon [18, 19] generalized the previous approach to the case of multi–level production systems by leveling the production rates of the corresponding sub–assemblies.

Apart from the aforementioned priority–rule–based heuristics, various solution techniques for the sequencing problem on mixed–model assembly lines are discussed in the open literature.

Bard et al. [2] suggest a tabu search algorithm which seeks to minimize the total line length and to level the component usage by a multi–criteria objective. Branch and bound techniques have been used by Scholl [23] to minimize work overload, as well as by

Bolat [3], who additionally considered setup costs. Kim et al. [11] present a genetic algorithm for the minimization of the total line length and Rachamadugu & Yano [22] propose a Markov process approach to minimize work overload.

Moreover, Steiner & Yeomans [24, 25] investigated a graph–theoretic procedure to minimize the deviation between actual and expected production rate of models. McCormick et al. [15] devise a transformation to a network flow formulation, whereas Kubiak & Sethi [14] introduce a transformation to the well–known assignment problem. Decker [6] considered a transformation to the traveling salesman problem.

As mentioned in Decker [6] and Domschke et al. [7], the sequencing problem of mixed–model assembly lines shows a close relationship to the permutation flow shop problem including earliest and latest start times. Nevertheless, this problem class cannot easily be adapted to the problem of sequencing mixed–model assembly lines because minimizing the makespan, as the most common objective in permutation flow shop, is not of crucial interest in the context of mixed–model assembly line sequencing.

In the following section, we describe the sequencing problem on mixed–model assembly lines. We discuss possible objectives, before we present a problem formulation. In the third section, we devise an algorithm for the assembly line order–sequencing (AOS) based on the leveling of workload. We illustrate the sequencing procedure by an example and present an extended assembly line batch sequencing algorithm (ABS) with respect to the requirements of automobile production. In the fourth section, we briefly describe experimental results concerning the performance of the AOS–algorithm and evaluate different policies of building sequences in automobile production. Finally, we give conclusions of this study, evaluate the operative usefulness of our algorithms, and give an outlook to further developments of the suggested approach.

2 Problem formulation

In this section we consider the effect of order–sequencing on the performance of mixed–model assembly lines. We discuss different criteria and objectives to evaluate the performance of an assembly line. Then, we briefly classify procedures for the assembly line sequencing problem described in the literature and give a motivation for the concept of workload leveling. Finally, we provide an integer programming formulation for the sequencing problem in question.

In the context of planning and running assembly lines we can distinguish between two main problem types [28]:

- Assembly line balancing
- Determination of an order–sequence

In what follows, we deal with the latter problem, the determination of a "good" order–sequence. In the open literature, the sequencing problem is either considered in a rather short–term context [23] or as part of the process of line balancing [16, 28].

We consider the sequencing problem in relation to the balancing problem for the following reason: the optimization of the line balance as well as the optimization of the order–sequence should improve the efficiency of an assembly line. In order to evaluate the performance of an assembly line in process, we suggest the following criteria:

- Utility work
- Labor utilization
- Component usage

Utility work denotes the work overload which cannot be performed by the regular operators of each station. We distinguish between the distribution and the maximum of utility work per station. The distribution of utility work to the stations and over the time determines the number and allocation of necessary "utility workers". Reducing the maximum utility work is considered to be equivalent to reducing the risk of stopping the conveyor [21]. A high labor utilization obviously corresponds to a high productivity of the assembly line. The labor utilization UT_l at station l

$(l = 1, \ldots, s)$ can be defined as

$$UT_l := \frac{T_{nl} - \sum_{k=1}^{n} U_{kl}}{\tau\, n\, w_l\, s_l}$$

where T_{nl} denotes the accumulated workload at station l for a sequence of length n. U_{kl} denotes the utility work incurred at station l by the unit in sequence position (stage) k. The labor capacity of station l can be computed by $\tau\, n\, w_l\, s_l$, where s_l denotes the number of units in station l. We assume, that station l has a fixed rate launch interval τ and that each unit in the station is processed by w_l operators. The labor utilization UT_l can be increased by reducing the labor capacity $\tau\, n\, w_l\, s_l$ or by decreasing the accumulated utility work $\sum_{k=1}^{n} U_{kl}$ of the sequence. Determining the labor capacity of each station l is part of the line balancing. The accumulated utility work $\sum_{k=1}^{n} U_{kl}$ depends on the sequence and the labor capacity. Thus, the utilization UT_l of each station can only be determined, if the line balance is known and a sequencing algorithm is available for a given order set. Therefore, sequencing does not only represent a short–term problem, but has to be considered in the context of line balancing, too. Obviously, the same sequencing procedure should be used to evaluate the line balance as to determine the order–sequence. This fact is not appropriately covered in recent research. Finally, with respect to material supply, a constant rate of component usage reduces the effort of material supply and leads to a smooth production in the underlying sub–assemblies.

We now introduce three objectives for the order–sequencing problem related to the evaluation criteria mentioned above. Leveling the deviation of the actual from the expected accumulated workload until stage k leads to uniform workload over the sequence. Due to that, the displacement of operators in their stations is reduced and utility work becomes less probable. Thus, we first consider the objective of leveling the workload in the different stations over the sequence. Let order i possess options $o \in O_i$. Then, the workload

$$t_{il} := \sum_{o \in O_i} p_{ol}$$

is required to assemble order i at station l, where p_{ol} denotes the workload caused by option o at station l. The order i assigned to sequence position k is denoted by unit i_k. The accumulated workload

$$T_{kl} := \sum_{q=1}^{k} t_{i_q l}$$

performed at station l until position (stage) k depends on the currently scheduled units i_q $(q = 1, \ldots, k)$. The average workload $\bar{t}_l$ performed at station l per unit is calculated by

$$\bar{t}_l := \frac{\sum_{i=1}^{n} t_{il}}{n},$$

such that the objective

$$\text{Min.} \sum_{k=1}^{n} \sum_{l=1}^{s} (k\bar{t}_l - T_{kl})^2 \tag{2.1}$$

represents the total squared deviation of the accumulated workload T_{kl} from the expected accumulated workload $k\bar{t}_l$ of each station l and of each stage k.

The second objective to evaluate the quality of a sequence considers the minimization of total utility work. In order to determine the total utility work some details of the underlying assembly line are required. According to [16, 28] we assume a paced assembly line with a fixed rate launch interval τ and open stations with *upstream allowance time* t_l^u and *downstream allowance time* t_l^d. Concurrent work is assumed to be not allowed, which means that two options assigned to different stations cannot be assembled simultaneously. The length s_l of station l is indicated by the number of units assigned to the station at the same time. Furthermore, the number of operators w_l assigned to each unit at station l may differ from station to station. The arrival time a_{kl} and the departure time d_{kl} of the unit in position k at station l $(l = 1, \ldots, s)$ are given by

$$a_{kl} := \tau(k-1) + \tau \sum_{q=1}^{l-1} s_q$$

and

$$d_{kl} := a_{kl} + \tau s_l.$$

The earliest possible start time to process the unit of stage k at station l is equal to $a_{kl} - t_l^u$, i.e., the point in time when the unit of stage k reaches the upstream limit. The areas between station boundary and upstream limit as well as downstream limit are called *overlap areas* of the station. The station length enlarged by these overlap areas is termed working area. With respect to utility work, we assume that the latest possible finish time of the unit in position k at station l is equal to the time $d_{kl} + t_l^d$ when the unit in question reaches the downstream limit. The configuration of a station l including the working area as well as the corresponding overlap areas is depicted in Fig. 1.

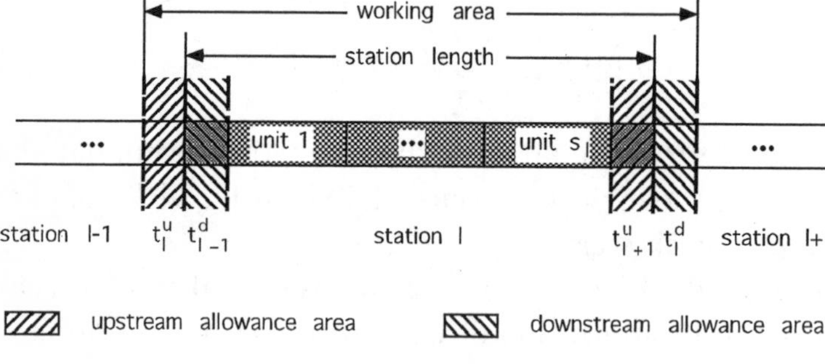

Figure 1: Configuration of station l

The part of the workload that cannot be performed within the working area is assumed to be utility work. Therefore, utility work occurs, iff the expected finish time $s_{kl} + t_{i_k l}/w_l$ exceeds the latest possible finish time $d_{kl} + t_l^d$ and is defined by

$$U_{kl} := \max\{0, s_{kl} + \frac{t_{i_k l}}{w_l} - (d_{kl} + t_l^d)\}.$$

Here, the start time s_{kl} of unit i_k at station l depends on the earliest possible start time $a_{kl} - t_l^u$ when unit i_k enters the working area, the finish time $f_{k,l-1}$ of unit i_k at station $l-1$ and the availability of operators at station l. In order to determine whether operators are available to process the unit in question, we assign the operators to the s_l units at station l following this policy: the $s_l w_l$ operators available at station l are divided into s_l teams of w_l operators, which are assigned to one unit each. Within the station preemption of workload is assumed to be allowed. Let i_k be the unit entering station l next. If the w_l operators of team s_l assigned to unit i_{k-s_l} have finished their workload, they continue with unit i_{k-s_l+1}. Simultaneously, team $s_l - 1$ shifts to unit i_{k-s_l+2}, and so on. Team 1 of station l is now without any unit and ready to process the unit i_k entering the station. In consequence, we consider the finish time $f_{k-s_l,l}$ of the unit i_{k-s_l} in order to determine the availability of operators to process the unit i_k that enters station l. Start time s_{kl} and finish time f_{kl} of unit i_k at station l can be calculated by

$$s_{kl} := \max\{a_{kl} - t_l^u, f_{k-s_l,l}, f_{k,l-1}\}$$

$$f_{kl} := \min\{s_{kl} + t_{i_k l}/w_l, d_{kl} + t_l^d\}$$

where we set $a_{11} := 0$, $f_{0l} := 0$ $(l = 1, \ldots, s)$, and $f_{k0} := 0$ $(k = -s_1 + 1, \ldots, n)$. The minimization of total utility work can be formulated as

$$\text{Min.} \sum_{k=1}^{n} \sum_{l=1}^{s} U_{kl}. \tag{2.2}$$

The start and finish times at station l are illustrated by Fig. 2, which depicts station l with a station length of $s_l = 2$ units. The horizontal bars represent the unit i_k of that stage k, given on the ordinate. The length of station l is represented by the gray area; dotted lines parallel to the station boundary mark the allowance limits of the stations as mentioned in the legend. The intersections of the boundary lines of station l and the bottom line of unit–bar i_k represent arrival time a_{kl} as well as departure time d_{kl} of unit i_k at station l; the start time of unit i_k at station l is determined by the finish time $f_{k-2,l}$ of unit i_{k-2} at station l. The handling of utility work is shown by unit i_{k-1}. That part of the workload exceeding the downstream limit $d_{k-1,l} + t_l^d$ is eliminated and is assumed to be performed by utility workers. The start time $s_{k+2,l}$ of unit i_{k+2} at station l is determined by the finish time $f_{k+2,l-1}$ of unit i_{k+2} at the previous station $l - 1$.

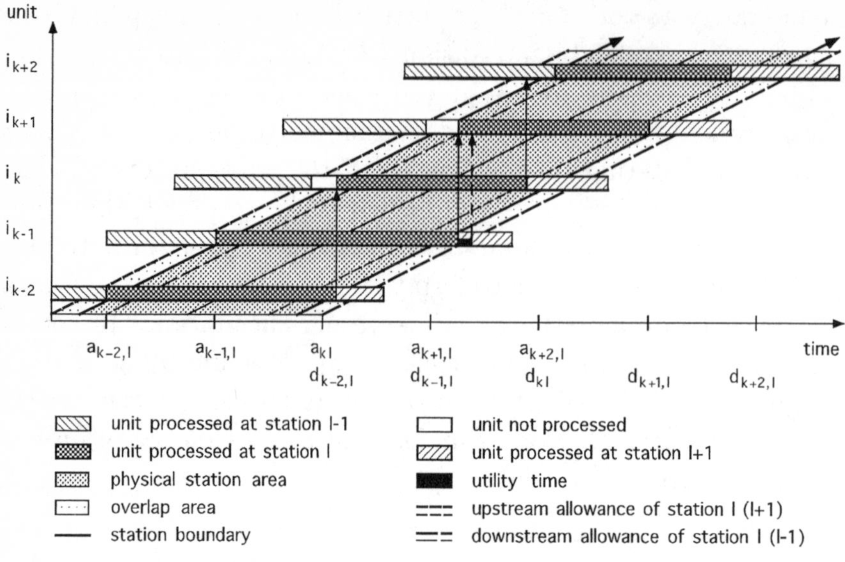

Figure 2: Start times of orders at station l

The third criteria to evaluate a sequence is a smooth component usage. In the case of automobile production a product option consists of several components, whereas each component belongs to exactly one product option. A uniform distribution of options over the sequence is considered to be equivalent to a uniform usage of components. Different options show different frequencies in the order set. Thus, a measure to make different options comparable with respect to their distribution over the sequence has to be defined. Therefore, we calculate the variation coefficient $\bar{s}_o/\bar{x}_o$ of the distances $k_2 - k_1$ between two consecutive orders i_{k_1} and i_{k_2} with option o. Mean distance $\bar{x}_o$ and standard deviation $\bar{s}_o$ can be calculated by

$$\bar{x}_o := \frac{1}{n_o} \sum_{j_o=2}^{n_o} k_{oj_o} - k_{o,j_o-1}$$

and

$$\bar{s}_o := \sqrt{\frac{1}{n_o - 1} \sum_{j_o=2}^{n_o} \left(k_{oj_o} - k_{o,j_o-1} - \bar{x}_o\right)^2}$$

respectively, where n_o denotes the number of orders with option o in the order set and k_{oj_o} the position of the j_o–th order with option o. Given $\bar{x}_o$ and $\bar{s}_o$, we can formulate the third objective, i.e. the minimization of the mean variation coefficient of all N_o options as

$$\text{Min.} \frac{1}{N_o} \sum_{o=1}^{N_o} \frac{\bar{s}_o}{\bar{x}_o} \; . \tag{2.3}$$

In this paper the main emphasis is on the leveling of workload as done by Sumichrast [26], whereas most sequencing algorithms known from literature consider one of the following objectives to determine a "good" order–sequence:

- Minimization of work overload [23, 30]
- Leveling occurrence of models types [17]
- Leveling component usage [20]

As mentioned above, the minimization of total work overload is equivalent to the minimization of total utility work. However, the minimization of total work overload does not necessarily lead to an even distribution of workload, which is desired for reasons of ergonomics and continuity of work, as well as for quality reasons. In order to avoid this drawback, the leveling of workload has to be considered for each station.

The real objective of leveling the occurrence of models over the sequence or leveling the usage of components is to achieve an even material supply. Implicitly, these approaches intend to level the workload and to minimize the variation in displacement of the operators. In the case of build–to–order production, leveling of models is impossible, since the model demand is equal to one.

In practical applications the determination of an order–sequence is often done by leveling the usage of components. Hereby, components are classified with respect to their influence on material supply and the variation in workload on the line. We distinguish between:

- basic components
- optional components
- order–dependent components

Basic components are required for each unit in an identical manner and lead to a constant workload and a constant effort of material supply at the stations. Optional components are used only in some units, depending on the order configuration, and cause additional workload and supplementary effort of material handling. Components required for each unit in an order–dependent configuration differ in workload, but do not cause an additional effort of material handling. Thus, only optional components need to be considered, if the focus is on the additional effort of material handling.

Different components lead to different workloads at the stations, such that the leveling of components does not necessarily lead to a uniform workload. Furthermore, different components, that incur workloads at the same station, may result in a high utility work at that station. Thus, high utility work is possible, even if each component shows a uniform distribution over the sequence. Due to the relationship between the usage of a component and the workload p_{ol} at a station l, we expect

1. an even distribution of components causing intensive workload

2. an even distribution of a modest total utility work over the time horizon

in leveling the workload.

The above considerations indicate that leveling the workload is a reasonable objective to order–sequencing on mixed–model assembly lines. Using (2.1) we present the workload leveling problem (WLP) for order–sequencing on a mixed model assembly line. The objective of WLP is to minimize the mean squared deviation of the actual accumulated workload from the expected accumulated workload of each stage at each station. The decision variable x_{ik} indicates whether an order i has already been scheduled at position $k_i \leq k$ of the sequence and is given by

$$x_{ik} = \begin{cases} 1, & \text{if order } i \text{ is scheduled in position } k_i \leq k \\ 0, & \text{otherwise} \end{cases}$$

Thus the WLP can be formulated as

$$\text{Min.} \quad \sum_{k=1}^{n} \sum_{l=1}^{s} (k\,\bar{t}_l - x_{i_k k} t_{i_k l})^2$$

$$\text{s.t.} \quad \sum_{i=1}^{n} x_{ik} = k \qquad (k = 1, \ldots, n) \qquad (2.4)$$

$$0 \le x_{i,k-1} \le x_{ik} \le 1 \qquad (i = 1, \ldots, n; k = 2, \ldots, n) \qquad (2.5)$$

$$x_{ik} \in \{0, 1\} \qquad (i = 1, \ldots, n; k = 1, \ldots, n) \qquad (2.6)$$

Restriction (2.4) and (2.6) ensure that there are exactly k units scheduled until stage k and restriction (2.5) guarantees that each order i is only scheduled once.

3 Sequencing algorithm

In this section we introduce an algorithm for the order–sequencing problem WLP on mixed–model assembly lines. We describe the basic ideas of the approach and give a formal representation of the basic workload leveling algorithm. Then, we illustrate the application of the algorithm by an example. Next, we discuss problems that arise in the practical application of sequencing algorithms in automobile production. With regard to that case, we suggest an extended workload leveling algorithm computing a batch–sequence.

The algorithm to be proposed is an iterative greedy heuristic. The two characteristic features are the leveling of workload and the determination of an order–sequence. Recent research in the field of sequencing on mixed–model assembly lines emphasizes the leveling of the production rate of outputs and the determination of model–sequences [13]. A workload based sequencing algorithm was proposed by Sumichrast [26], who used, as the rate to level, the accumulated workload T_{nl} of the order set at station l divided by the total workload $\sum_{l=1}^{s} T_{nl}$ of the order set at all stations. In contrast to Sumichrast, we consider the average workload $\bar{t}_l$ at station l over the sequence as the expected workload of station l at each stage. Thus, the accumulated workload expected to be performed until stage k at station l is given by $k\,\bar{t}_l$. The algorithm consecutively schedules order $i_1, \ldots, i_n$ where at each stage k the eligible order i^* with minimal priority value v_{i^*k} is scheduled. The priority value v_{ik} of each eligible order $i \in E$ at stage k is given by the minimal squared deviation of the expected accumulated workload $k\bar{t}_l$ from the actual accumulated workload $T_{k-1,l} + t_{il}$. If there is more than one eligible order i^* with minimal priority value v_{i^*k}, the order with smallest index is chosen. A representation in pseudo code of the algorithm, approximately solving the WLP with time complexity $O(n^2 s)$, is given in the following:

Algorithm [AOS]

Step 1: Initialization

$k := 1$

$E := \{1, ..., n\}$

For all $l \in \{1, \ldots, s\}$: $T_{kl} := 0$ and $\bar{t}_l := \frac{\sum_{i=1}^{n} t_{il}}{n}$

Step 2: Sequencing the orders

While $E \neq \emptyset$ Do

For all $i \in E$: $v_{ik} := \sum\limits_{l=1}^{s} (k\bar{t}_l - T_{k-1,l} - t_{il})^2$

$i_k := \min\{i \in E | v_{ik} = \min\limits_{j \in E} v_{jk}\}$

$E := E \setminus \{i_k\}$

For all $l \in \{1, \ldots, s\}$: $T_{kl} := T_{k-1,l} + t_{i_k l}$

$k := k + 1$

End (While)

In order to illustrate the proceeding of the AOS–algorithm, we consider an example with five stations and six orders. Again, each order consists of several options and each option may incur workload at several stations. Therefore, the set of options O_i determines the workload t_{il} of order i at station l. Table 1 shows the workloads t_{il} of orders $i = 1, \ldots, 6$ at stations $l_1, \ldots, l_5$. These quantities lead to a constant average workload of $\bar{t}_l = 2.7$ time units at each station l, depicted in the bottom line of Table 1. Assuming a fixed–rate launch interval τ of three minutes, we obtain an expected labor utilization of 90% at each station l.

order i \ station l	1	2	3	4	5
1	1.4	4.2	1.3	4.3	1.7
2	1.4	1.8	1.3	3.1	1.7
3	5.2	2.4	4.9	3.1	4.5
4	3.4	1.8	3.7	3.1	3.3
5	1.4	4.2	1.3	1.9	1.7
6	3.4	1.8	3.7	0.7	3.3
$\bar{t}_l$	2.7	2.7	2.7	2.7	2.7

Table 1: Allocation of workload t_{il}

Applying the AOS–algorithm, we compute the priority values v_{i1} of each order i. The priority values v_{i1} $(i = 1, \ldots, 6)$ are shown in the second row of Table 2.

stage k \order i	1	2	3	4	5	6
1	9.46	5.62	14.58	2.82	7.54	6.66
2	5.04	4.56	28.32		1.20	13.2
3	14.66	8.90	8.90			6.26
4	4.80	8.16	22.96			
5		14.58	5.62			
6		0				

Table 2: Priority value v_{ik} of orders

For example, with $T_{0l} := 0$ for each station l, the priority value v_{11} of order $i = 1$ at the stage $k = 1$ can be calculated by

$$(2.7 - 1.4)^2 + (2.7 - 4.2)^2 + (2.7 - 1.3)^2 + (2.7 - 4.3)^2 + (2.7 - 1.7)^2 = 9.46$$

Since order $i = 4$ has the minimal priority value v_{i^*1}, we set $i_1 = 4$. At every further stage k, the priority values v_{ik} of units, that are not yet scheduled, are calculated. For example, the priority value v_{12} of order $i = 1$ at stage $k = 2$ can be computed by

$$(5.4 - 3.4 - 1.4)^2 + (5.4 - 1.8 - 4.2)^2 + (5.4 - 3.7 - 1.3)^2 + (5.4 - 3.1 - 4.3)^2 + (5.4 - 3.3 - 1.7)^2 = 5.04$$

Continuing with the AOS–algorithm, we obtain the order–sequence $(4, 5, 6, 1, 3, 2)$ with an objective function value equal to 20.7, whereas an optimal sequence is $(4, 1, 6, 5, 3, 2)$ with an objective function value equal to 18.78.

With regard to automobile production, an algorithm for the determination of an order–sequence has to consider, additionally, the structure of the production system and the constraints of material supply. Before we investigate sequencing policies in context of automobile production, we give a short overview of the organization of the production system considered.

In automobile production the three sub–systems *body shop*, *paint shop* and *assembly shop* are distinguished. Each sub–system has different production and scheduling restrictions. Body shop as well as assembly shop are organized as a mixed–model–system, whereas the paint shop is typically a multi–model line. With respect to sequencing, the "models" are defined by specific shop–related options, which differ from shop to shop. For instance, in the body shop the number of doors and the sunroof may determine the model type of an order, whereas in the paint shop models are defined by the color, in general. The basic idea of the new approach is to define models in the assembly shop with respect to the options of an order. Since it is unlikely that two cars possess the same set of options, a model demand equal to one has to be assumed in the assembly shop. Since in different shops different options are considered to define model types, the model types differ from shop to shop. Therefore, an optimal sequence for the paint shop in general does not correspond to an optimal sequence for the other shops.

In what follows, we investigate the problem "how to provide the assembly shop with a good order–sequence". Here, the procedure of order–sequencing for the assembly shop depends on:

- the sequencing policy
- the quality of the painting process with respect to sequencing
- the performance of the sorting buffer providing the assembly shop

Sequencing policies differ in the point in time at which the sequence is determined, the location in the production system where the sequence is built physically, the orders eligible for each position of the sequence, and other technological constraints. The quality of the paint process with respect to the sequencing problem depends on the probability of rework and the length of rework cycles. The order–sequence can be changed in *sorting buffers* between the shops. The performance of a sorting buffer depends on its size, the type of accessing stored units and the velocity of providing an expected unit. We now investigate three sequencing policies with respect to the resulting order–sequences in the assembly shop:

1. resorting a batch–sequence disturbed in the painting process
2. scheduling a buffer–sequence on the basis of the units available in the sorting buffer
3. scheduling an "optimal" sequence in the assembly shop, assuming that each unit can be provided by the sorting buffer in time

Applying the first policy, we suppose that the sequences in the paint shop and in the assembly shop are identical. In determining the batch–sequence, we have to consider the size of color batches in the paint shop as well as the workload of orders at the assembly stations. The advantage of this policy is a long–term look ahead of the order–sequence to assemble. In the paint shop the sequence of units is disturbed by rework cycles. The ability to resort all units disturbed in the paint process depends on the performance of the sorting buffer. The consideration of the color batch restriction, generally, results in a reduced quality of the sequence with respect to the leveling of workload.

Providing a *buffer–sequence* according to the second policy results in a short look ahead of the orders entering the assembly shop next. That is unfavorable with regard to material supply.

The third policy entails a long look ahead of the sequence in the assembly shop. However, the sequence of units leaving the paint shop is not deterministic due to the possibility of rework in the paint shop. Therefore, this policy is of more theoretical significance but it can be used as a reference for the quality of the sequences according to the policies 1 and 2, respectively.

In order to determine the batch–sequence of policy 1, we propose an extended workload leveling algorithm termed as assembly line batch sequencing algorithm (ABS). For the determination of order–sequences according to policy 2 and 3 we use the AOS–algorithm. Here, we can apply the AOS–algorithm for policy 2, if we consider the orders available in the sorting buffer to be the set of eligible orders at each stage.

The basic idea of the ABS–algorithm is to determine a sequence of batches and then to schedule orders within each batch with respect to the leveling of workload. In doing so, we first choose the color c^* of the orders of the next batch. Color c^* is the color with the maximum positive deviation of the actual from the expected amount of scheduled orders with color c. If there is more than one color c^* with that property, the color with smallest index is chosen. The actual size of the color batch AB is initially set to the minimum of batch size B and the amount of eligible orders with color c^*. Then, we schedule AB orders of the current color c^* according to the AOS–algorithm where c_i denotes the color of oder i. This procedure is repeated until all orders are scheduled.

A formal representation of the ABS–algorithm with time complexity $O(n^2 s)$ is given as follows:

Algorithm [ABS]

Step 1: Initialization

$k := 1;\ AB := 0$

$E := \{1, ..., n\}$

For all $c \in \{1, \ldots, N_c\}$ Do

$E_c := \{i | c_i = c\}$

$AC_c := \frac{|E_c|}{n}$

$S_c := 0$

For all $l \in \{1, \ldots, s\}$: $T_{0l} := 0$ and $\bar{t}_l := \sum_{i=1}^{n} t_{il}/n$

Step 2: Sequencing the orders

While $E \neq \emptyset$

If $AB = 0$ Then Do

$$c^* := \min\{c | E_c \neq \emptyset \wedge (k\,AC_c - S_c) = \max_{\gamma \in \{1,...,N_c\}} (kAC_\gamma - S_\gamma)\}$$

$AB := \min(B, |E_{c^*}|)$

For all $i \in E_{c^*}$: $v_{ik} := \sum_{l=1}^{s} (k\bar{t}_l - T_{k-1,l} - t_{il})^2$

$i_k := \min\{i \in E_{c^*} | v_{ik} = \min_{j \in E_{c^*}} v_{jk}\}$

$E := E \setminus \{i_k\}$

$E_{c^*} := E_{c^*} \setminus \{i_k\}$

$S_{c^*} := S_{c^*} + 1$

For all $l \in \{1, \ldots, s\}$: $T_{kl} := T_{k-1,l} + t_{i_k l}$

$k := k + 1;\ AB := AB - 1$

End (While)

In general, the ABS–algorithm can be used if the workload t_{il} of each order i at each station l is given and if a model type can be assigned to each order.

4 Experimental performance analysis

We briefly report on an experimental analysis of the algorithms introduced in Section 3. First, details of the underlying assembly line are presented. Then, the AOS–algorithm is compared with the two workload leveling heuristics of [13] and [26]. Finally, an evaluation of the sequencing policies proposed in the previous section is given and further experiments are briefly discussed. For a detailed view, we refer to Engel [8].

The experimental analysis was part of a recent research, initiated by IBM Informationssysteme GmbH, Germany, concerning production planning of mixed–model assembly lines in automobile production. The described sequencing policies have been used for the evaluation of mixed–model assembly lines in automobile production. Sequences are evaluated by the leveling of workload WL, the total amount of utility work U and the leveling of options OL according to the objectives (2.1), (2.2) and (2.3), respectively.

With respect to the characteristics of automobile production, we consider a paced assembly line with $s = 30$ stations, where the station lengths s_l of three specific stations are equal to 6, 3, and 4 units. The remaining stations obtain a station length of $s_l = 2$ units. $w_l = 2$ operators are allocated to each unit at each station l $(l = 1, \ldots, s)$. The fixed launch rate τ is given by three minutes. Upstream allowance time t_l^u and downstream allowance time t_l^d of each station are uniformly set to 50% of the fixed launch rate τ. Frequencies of $N_o = 20$ options and workload p_{ol} incurred by option o in station l are given similar to those used in practical applications.

The AOS–algorithm can be applied to the case of order–sequencing as well as to the case of model–sequencing. For the case of order–sequencing, we generated 100 sets containing 100 orders where each order possesses an individual configuration of options. The set of orders was generated by a random procedure, so that the set of orders contains the fixed frequencies N_o for each option o. The order–sequence determined by the AOS–algorithm was compared with a random sequence and with the sequence determined by the *Time Spread* heuristic devised in [26].

For the case of model–sequencing, we generated 10 sets of 100 orders, where 5 or 10 model types are distinguished and all orders of a model type possess the same configuration of options. The model–sequence determined by AOS was compared with a random sequence and, additionally, with an algorithm described by Kubiak [13]. Table 3 and 4 show the mean objectives WL, U, and OL over all sequences of the test set.

Considering Table 3, we see that the AOS–algorithm is markedly superior to the Time Spread heuristic with respect to all objectives. The poor performance of the Time Spread heuristic can be explained by its scheduling criteria. The Time Spread heuristic prefers orders that have a small total workload and show a very even distribution of workload over the stations. With regard to objective (2.1) of workload leveling WL this leads to a high deviation of the accumulated workload from the expected workload for stages in the middle of the sequence. Since the Time Spread–heuristic, obviously, does not lead to a leveling of workload, it seems to be inadequate for the case of order–sequencing.

sequence\objective	WL	U	OL
Random	300147.19	443.19	0.74
Time Spread	5667728.76	551.09	0.80
AOS	15846.04	263.31	0.54

Table 3: Comparison between Random,Time Spread, and AOS

Considering the case of model–sequencing the AOS–algorithm outperforms the algorithm of Kubiak with respect to the objectives WL and U, whereas the latter one shows a better performance in OL (see Table 4). Kubiak determines the average workload for each station and for each model. At each stage the model is scheduled, which obtains the maximum deviation between expected and accumulated workload at the previous stage.

In order to evaluate the three sequencing policies for the assembly shop, mentioned in Section 3, we generated 10 sets containing 300 different orders. A specific color is assigned to each order. The number of colors available was given by $N_c = 10$. With respect to sequencing policy 1, we generate a *batch–sequence* with

sequence\objective	WL	U	OL
Random	274355.18	641.41	0.71
Kubiak	42836.45	586.07	0.30
AOS	32766.75	486.77	0.33

Table 4: Comparison between Random, Kubiak, and AOS

color batch size $B = 5$ by applying the ABS–algorithm.

The determined batch–sequence is assumed to be the input sequence of the paint shop. With no rework cycles in the paint shop, this sequence theoretically passes to the assembly shop. In general, the input sequence of the paint shop is disturbed by rework cycles. Therefore, the batch–sequence was randomly disturbed with a disturb factor of 10%, which means that on the average 90% of the units leave the paint shop without running through a rework cycle. The length of the rework cycle was uniformly chosen between 10 and 60 units per delay. The disturbed batch–sequence is called *disturbed sequence.* Since a long look ahead of the assembly sequence is desired, we seek to resort the disturbed sequence to the original batch–sequence. The ability to resort the batch–sequence depends on size and accessibility of the sorting buffer. In the considered problems, a buffer size of 40 units and random buffer access are assumed. The result of resorting the disturbed sequence is denoted by *resorted sequence.*

Sequencing policy 2 provides a sequence on the set of units actually leaving the paint shop. Considering the units of the disturbed sequence that are in the sorting buffer, a *buffer–sequence* can be computed with the AOS–algorithm. Finally, the AOS–sequence is generated on basis of the total order set. The AOS–sequence is only of theoretical significance because, in general, this sequence can only be provided to the assembly shop, if the buffer is of size n. Table 5 shows the objectives of the different sequences. Loosely speaking, the batch–sequence is not as good as the buffer–sequence or the AOS–sequence, because the consideration of batches reduces the set of eligible jobs at each stage. Due to a relative small buffer size the resorted sequence is generally not as good as the batch–sequence. Obviously, the AOS–sequence shows the best performance, but with respect to practical applications the buffer sequence outperforms each available sequence.

policy\objective	WL	U	OL
Batch–Sequence	114501.06	934.19	0.69
Disturbed Sequence	267942.29	1018.87	0.72
Resorted Sequence	129065.68	959.59	0.70
Buffer–Sequence	66324.83	732.38	0.60
AOS–Sequence	56247.26	697.56	0.58

Table 5: Sequencing policies

Further tests to determine the performance of AOS with regard to variations in the frequency of options, length of the sequence, size of the color batch or length of the allowance limits have been done. For further details we refer to Engel [8].

5 Conclusions

In this paper we discussed the performance analysis of mixed–model assembly lines with given line balance. We presented three objectives for the sequencing problem. We motivated a workload leveling approach and introduced an integer programming formulation for the workload leveling problem WLP. We devised two polynomial heuristics AOS and ABS with time complexity $O(n^2s)$ for the WLP. Thereby, AOS provides an order–sequence and ABS computes batch–sequences of orders. We proposed three sequencing policies for practical applications in automobile production. In an experimental performance analysis we compared the AOS–algorithm with two workload leveling algorithms for order– and model–sequencing, respectively. The AOS–algorithm outperforms the heuristic of [26] for the problem of order–sequencing as well as the heuristic of [13] for the problem of model–sequencing. Finally, we evaluated the proposed sequencing–policies of automobile production.

Important areas of further research are leveling the variation in workload at the stations over the sequence as well as resource constraints of options over time, which is important for practical applications.

References

[1] Bard, J.F., Dar–El, E.M., and Shtub, A. (1992), An analytic framework for sequencing mixed model assembly lines, *International Journal of Production Research*, Vol. 30, pp. 35–48

[2] Bard, J.F., Shtub, A., and Joshi, S.B. (1994), Sequencing mixed–model assembly lines to level parts usage and minimize line length, *International Journal of Production Research*, Vol. 32, pp. 2431–2454

[3] Bolat, A. (1994), Sequencing jobs on an automobile assembly line: objectives and procedures, *International Journal of Production Research*, Vol. 32, pp. 1219–1236

[4] Dar–El, E.M. (1978), Mixed–model assembly line sequencing problems, *OMEGA*, Vol. 6, pp. 313–323

[5] Dar–El, E.M. and Cother, R.F. (1975), Assembly lines sequencing for model mix, *International Journal of Production Research*, Vol. 13, pp. 463–477

[6] Decker, M. (1993), *Variantenfließfertigung*, Schriften zur quantitativen Betriebswirtschaftslehre, Vol. 7, Physica, Heidelberg

[7] Domschke, W., Scholl, A., and Voß S. (1993), *Produktionsplanung – Ablauforganisatorische Aspekte*, Springer, Berlin

[8] Engel, C. (1997), Belastungsnivellierung in der Variantenfließfertigung, *Diploma Thesis*, Institut für Wirtschaftstheorie und Operations Research, University of Karlsruhe

[9] Görke, M. and Lentes, H.–P. (1981), Modellfolgebestimmung bei gemischter Produktfertigung, *wt – Zeitschrift für industrielle Fertigung*, Vol. 71, pp. 153–160

[10] Inman, R.R. and Bulfin, R.L. (1992), Quick and dirty sequencing for mixed–model multi–level JIT systems, *International Journal of Production Research*, Vol. 30, pp. 2011–2018

[11] Kim, Y.K, Hyun, C.J. and Kim, Y. (1996), Sequencing in mixed–model assembly lines: A genetic algorithm approach, *Computers and Operations Research*, Vol. 23, pp. 1131–1145

[12] Köther, R. (1986), Verfahren zur Verringerung von Modell–Mix–Verlusten in Fließmontagen, *IPA–IAO Forschung und Praxis*, Vol. 93, Springer, Berlin

[13] Kubiak, W. (1993), Minimizing variation of production rates in just–in–time systems: A survey, *European Journal of Operational Research*, Vol. 66, pp. 259–271

[14] Kubiak, W. and Sethi, S. (1991), A note on level schedules for mixed–model assembly lines in just–in–time production systems, *Management Science*, Vol. 37, pp. 121–122

[15] McCormick, S.T., Pinedo, M.L., Shenker, S., and Wolf, B. (1989), Sequencing in an assembly line with blocking to minimize cycle time, *Operations Research*, Vol. 37, pp. 925–935

[16] Macaskill, J.L. (1973), Computer simulation for mixed–model production lines, *Management Science*, Vol. 20, pp. 341–348

[17] Miltenburg, J. (1989), Level schedules for mixed–model assembly lines in just–in–time production systems, *Management Science*, Vol. 35, pp. 192–207

[18] Miltenburg, J. and Sinnamon, G. (1989), Scheduling mixed–model multi–level just–in–time production systems, *International Journal of Production Research*, Vol. 27, pp. 1487–1509

[19] Miltenburg, J. and Sinnamon, G. (1992), Algorithms for scheduling multi–level just–in–time production systems, *IIE Transactions*, Vol. 24, pp. 121–130

[20] Monden, Y. (1983), *Toyota Production System*, Industrial Engineering and Management Press, Atlanta

[21] Okamura, K. and Yamashina, H. (1979), A heuristic algorithm for the assembly line model–mix sequencing problem to minimize the risk of stopping the conveyor, *International Journal of Production Research*, Vol. 17, pp. 233–247

[22] Rachamadugu, R. and Yano, C.A. (1994), Analytical tool for assembly line design and sequencing, *IIE Transactions*, Vol. 26, pp. 2–11

[23] Scholl, A. (1995), *Balancing and Sequencing of Assembly Lines*, Physica, Heidelberg

[24] Steiner, G. and Yeomans, S. (1993), Level schedules for mixed–model just–in–time processes, *Management Science*, Vol. 39, pp. 728–735

[25] Steiner, G. and Yeomans, S. (1996), Optimal level schedules in mixed–model, multi–level JIT assembly systems with pegging, *European Journal of Operational Research*, Vol. 95, pp. 38–52

[26] Sumichrast, R.T., Russell, R.S., and Taylor, B.W. (1992), A comparative analysis of sequencing procedures for mixed–model assembly lines in a just–in–time production system, *International Journal of Production Research*, Vol. 30, pp. 199–214

[27] Sumichrast, R.T. and Clayton, E.R. (1996), Evaluating sequences for paced, mixed–model assembly lines with JIT component fabrication, *International Journal of Production Research*, Vol. 34, pp. 3125–3143

[28] Thomopoulos, N.T. (1967), Line balancing–sequencing for mixed–model assembly, *Management Science*, Vol. 14, pp. 59–75

[29] Tsai, L.H. (1995), Mixed–model sequencing to minimize utility work and the risk of conveyor stoppage, *Management Science*, Vol. 41, pp. 485–495

[30] Yano, C.A. and Rachamadugu, R. (1991), Sequencing to minimize work overload in assembly lines with product options, *Management Science*, Vol. 37, pp. 572–586

Chapter 5

Production Control

Automatic Production Control

- A New Approach in Production Planning and Control Based on Methods of Control Theory -

Hans-Peter Wiendahl and Jan-Wilhelm Breithaupt
Institute of Production Systems
University of Hanover

Abstract: Control theory provides excellent tools to analyse and control dynamic systems. Using the funnel model and the theory of the logistic operating curve a continuous model of a single production system has been developed. Based on this model an extended model for several work systems connected via the material flow was designed. For a single work system a backlog as well as wip controller has been developed. The controller interacts to adjust the capacity of the work system to eliminate the backlog as soon as possible. Simulation experiments confirm that this concept ensures the synchronisation of capacity and work. A suggestion to integrate the strategy into production planning and control (PPC) on the planning level has been generated. The objective of this approach is to develop the present open loop control realised in PPC into a closed loop control with defined control and reference variables.

1 Global Competition

The continuing structural changes in society and, thereby, in the production environment contain new challenges which companies have to face with innovative approaches in order to remain competitive in the future. These changes affect the categories: product, market, production location and process.

In a long-term study published in 1993 McKinsey investigated the features with which internationally successful companies assert themselves in the battled world market [1]. The study illustrates that international competition is based upon three essential pillars: *price*, *quality* and *time*. The successful companies dominate in all three dimensions. In addition to these process related criteria the company related features *innovation rate*, *agility* and the *ability to learn* appear increasingly as decisive points on the global markets [1,2,3]. (Fig. 1) Companies which are able to adapt their production struc-

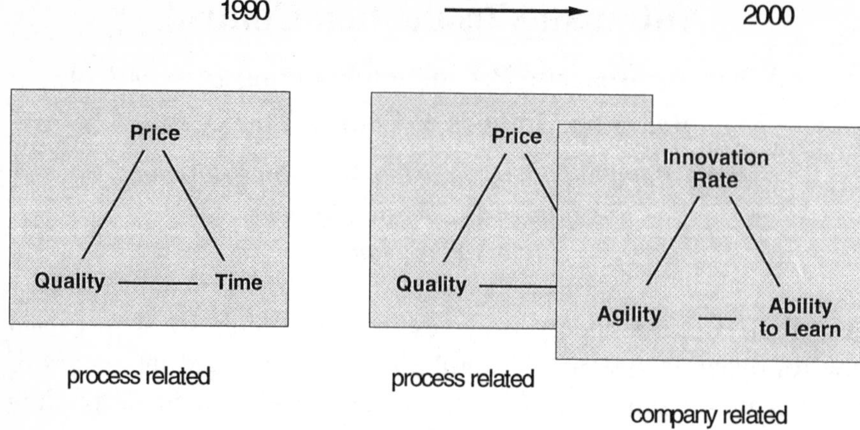

©IFA D1662BRE

Figure 1: Pillars of the global competition

tures according to fast changing market requirements and technologies will win market shares. Thereby it is not very helpful "..to search for the solutions to problems by the competitors because the potentials in a company are then concentrated on copying existing solutions and the company always remains in second place" [2]. Companies and research are therefore called upon to develop their own innovative ideas and to expand them to new tools and procedures. Within the scope of this article a new perspective is pointed out under the catchword *automatic production control* (APC), which can be helpful in improving companies competitiveness in the future.

2 The Model

2.1 The Logistical Objectives

The quality of the logistical performance is determined by lead time and schedule performance in the job shop as outlined in Fig. 2. On the other hand, loading of the production facilities and work-in-process (wip) influence the profitability of the manufacturing process. From these partly conflicting requirements concerning the PPC the four main objectives of the production process can be derived. Short lead time and low schedule deviation represent the market-related objectives, whereas low inventory and high and steady loading of the work systems are the factory-related objectives.

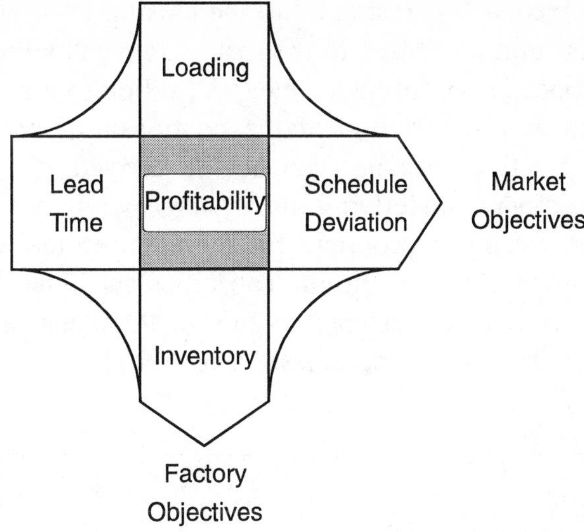

©IFA C0943E

Figure 2: Logistical objectives in manufacturing [4]

In order to adjust the logistical objectives under the consideration of their mutual dependencies and the aspect of competitive production, there exists the demand to design a controllable production and order processing. While conventional PPC systems predominantly control (open-loop) and therefore a feedback is missing, a self-controlling process can be achieved by a (closed-loop) control by defining appropriate reference and correcting variables [5,6].

2.2 An Elementary Continuous Production Model

Present day PPC systems differentiate between the planning level and the operational level. These views can also be applied to the production control. Due to the different nature of the levels, different models are needed. Single events are of interest on the operational level. A more global view on the planning level makes the utilisation of a continuous model possible. This is desirable because for continuous processes control theory has much more methods available than for discrete. In addition, the planning level is responsible for the definition of the side constraints for attaining the goals on the operational level. The planning level of the production control is therefore to be the essential object to be looked at in the following [5,6].

Various models for production control have already been developed. Most of them are based on a simple control loop. The feedback in today's MRP II systems can be mentioned here as an example

(Fig.3). The feedback is restricted to the closing of information circuits. The reaction according to the results remains with the system user. Such concepts are missing a clear definition of the control variables as well as a description of the relationship between effects. Thus, the correcting variables can not be derived effectively. The deterministic models underlying this system merely describe reality statically and are not appropriate for representing the dynamics of reality (i.e. assumptions of infinite capacities and fixed lead times). Most of the systems are designed to enforce the plans generated and show a strong static nature because of this [5,6,7].

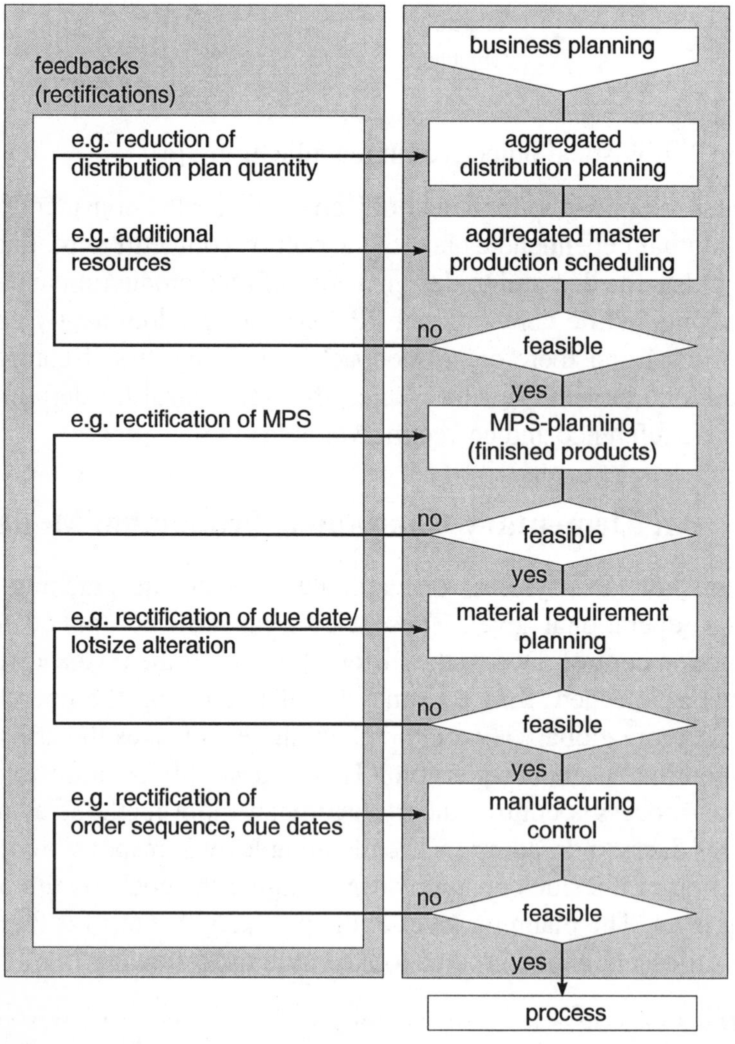

Figure 3: Realised feedback of MRP II concepts [5,6]

As mentioned above the application of control theory requires a continuous production model. Such a model for a single work system was developed and evaluated in 1995 by Petermann with the aid of the funnel model and the logistic operating curve [5,6]. An extended model for modelling several work systems connected via the material flow was developed within a running research project. This model is described in the following.

The underlying funnel model [4,8,9] describes the individual events at a work system and represents therefore a discrete model which has to be substituted far-reaching with a continuous model. This is successful only when the microscopic behaviour of the discrete individual process disappears behind the macroscopic of the system. This is to be assumed for controlling on the planning level, because here the individual event is not of interest, but rather mean variables, like loading, work-in-process or lead time observed over a period of time.

With the aid of the funnel model the order processing of a work system can be described as the flow of a liquid through the funnel (Fig. 4). The incoming orders, measured in hours of work content, form a stock on pending lots, which have to flow through the funnel outlet. The diameter of the outlet can be described as the capacity of the work system which is adjustable within limits (actual performance).

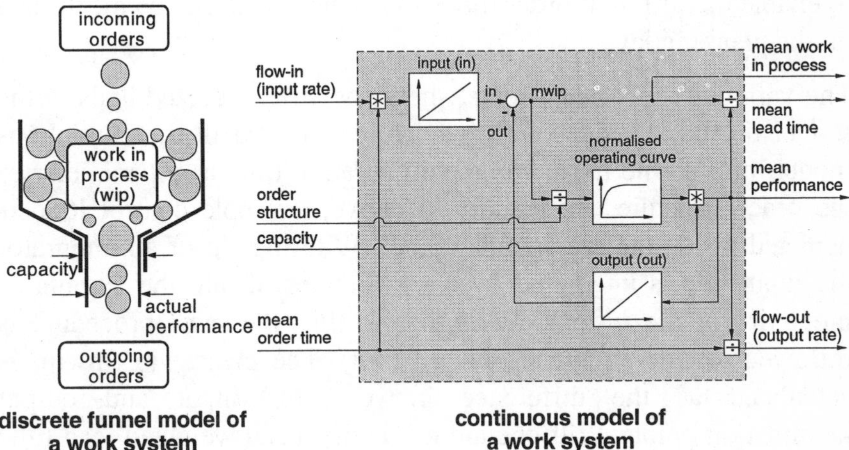

©IFA D1654E

Figure 4: Comparison of the Input and Output Parameter of the Discrete and the Continuous Model

It is obvious that the mean lead time for incoming orders changes with the stock on pending orders and the system's performance [4]. These interdependencies between mean lead time (mlt), mean work-in-process (mwip) and mean performance (mper) are stated through the funnel formula below:

$$mlt = \frac{mwip}{mper} \tag{1}$$

Detailed derivations of funnel model and formula are specified in [4,8,9].

The input and output variables as well as a simple control loop of the continuous model are shown in Fig. 4 on the right side. The incoming orders, measured in hours of work content in the funnel model, are converted into the input rate (dimension: number of orders per unit of time). Multiplied by the mean order time the dimension of the input rate is changed into work content per time unit.

Similar to the input rate, the output of the work system is converted into the output rate by dividing the mean performance through the mean order time. These transformations are required because the material flow between two work systems is measured in number of orders per unit of time, whereas within the work system the work content is of interest [4]. In reality the order time of several orders processed by a specific work system differ. Therefore, the transformation leads to practicable result only over time periods long enough to enable the different order times to balance. This can be mentioned on the planning level.

The various sizes of the orders, shown as different sized balls in the funnel on the left side of Fig. 4, are considered in the continuous model through the mean weighted operation time as a parameter of the order structure. As mentioned above, a simple control loop is depicted inside the continuous model. With the aid of an integrator the input rate is integrated over a time interval into the cumulated input (in) of the system. Analogous to this, the same procedure is followed for the cumulated output (out). The mean wip (mwip) is calculated as the difference between the input and output (summation point). With the aid of (1) the mean weighted lead time results can be calculated by dividing the mean wip through the output rate. The four variables, output rate, mean performance, mean weighted lead time and mean wip, are the output variables of the system.

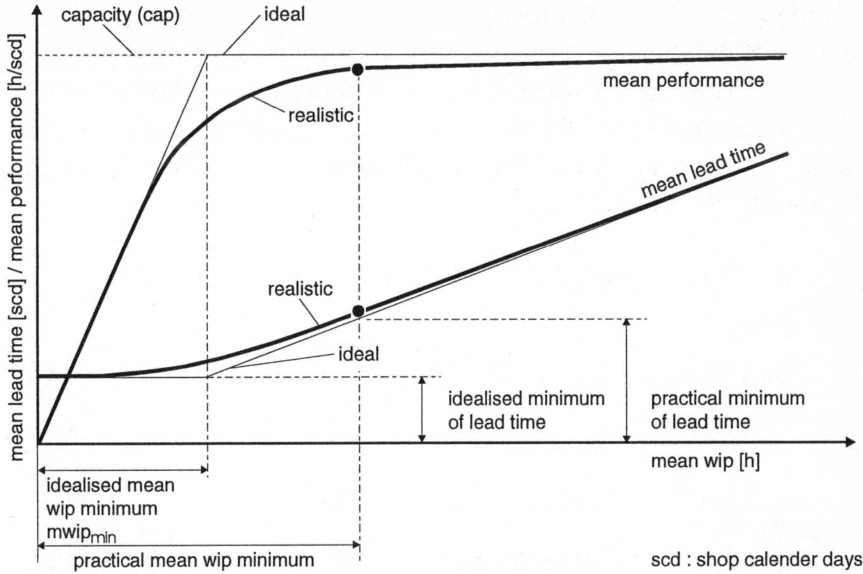

©IFA D4172E

Figure 5: Interdependency between output, lead time and work-in-process (wip) [10]

Within a control circuit each straight line requires a transmission function. For this reason, several former attempts to design a closed-loop control for the PPC failed, because clear definitions of the interdependencies between control and reference variables were missing.

The logistic operating curve forms the connection between the mean wip, the mean weighted lead time and the mean performance with the aid of the capacity and the ideal minimal wip (order structure) as input parameters (Fig. 5) [4,10,11].

The curve states that the production of a work system is independent of the wip as long as every work system has a stock on pending orders at all times. Then, the performance of the system is almost equal to its capacity. Only if the inventory is further reduced, losses in production will occur due to interruptions in the material flow. On the other hand, the lead time decreases in proportion with the wip until the physical minimum is reached: Beyond this point the lead time cannot be reduced further, because it is limited by the sum of the operation time and minimum transport time (idealised minimum of lead time).

The idealised mean wip minimum ($mwip_{min}$) represents the wip level that is necessary to run the system under idealised conditions, assuming that no arriving order has to wait and, nevertheless, no interruption in the material flow occur. This cannot be found in practice, so the realistic curves differ from the idealised ones shown in (Fig. 5). Nyhuis found out, that it is possible to calculate these realistic curves for most of the job shop productions [10] with the following equations (2) and (3):

$$mwip(t) = mwip_{\min} \cdot \left(1 - \left(1 - \sqrt[4]{t}\right)^4\right) + mwip_{\min} \cdot \alpha_1 \cdot t \qquad (2)$$

$$mper(t) = per_{\max} \cdot \left(1 - \left(1 - \sqrt[4]{t}\right)^4\right) \qquad (3)$$

with

mwip(t)	*: mean work-in-process [h]*
mper(t)	*: mean performance per shop calendar day [h/scd]*
$mwip_{min}$	*: idealised mean work-in-process minimum [h]*
per_{max}	*: maximally available performance [h/scd]*
α_1	*: streching parameter [-] (default-value: 10)*
t	*: running parameter* $(0 < t < 1)$

A detailed derivation of these formulas is specified in [10]. With the aid of (2) and (3) a pair of values for wip and performance can be calculated dependent to the running parameter t. With these pairs, the course of the performance is defined. Using the funnel formula, the course of the lead time can also be determined.

A major problem for the production control occurs due to the structural parameters, $mwip_{min}$ and per_{max}, of the logistic operating curve which vary over time. This behaviour leads to a variable transmission function which is unacceptable for a control task. Therefore it is necessary to normalise the logistic operating curve. The normalised version describes the interdependence between utilisation (ut) and relative mean work in process ($mwip_{rel}$) which is the relation between mean wip (mwip) and idealised minimum mean wip ($mwip_{min}$) [10,11].

The normalised logistic operating curve is the core of the continuous model and forms the connection between the input and output variables.

2.3 The Connection of Several Work Systems

In the previous chapter a continuous model for a single work system was presented. In principle, this model is suitable for the connection of several work systems via the material flow, but the connection between the systems itself is still missing. To fill the gap two steps are necessary: Firstly, the different types of connections have to be examined and secondly, this has to be integrated into a model based on control theory.

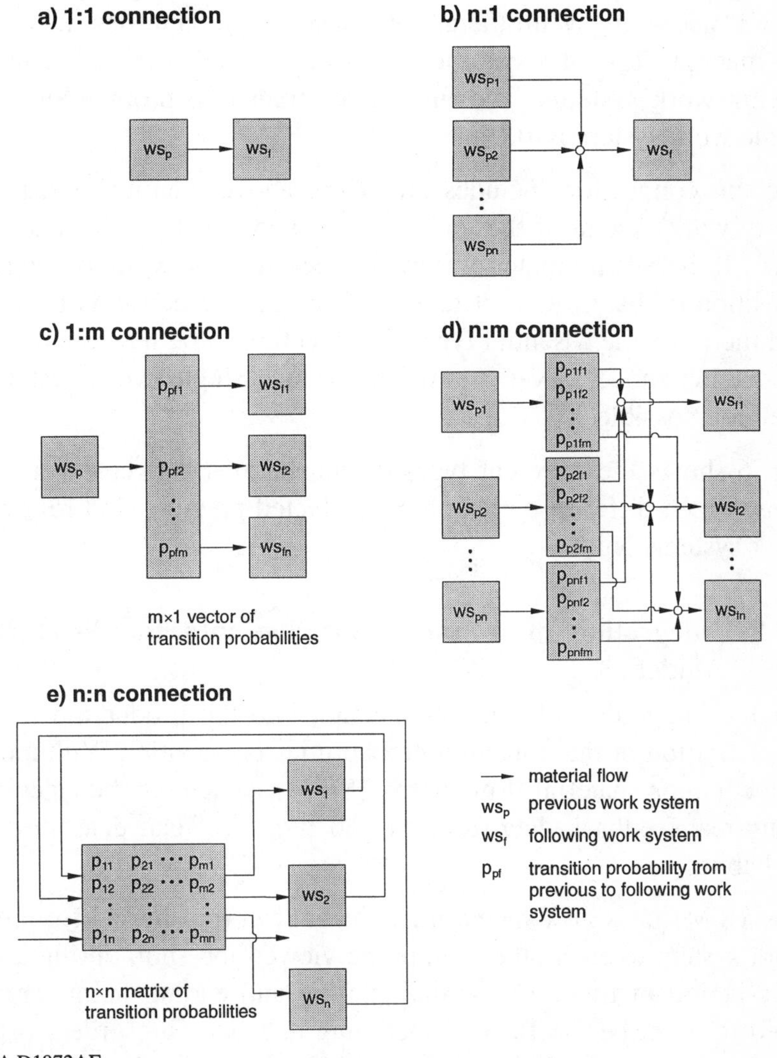

©IFA D1972AE

Figure 6: Types of connections between work systems via the material flow

2.3.1 Different Types of Connections

Five different types of connections can be distinguished. In a 1:1-connection all the orders processed by a work system flow to only one following work system, which itself does not process any other orders. If the output of n systems is processed by only one following work system, the connection is called n:1. The input of this following system is the summarised output of the previous ones.

More complex are models for 1:m respectively n:m connections, because the previous work systems deliver to more than one following system. Therefore a distribution of the outgoing material flow is necessary. With the aid of transition probabilities the outgoing material flow of a specific work systems is distributed to m following work systems. The sum of the transition probabilities of a single work system is 100%.

The n:n connection includes all of the above mentioned features. Every work system of the job shop can theoretically deliver to each other. If there is no material flow between specific work systems the transition probability is set to zero. The n:n connection is universal and therefore the essential type of connection to be integrated in the production model. Even transitions from a single work system to itself are possible.

Fig. 6 shows the different types of connections between work systems, which differ in the number of affected previous and following work systems only.

2.3.2 Integration of a n:n Connection into the Production Model

For the integration of a n:n connection into the production model, the definition of the transition probabilities is essential. With the aid of a common material flow matrix (MFM), which can be calculated using real feedback data from the job shop, one can determine the probabilities.

The MFM states how many orders have been transported from each work system to each other within the viewed job shop during a specific period of time. The virtual starting and ending points are defined to describe the flow of incoming and outgoing orders [4]. In the upper left part, Fig. 7 shows a MFM for a job shop of 6 work systems.

material flow diagram

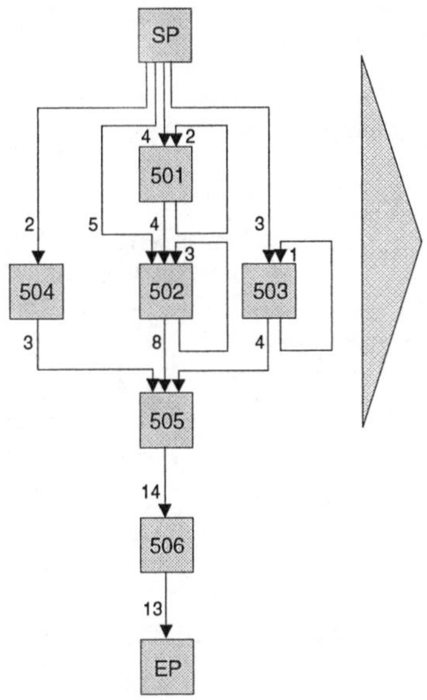

material flow matrix

following work system (columns) / previous work system (rows)

	SP	501	502	503	504	505	506	EP
SP	0	4	5	3	2	0	0	0
501	0	2	4	0	0	0	0	0
502	0	0	3	0	0	8	0	0
503	0	0	0	1	0	4	0	0
504	0	0	0	0	0	3	0	0
505	0	0	0	0	0	0	14	0
506	0	0	0	0	0	0	0	13
EP	0	0	0	0	0	0	0	1

$$p(i,j)=\frac{nt(i,j)}{\sum_{x=1}^{n} nt(i,x)}$$

normalised material flow matrix

following work system (columns) / previous work system (rows)

	SP	501	502	503	504	505	506	EP
SP	0	0,3	0,4	0,2	0,1	0	0	0
501	0	0,3	0,7	0	0	0	0	0
502	0	0	0,3	0	0	0,7	0	0
503	0	0	0	0,2	0	0,8	0	0
504	0	0	0	0	0	1,0	0	0
505	0	0	0	0	0	0	1,0	0
506	0	0	0	0	0	0	0	1,0
EP	0	0	0	0	0	0	0	1,0

SP : starting point
EP : ending point
501 : work system number
p(i,j) : transition probability from work system i to work system j
nt(i,j) : number of transitions from work system i to work system j

©IFA F5817E

Figure 7: Derivation of the normalised material flow matrix

For the calculation of the transition probabilities the absolute numbers of orders flowing from previous work systems to following are not of interest, but rather the percentage relation of the distribution of the system's output as mentioned above. Therefore it is necessary to normalise the material flow in relation to the summarised output of each work system (Fig.7, lower part).

Apart from this, the content of the normalised MFM is time independent as long as the structure of the production program remains the same. Only if the production program varies over a longer period of time, the model has to be adapted by using actual feedback data to calculate a new MFM.

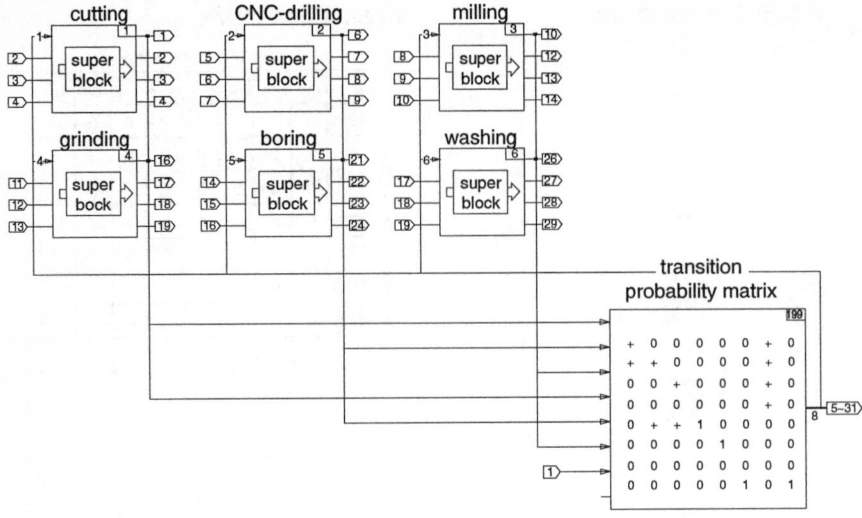

©IFA F5816E

Figure 8: Example of a continuous job shop model (with 6 work systems)

The normalised MFM can be integrated into the model based on control theory as the transition probability matrix. In Fig. 8 shows a simulation model based on the job shop mentioned before. Each work system is represented through a „superblock“, which contains the elementary work system model. Analogous to the elementary work system model, each superblock has four inputs and outputs. Via the input and the output rate the superblocks are connected with the transition probability matrix block, which has the task to distribute the output of each work system to other work systems. The input as well as the output of the whole job shop is directly connected with this block.

2.4 Model Evaluation

The transition from a discrete process to a continuous process is difficult. Is this transition fundamentally allowed? When does the microscopic behaviour of the single event (here: the processing of an operation) stands behind the macroscopic behaviour of the system with reference to the viewed input and output variables? To be able to answer these questions, extensive model evaluations are necessary.

The basic idea of the model evaluation is in principle very simple. (Fig. 9). The input and output variables described in the model are

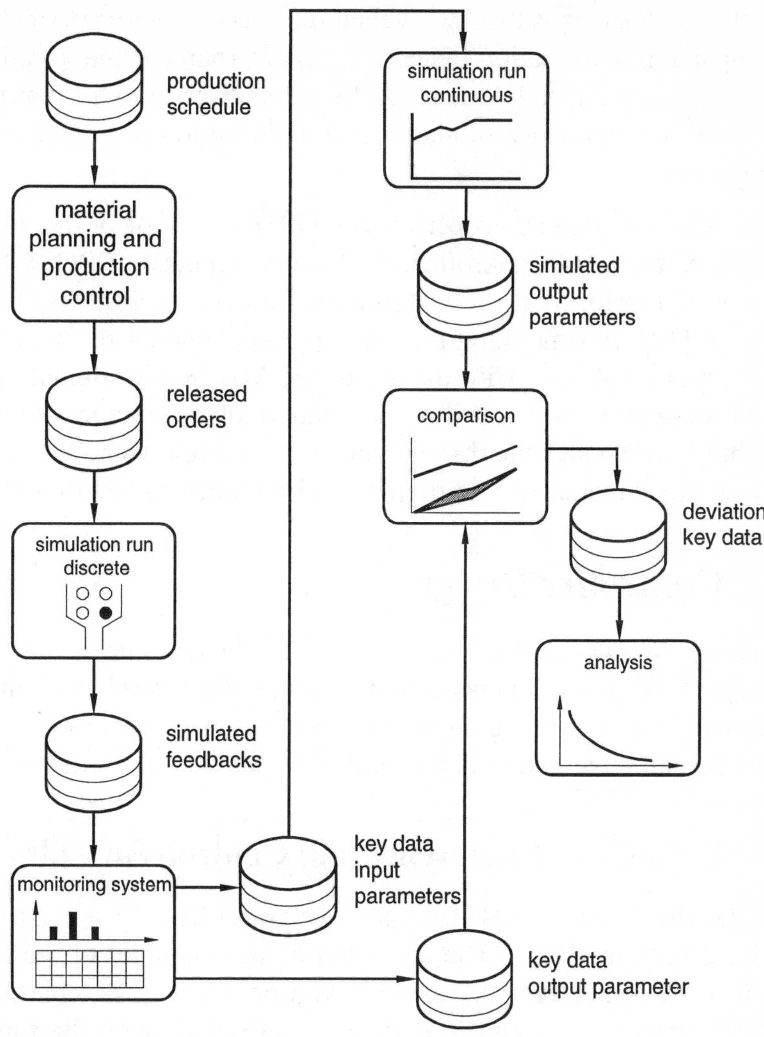

©IFA F1462E

Figure 9: The order of events of the experiments for the evaluation of the continuous model

measured in an existing job shop. Then the measured input variables are used as an input signal of the model and the system's reply is recorded. Finally the output signal curves from the model are compared to those output variables measured in reality. The question of under which conditions the input and output variables measured in reality agree with those of the model can be answered with this method. In reality though, the problem of measurement error (e.g. faulty feedback) arises. Therefore a possibility to reduce or even exclude the disturbing influence of such measurement errors has to be examined. The conditions necessary for this can be found in the laboratory. This means that the evaluation tests can not be performed in comparison with reality. Because the background of the questions asked exists mostly in the transition from the discrete to the continuous model, the questions should also be reduced to this central issue of transition.

For the discrete simulations performed within the scope of the model evaluation, the event-controlled production simulator PROSIM II, which was developed at the Institute for Production Systems (IFA), was used [12]. A simulation for the continuous model for the comparison was performed with the aid of the MATRIXx simulation, a special simulation software for the support of cybernetic systems. The ability to transfer the discrete model of a single work system to the continuous model was verified based on numerous test runs [5].

3 Controller Design

The design of a controller concept has to be divided into two different steps. First of all, it is necessary to define the controllable output variables under consideration of the input variables. Afterwards, the connection and interaction of the controller has to be designed.

3.1 Definition of Reference and Control Variables

With the aid of the models presented two controllers for a work system have been modelled. The input and output variables of the continuous model have been discussed in section 2.2. The variables output performance, wip, and lead time are linked through the funnel formula as mentioned above. Therefore only two of the variables are controllable simultaneously.

The essential task of a work system is to allocate the required performance to process the system load (production schedule). For this reason the output performance, respectively, the output rate attains importance. On a closer examination of the system's output performance, the actual output rate is of less importance. The question of interest is whether the planned work is finished by a certain date. The difference between the planned sum and the actual output is defined to be the backlog of the system. The backlog of the system thus becomes the most important control variable for monitoring the output performance of the production system. In order to utilise a clear definition, the newly developed controller is therefore named

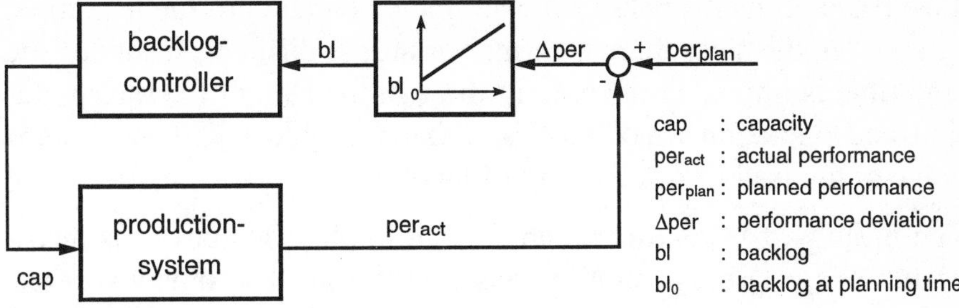

©IFA D1794E

Figure 10: Concept of an automatic backlog controller

backlog controller instead of output controller. The capacity is used here as a correcting variable of the system. Fig. 10 shows the concept of a backlog controller.

The planned performance is the reference variable. The difference between the actual and the planned performance is integrated over a time interval. The result is the above mentioned backlog. Referring to the actual backlog the backlog controller adjusts the required capacity of the work system. Because in reality it is impossible to adjust the capacity immediately, a reaction time between the request for capacity and the following allocation was introduced [5,6].

After defining the backlog as a control variable, a decision whether the lead time or the wip shall be controlled is necessary. A controllable order processing based on production scheduling supports the implementation of the lead time as control variable.

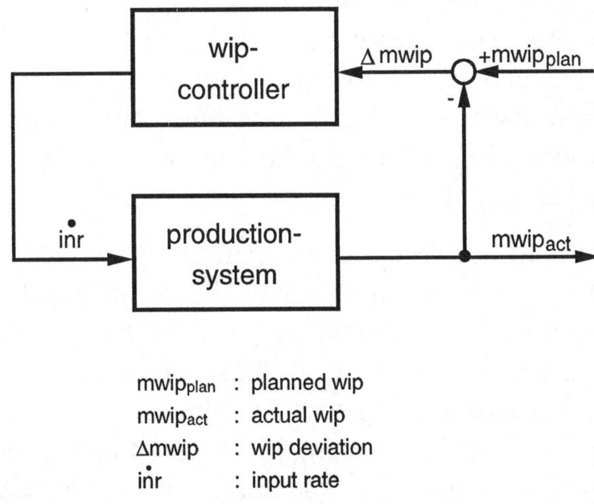

©IFA D1795E

Figure 11: Concept of an automatic wip controller

Two basic problems result from an implementation of the lead time: First of all, the measurement of the variable is difficult; secondly, the lead time is limited at the bottom through the sum of the transportation and operation time. This is a major problem if the reference value of the lead time falls short of this limit.

The mean wip as control variable is not limited. Moreover its measurement is easier and more precise. Therefore it is obvious to use wip control. The main task of the wip controller (Fig. 11) is to set the system to an operating point on the operating characteristic curve that was defined within the scope of production planning. The planned wip is the reference variable. Referring to the difference between planned and actual wip the wip controller adjusts the input rate of the production system [5,6].

3.2 Combination of the Backlog and the wip Controller

The calculation of the required capacity for the next period depends on the operating point on the operating characteristic curve for the work system. The developed concept for the backlog controller only functions, when the planned utilisation of the system is reached since otherwise backlog does not arise. The wip controller is suitable for this task. The basic functionality of both controllers can be compared with the conventional production control methods. In the case of increasing backlog in a production system, it is useful to increase the capacity. If the lead time keeps growing, the line-up (queue) in front of the work system can be diminished through reducing the input rate of the system. The logistic operating curve is a qualified tool for combining both concepts with each other. Simulation experiments have confirmed that this approach guarantees the synchronisation of capacity and work. Fig. 12 shows the integration of both controllers in a controller concept [5,6].

The first step is to decide in which operating state on the characteristic curve the system should be driven. This can be done by deciding which utilisation the system ought to reach. For important or expensive systems this value must be higher than for the other systems. So the backlog control loop is enhanced by calculating the planned output first and then using the planned utilisation for determining the necessary capacity. The relative wip is multiplied by the

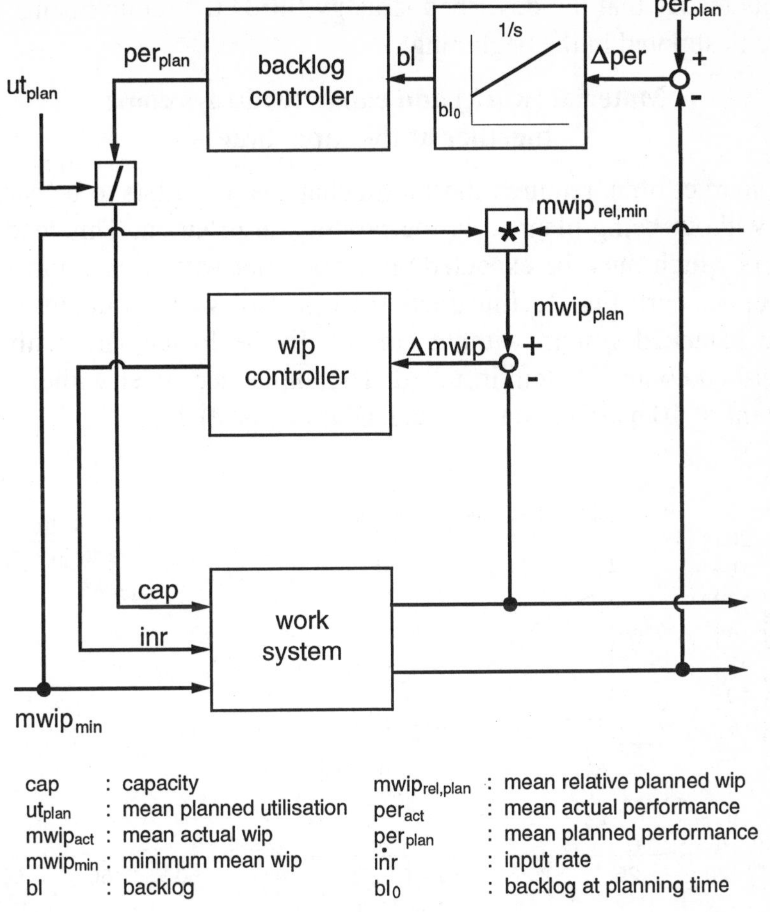

©IFA C0106E

Figure 12: Concept of a combined wip and backlog controller

mean wip minimum ($mwip_{min}$). This results in the planned mean wip. Deviations between the planned and realised performance of the system are integrated over a time interval and defined to be the backlog of the system. The backlog controller calculates the planned performance for the next period which is divided by the planned utilisation. The result represents the corrected capacity of the system. The planned relative mean wip is multiplied by the actual minimum mean wip. This delivers the planned mean wip as the reference value of the wip controller for the next period. This value is compared with the actual mean wip in the system. If deviations occur the wip controller corrects the input rate.

The whole concept and the control design were created by using control theory methods and simulations. Furthermore simulations were used to evaluate the control design. Those simulation experiments testify that the described strategy fulfils the requirements that were postulated in the beginning:

Material (work) and capacity always come together at the same time.

The market often requires short term changes for orders to be carried out without being planned in the production program. This leads to orders which must be executed unplanned but with high priority. In the upper part, Fig. 13 illustrates the effect of such an urgent order on a balanced system without control; in the lower part, with the described control system installed. The unplanned order with a work content of 10 h arrives on shop calendar day (scd) 26.

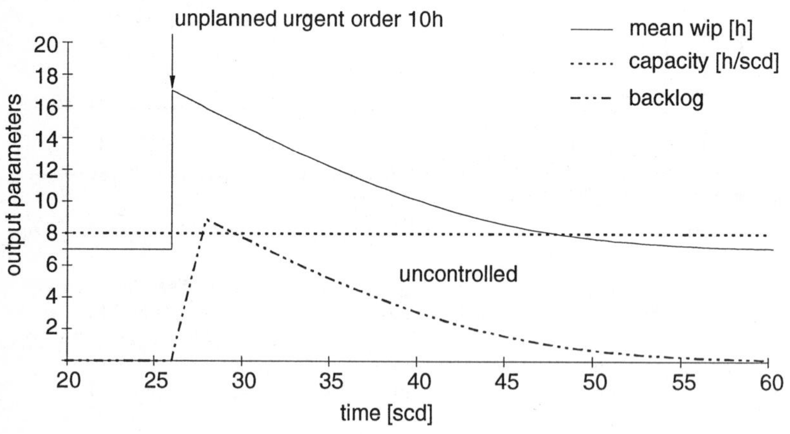

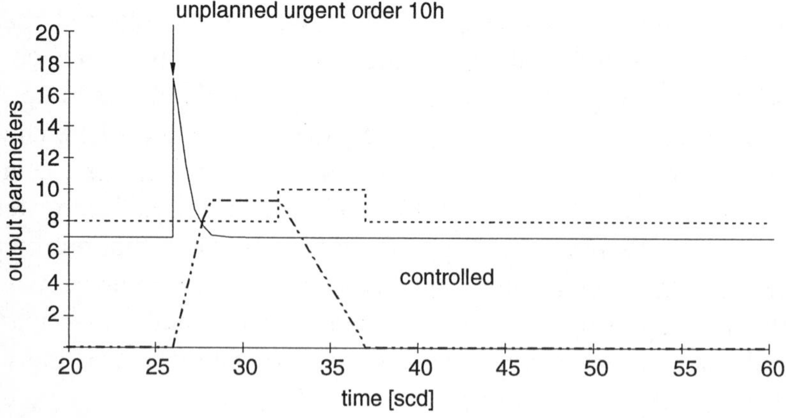

Figure 13: Impact of an unplanned urgent order on the control parameters

As it is called an urgent order it is processed immediately after its arrival. Due to this, wip increases by 10 h to 17 h and backlog comes up because planned work cannot be carried out. Because of its low time constant the system is able to reduce both backlog and wip to the initial level relatively fast (34 scd).

If the balanced system was driven with a higher utilisation this would take much more time. For example, running with 98 % utilisation the same system needs approximately 200 scd to balance the disturbance caused by one single unplanned order of 10 h work content.

The controlled system reacts completely different (Fig. 13, lower part). As the first measure the wip controller reduces the input rate to decrease wip to the planned level. The backlog controller works periodically every 5 days and corrects the capacity after 2 days reaction time at scd 32 exactly to that value that is necessary to decrease the backlog to zero during the following period.

At the same time the wip controller increases the input rate. So there is enough work in the system. This demonstrates that work that cannot be performed is not released until there is sufficient capacity available to carry the work out. Capacity and work come together at the same time keeping lead times at the planned level and compensating disturbances between load and capacity. The quality of this process in a system with this control strategy installed is independent from the initial operating state of the production system. The behaviour of the uncontrolled system becomes worse in proportion with a higher utilisation as previously mentioned [5,6].

4 Integration of the Developed Controllers in a PPC Environment

The PPC system of tomorrow will certainly not be based on an integrated circuit. From today's view, three areas can be identified in which changes in PPC could arise from the previously mentioned approach.

To begin with, it is to be expected that new elements enlarge the understanding of the process and its rules of behaviour as it was shown in the example above. Beyond this, it is quite imaginable that improved techniques and algorithms for the planning and enforcing of order processing can be developed on the basis of these models. It can be determined quite definitely, that it is necessary to think over the functional architecture in PPC if a controlled process is to be attained. The following are new characteristics of a process governed by controlling:

- short-term realisation of changing goals with respect to time and place (guidance behaviour)
- compensation of occurring disturbances (disturbance sensitivity)
- quickest possible attainment of a state of balance (stability)

A proposal for such a functional concept is shown in Fig. 14. The levels, planning and operation, are kept in the usual form [13]. Main tasks are named within the levels with respect to production control.

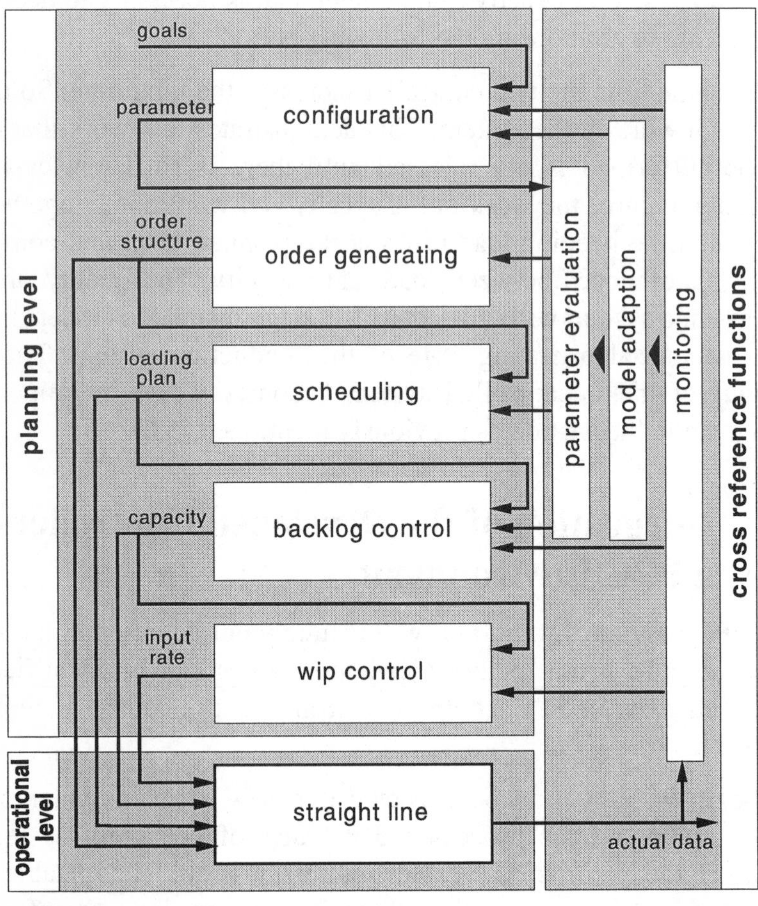

©IFA D1348EBr

Fig. 14: Functional architecture of an APC system

Three cross reference functions are mentioned alongside. The entire architecture must demonstrate a time independent transparency between the control variables and the parameters used in the functions. This can be realised most consistently when all implemented procedures are based as much as possible on the same models.

The wip controller affects the input side through the order release; the backlog controller has to guarantee the capacity. The higher the capacity flexibility of a production is, the more the output performance control attains importance. Capacity flexibility can be described by the length of time, in which a change of capacity can be realised and by the limits within which this change can range. The scheduling and generating of orders are to be understood as correcting variables. The order generation results in the corresponding order structure which most likely arises in the work system. The scheduling sets up a dynamic loading plan for the work system and determines, beyond this, the planned start dates.

The configuration represents a function, which is largely unknown in the classical field of PPC. Its task is to secure the commitment of consistent and also reachable goals. On top of this, it represents the connecting link between goals and process parameters which brings about the transparency demanded above.

The total system does not work automatically, but rather with a strong linkage to the human being. Other important cross-section functions are the model adaptation and the parameter evaluation. The task of the latter is the monitoring of the parameter adjustments during operation and, if necessary, to point out inconsistencies that arise. These inconsistencies can occur from disturbances or from the process itself. The model adaptation has to guarantee a continual and highest possible automatical adjustment of the models used, because even controlling can not function correctly when the underlying control path model does not comply with reality anymore. All three cross-section functions work closely together and can mutually support each other (Petermann, 1993).

The controller concept will be adapted to the extended job shop model. As a first step a backlog controller has been designed. A presentation of this controller will be made in the near future.

References

[1] McKinsey & Company, Inc.(Editor); Rommel, G.(1993), *Einfach überlegen*, Stuttgart, Schäffer-Poeschel

[2] Warnecke, H.-J.; Hüser, M. (1993), *The Fractal Company – a Revolution in Corporate Culture*, Berlin, Springer

[3] Wildemann, H. (1994), *Die modulare Fabrik - Kundennahe Produktion durch Fertigungssegmentierung*, St. Gallen, gfmt

[4] Wiendahl, H.-P. (1995), *Load-orientated manufacturing control*, New York, Springer

[5] Petermann, D. (1996), *Modellbasierte Produktionsregelung*, Fortschritt-Berichte VDI, Reihe 20, Nr. 193, Düsseldorf, VDI

[6] Wiendahl, H.-P.; Breithaupt, J.-W. (1997) Production Planning and Control based on Control Theory – A new Approach in PPC, in: *Conference Proceedings "The second World Congress on Intelligent Manufacturing Processes & Systems"*, June 10th-13th 1997, pp.199-204, Budapest, Springer

[7] Hopp, W. J., Spearman, M. L. (1996), *Factory Physics – Foundations of Manufacturing Management*, Chicago, Irwin

[8] Kettner, H.; Bechte, W. (1981), Neue Wege der Fertigungssteuerung durch belastungsorientierte Auftragsfreigabe, *VDI-Z*, 123 (1981) 11, Düsseldorf, VDI

[9] Wiendahl, H.-P. (1988), The Throughput Diagram, *Annals of the CIRP*, Vol. 37/1/1/1988, pp. 465-468, CIRP

[10] Nyhuis, P. (1991), *Durchlauforientierte Losgrößenbestimmung*, Fortschritt-Berichte VDI, Reihe 2, Nr. 225, Düsseldorf, VDI

[11] Nyhuis, P. (1994), Logistic operating curves – a comprehensive method of rating logistic potentials, *paper at the conference „EURO XII / OR36"*, University of Strathclyde, Glasgow

[12] Wiendahl, H.-P.; Scholtissek, P. (1994), A Simulation Based System to Evaluate the Performance of Production Management Systems, in: *Production Management Methods*, P. 187-194, Oxford, Elsevier Science

[13] Hackstein, R. (1989), *Produktionsplanung und -steuerung (PPS)-Ein Handbuch für die Betriebspraxis*, Düsseldorf, VDI

An Interactive MRP II – Scheduling System

Miguel Nussbaum
Dpto. de Ciencias de la Computación
Pontificia Universidad Católica de Chile

Marcos Singer[1]
Escuela de Administración
Pontificia Universidad Católica de Chile

Gilda Garretón, Olivar Hernandez
Dpto. de Ciencias de la Computación
Pontificia Universidad Católica de Chile

Abstract: We implement a Manufacturing Resources Planning (MRP II) system that considers the process that begins when an order is placed by a customer and ends with the production of the corresponding item. Two implementations, in a printing shop and in an appliance factory, show that the success of such a system depends on its capability to complement, rather than replace, the human planner. Such a capability is based on the system's architecture: The data structure, the scheduling heuristic, and the graphic interface required by the user in the shop floor.

1. Introduction

The classic MRP II literature defines this type of system as a software tool designed to ensure that materials and components are available in time for production [1,2]. More advanced

---

[1] Partial support for this work was provided by Fundación Andes, project C-13222/7

versions also optimize the utilization of the resources of the plant [3,4]. These systems are focused on the production model and scheduling algorithms, leaving to a secondary level the requirements of the end user and his industrial environment. As a prove of that, most surveys that compare different MRP II systems focus on the complexity of their model (alternative routing, tooling, step overlapping, etc.) and on scheduling capabilities (rule based, bottleneck utilization, what if support, etc.) [5]. Such an approach makes the implicit assumption that the working model of the plant must be adapted to the system structure. However, our experience shows exactly the opposite: The logic of the system must match the way things are thought and done on the shop floor.

Based on this principle, we design our MRP II system by first defining what the end user wants to see, working backwards along the system's architecture to make sure that the graphic interface required can actually be implemented [6]. Section 2 presents the main modules of the system, which follows the production process across a number of functional units in the company, rather than the Production Plant only. Section 3 presents the data structure that includes commercial, machine and product data, showing the graphic user interface that allows its definition. Section 4 explains the scheduling heuristic that interacts with the MRP II, which is based on detecting and trying not to overload heavily used resources. Section 5 explains the main reports that are generated, which include multiple Gantt Charts and inventory reports. Section 6 presents a number of issues pertaining to the implementation of this system in two Chilean factories, including some process reengineering, interaction with the human scheduler, information feedback from the plant, etc. An extension of this system is briefly explained in Section 7, which allows the simulation of different scenarios for the evaluation of possible changes in the production plan or the structure of the plant. Finally, Section 8 presents the main conclusions of this work.

2. Description of the System

We conceptualize the MRP II system as an interface between the Commercial Unit and the Plant, in the process of manufacturing the products required by the clients. Figure 1 shows the flow diagram that corresponds to such a process, which is triggered by an order placed by a customer and ends with the production of the corresponding item [7]. As will become clear later in this paper, our system aims to encompass each one of the tasks and documents that are part of this process.

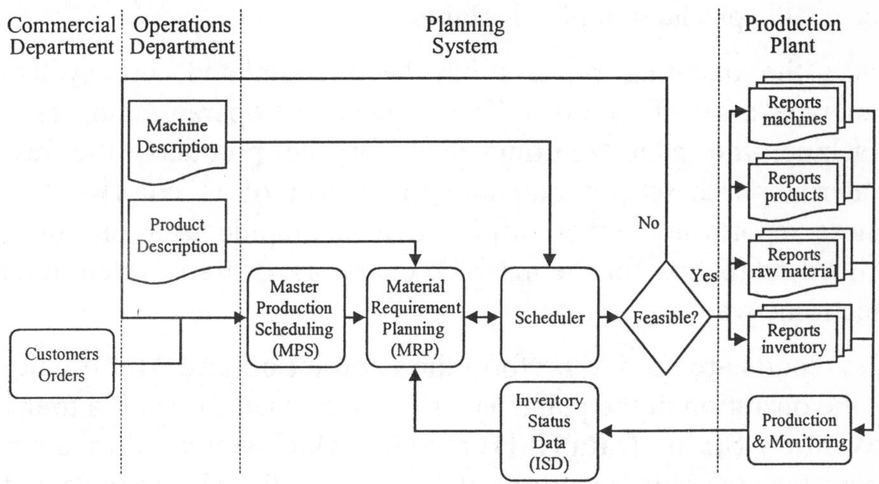

Figure 1: MRP II System's Structure

The triggering component is the **Master Production Scheduling** (MPS) module that is based on commercial information such as business and sales planning, demand forecasting and orders placed by the Clients. This data is then complemented with operational information such as the **Product Description** that includes the Bill of Materials (BOM), which is a hierarchical description of the sub-products and raw materials that are necessary to produce the final product, assembly, sub-assembly or sub-part.

The above input is processed by **the Material Requirements Planning** (MRP) Module that generates the corresponding purchase and production orders, without considering the capacity of the plant. Then the **Scheduler** module, with the information from the **Machine Description**, assigns the production tasks to the different machines in the plant. This process may result in changes in the production orders (due to maintenance for example) that are used by the MRP module to adjust the timing of the purchase and production orders. If a feasible schedule is not found, the MPS is modified, repeating this process until an acceptable production plan is found.

Once the planning process has been completed, the system generates a set of **Reports**: The sequence of operations on each machine, the manufacturing plan for the products, the raw material purchase program and the report of inventory level. These reports are either displayed in a graphic user interface, printed on reports for the machine operators, or transmitted in an electronic way.

The reports are used to perform the **Production and Monitoring** of the operation in the plant. The results are then forwarded to the **Inventory Status Data** (ISD) module, which stores information about the available inventory of final products, sub-products and raw material. The information is then fed back into the MRP module in order to make the proper adjustments, which are later given to the Scheduler to update the production plan for the plant.

The simplified description of the information flow of the system is displayed to the user by the Control Panel that is shown in Figure 2. The user can select different modules depending on whether the input required is available. For instance, until the plant information has been properly defined, it won't be possible to use the Planner.

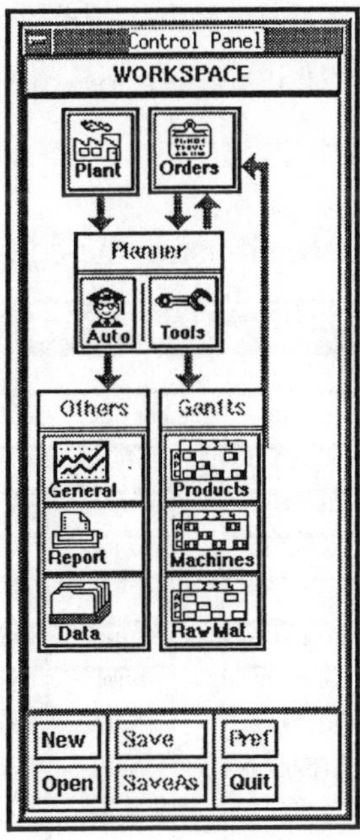

Figure 2: Control Panel

3. Commercial and Operational Data Structures

The commercial input of our system is limited to the customers' orders that are described by the following parameters:

3.1 External Order

Order ID (**OID**): Order Identifier.

Product ID (**PID**): Product Identifier.

Quantity (**Q**): Amount of the product required in order to satisfy the corresponding order.

Early Start (**ED**), *Due Date* (**DD**): Defines the time range within which the product has to be delivered.

Delayable (**DL**): TRUE if the product delivery can be delayed until after the due date if it cannot be planned within the defined time range.

These parameters can be defined using the editor of Figure 3:

Orders Editor

Order	Product	Quantity		Early Start	Due Date	Delay	Status	Planned I
1-1 oil-HKO	oil-HKO	900000.00	kg.	29/7/1992	17/8/1992			
2-1 soft detergent	softBox	180000.00	kg.	29/7/1992	9/8/1992			
3-1 low density HPO	oil-HPO	800000.00	kg.	29/7/1992	11/8/1992			
5-1 detergent 180g	deterg-box	150000.00	kg.	29/7/1992	20/8/1992			
6-1 B wine-label	wine-label	300000.00	un.	29/7/1992	18/8/1992			
6-2 A wine-label	wine-label	250000.00	un.	6/7/1992	1/8/1992			
7-1 medium shoes box	shoes-GBox	300000.00		26/7/1992	7/8/1992			
1-2 oil-HKO	oil-HKO	900000.00	kg.	10/9/1992	30/9/1992			
2-2 soft detergent	softBox	180000.00	kg.	10/9/1992	21/9/1992			
3-2 low density HPO	oil-HPO	800000.00	kg.	10/9/1992	25/9/1992			
5-2 detergent 180g	deterg-box	150000.00	kg.	10/9/1992	3/10/1992			
6-3 B wine-label	wine-label	300000.00	un.	10/9/1992	1/10/1992			
6-4 A wine-label	wine-label	450000.00	un.	16/8/1992	12/9/1992			

Add Order | Duplicate Order | Select All | Delete Order | Unselect All

Figure 3: Customers' Orders Editor

The operational input of our system describes in detail the products.

3.2 Product Descriptor

Product ID (**PID**): Production identification.

Product Level (**PL**): Relative position of the product inside the BOM. According to this parameter, the Basic MRP generates internal requirements hierarchically, maintaining consistency between due dates. In other words, products of the lowest hierarchy have to be produced earliest.

3.3 Inventory Information

Safety Stock (**SS**): Amount kept in inventory to protect against unforeseen events.

Available Inventory (**OH**): On hand inventory which changes according to the amount produced.

3.4 Production Variant

Production Variants (**V**): Each product may have different production alternatives that produce the same product but use a different machine sequence. This is determined by the product's particular recipe.

Lot Size (**LS**): There are cases where machine requirements define a production unit. For instance a barrel of 220 liters.

Minimum Batch Size (**BS**): Economic considerations sometimes make a small lot size unfeasible. For example, set up times may be considerably long with respect to the time required to produce one lot. Therefore a set of lots is defined as the minimum batch size.

3.5 Recipe Information

Step number (**SN**), *Step description* (**SD**): The recipe is defined by a sequence of steps, where each step has a descriptor or name.

Available Machines (**AM**): A given step may be performed by a number of machines. The scheduler selects one of them according to the heuristics described in Section 4.

Step Requirements (**SR**): Indicates which previous steps must be finished before beginning this step.

Product Requirements (**PR**): Indicates those products, and their amounts, which are required by this step (are of a lower level in the BOM) and are manufactured in the plant.

Raw material Requirements (**RR**): Lists those products and amounts purchased from a third party, which are required by this step.

For instance, consider the product "Label Oil-X23", which has a safety stock of 300 units and has an available stock of 2000 units. Two independent recipes are defined: For variant 1, the lot size is 50 units and the batch size is 100 labels. This variant has 4 steps, with the first two being in parallel. Step 1 can be performed in the machine Printer1 or machine Printer2, Step 2 either on machines Printer 3 or Printer 1. Step 3 requires that step 1 and 2 have been already completed, and can be performed either on machines PastingA or PastingW. Step 4 requires that step 3 has been finished, and can be performed either on machines Cut3 or CutD. The sub-product required for processing step 1 is Box AS, and for step 3 is LabelRed. Step 1 requires the raw materials PaintGreen and PaintBlue, while step 2 PaintWhite and step 3 requires Aniline. The data structure for this product can be seen below, while the graphic interpretation of its BOM is shown in Figure 4.

PID: "Label Oil-X23"
PL: Level 1
SS: 300 labels.
OH: 2000 labels.

Variant 1
LT: 50 labels.
BS: 100 labels.
Steps:

SN	SD	AM	SR	PR	RR
1	Printing	Printer1, Printer2	None	Box AS(1.0)	PaintGreen (2.0) PaintBlue(1.4)
2	Printing	Printer3, Printer1	None	None	PaintWhite (0.4)
3	Pasting	PastingA, PastingW	S1, S2	LabelRed(2.0)	Aniline (1.5)
4	cutting	Cut3, CutD	S3	None	None

Variant 2
LT: 75 labels.
BS: 150 labels.
Steps:

SN	SD	AM	SR	PR	RR
1	Printing	Printer3 PrinterA	None	Box AS(2.5)	PaintGreen (2.0) PaintBlue(1.4)
2	Printing	PrinterX, PrinterQ	None	None	PaintWhite (0.4)
3	Pasting	Pasting2, PastingA	S1, S2	LabelRed(3.0)	Aniline (1.3)
4	cutting	Cut3, CutD	S3	None	None

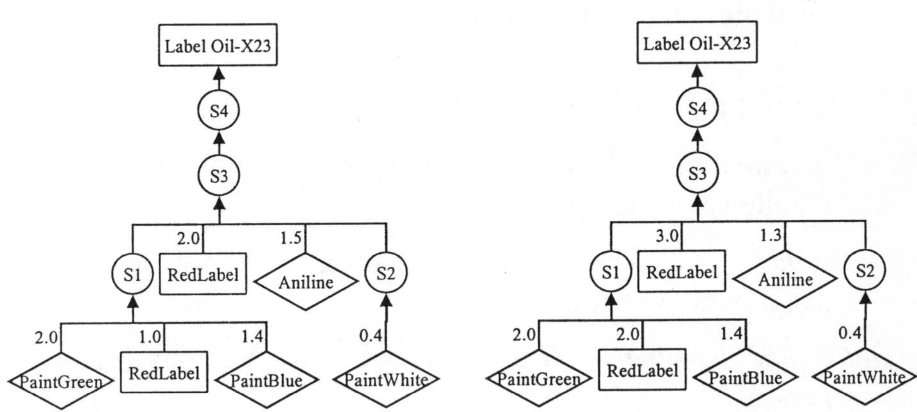

Figure 4: Bill of Material for Product Label Oil-X23

3.6 Raw Material Parameters

A raw material is an item necessary for the manufacturing of the product and is obtained from a provider external to the company. The following attributes define it.

Raw Material Name (**RN**): Material name.

Lot Size (**RLS**): Minimum amount of the purchase order.

Lead-Time (**RLT**): Time between the placement of the order and its delivery.

Safety Stock (**RSS**): Emergency Stock in the warehouse.

Available Stock (**ROH**): Actual stock at the warehouse.

3.7 Production Parameters

Set-up Time: Preparation time for the process. This may depend on the previous production process executed on the machine.

Process Time: Time to process the task on the machine for the corresponding step.

Cleaning Time: After using the machine to perform a given step, cleaning it may be required.

3.8 Maintenance Parameters

Next maintenance: The time remaining until the next maintenance period.

Maintenance Time: The time that the machine is not operating due to maintenance.

Maintenance Frequency: The time between two maintenance periods.

The above elements are defined using multiple editors that are shown in Figure 5.

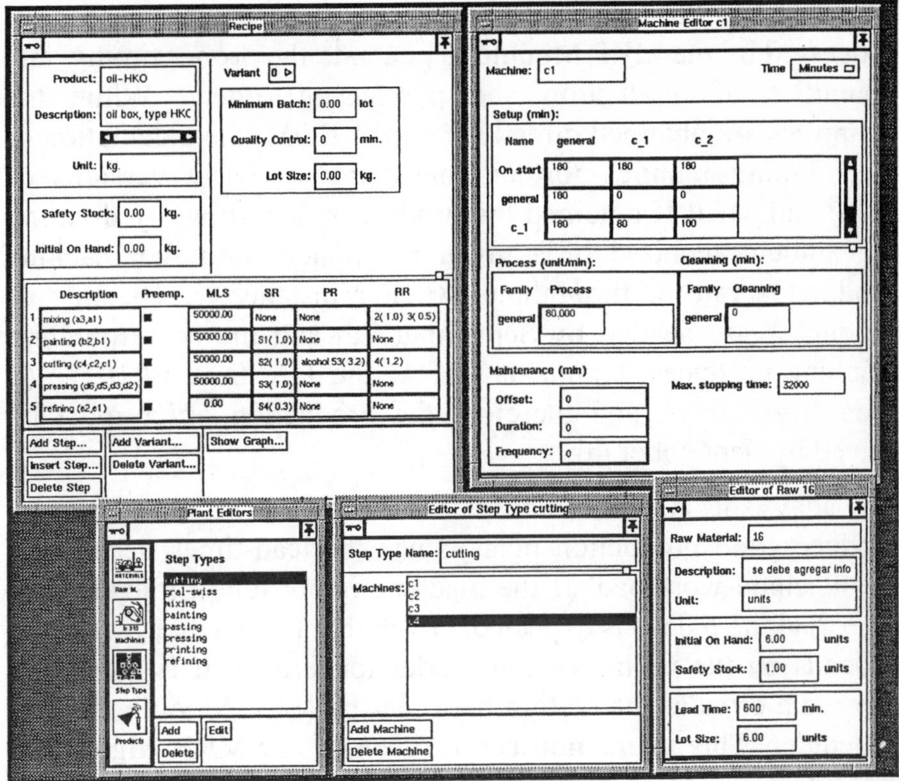

Figure 5: User Interface

4. Planning Heuristic

The first step of the planning process, performed by the MPS Module, is to generate a tentative production schedule that meets the requirements of the customers' orders. We say it is tentative, as it only considers the high level restrictions on the plant. For instance, the lead-time committed to a customer for a particular product cannot be shorter than the total time required by the machines that manufacture it (assuming that they do not work in parallel).

The Master Production Schedule of final products is then processed by the MRP Module to generate the proper timing, and quantities of production and purchasing orders. While the quantities are obtained directly from the BOM, the calculation of their timing requires further consideration. Early versions of MRP and MRP II obtained the production lead-time by dividing the amount required of a given component into the machine production rate or throughput [8]. For instance, if 10 units of product X are needed by September 9th and the corresponding machine produces 2 units/day of X, the lead-time is $10X$ / (2 X/day) = 5 days, and therefore the production order must be placed by September 4th.

Nowadays, it is well known that such a procedure fails to produce good production plans, since the lead-times depend on the machine workload: If the machine is not idle a longer lead-time than the 5 days is expected. Given that the machine schedule depends on the timing of the production orders, it is clear that there is a circular interaction between the MRP Module and the Scheduler. This interaction can be seen in our MRP Algorithm, presented in the Appendix, which successively calls the Scheduler as a subroutine.

The function of the Scheduler Module is both to select and to sequence the machine where a particular step is to be programmed [9]. The selection operation favors those machines that are less required by other orders. We define the *machine factor* as a measure of such usage of a machine. It is calculated as follows: We define a *product cluster* as the set of product orders that may use a machine where the step that is currently being analyzed may be scheduled. As each one of those orders may have a number of possible production paths defined by its different variants, we count how many of those paths use the machine for which we are calculating the machine factor. If all the paths of a given order use the machine, then we increase the machine factor by one. If half of the paths use it then we increase the factor by ½, and so on. We perform the same procedure for all the orders in the cluster, adding the factors weighted by the size and importance of each order. Higher factors mean busier machines, which should be avoided. The Appendix shows the Machine Factor algorithm in detail.

Once the machine has been selected it is necessary to sequence the operations on it. Recalling that the MRP scheduling is performed backwards, that is, steps in higher hierarchies of the BOM are considered first, we favor those sequences that produce the smallest "anticipation", a concept that we define as follows. Consider the example of Figure 6: To sequence A before B demands an anticipation of 5 days before the production is actually needed, and therefore it requires the storage of five days of inventory. To sequence B before A demands storage of 8 days of inventory. Therefore, assuming that the cost of storage is the same for A and B, the system will prefer the first sequence.

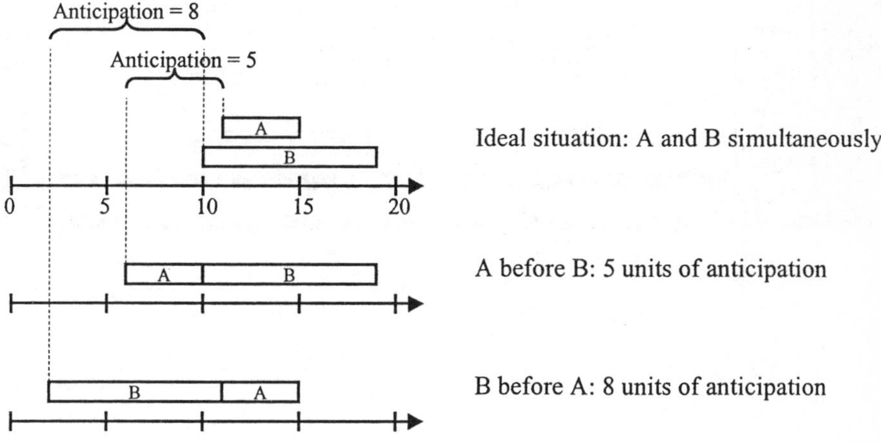

Figure 6: Anticipation Due to Different Step Sequences

5. Reports

The results that are obtained by the planning system are displayed through three Gantt Charts and an Inventory Report. Figure 7 shows the Product Gantt Chart that displays the products and sub-products in its vertical axis, indicating their planned starting and completion times, as well as their due dates. Figure 8 shows the Machine Gantt Chart that has as its vertical axis the different machines of the plant, illustrating their occupation by the different products. Figure 9 shows the Raw Materials Gantt Chart, which displays the corresponding raw material purchasing orders. These Gantt Charts are interconnected; when a product is selected with the mouse, all the machines where it is processed are highlighted as well as the raw material purchasing orders that are required. Analogously, when a machine is selected, the products that it manufactures are highlighted.

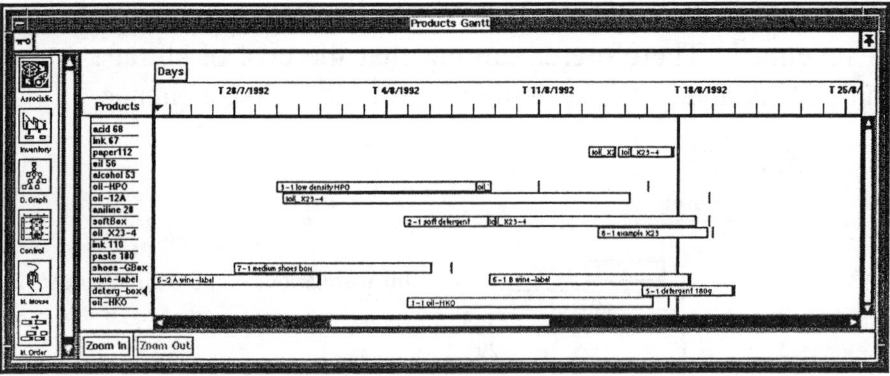

Figure 7: Gantt Chart for Products

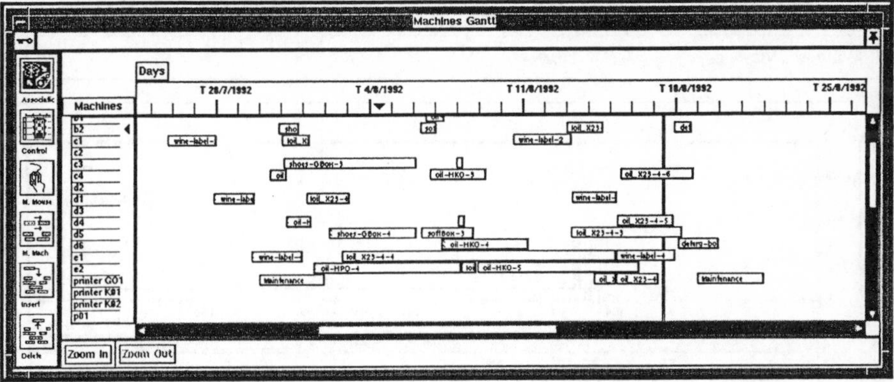

Figure 8: Gantt Charts of Machines

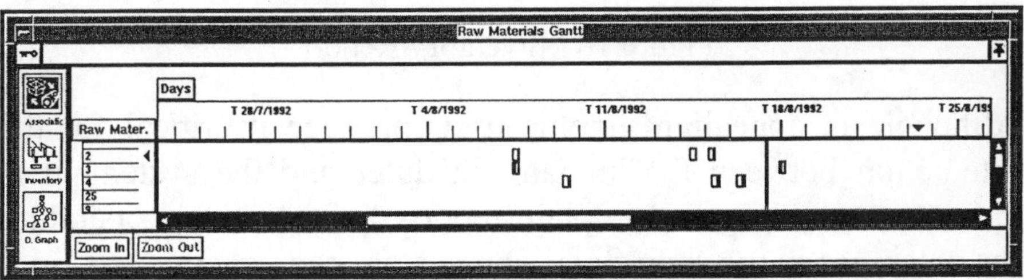

Figure 9: Gantt Charts for Orders of Raw Material

The system also generates the inventory report in Figure 10, which is monitoring the inventory level of the so-called "Prod 11." The vertical bars show the completion of a production order, which can be either from an internal or external customer. The report also shows the orders that decrease the inventory using a dotted bar. If a bar is selected with the mouse, the corresponding tasks in the Gantt Charts are highlighted, and vice versa.

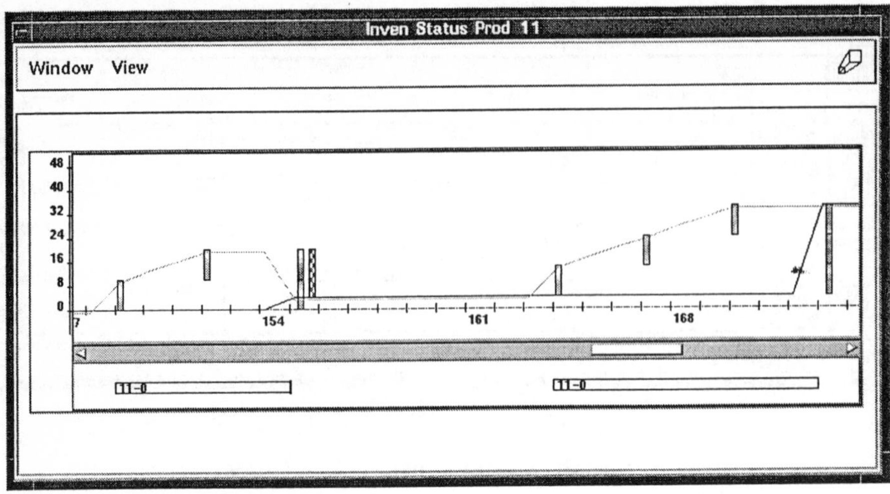

Figure 10 : Inventory Report

Although an appealing graphic user interface it critical for the interaction between the human scheduler and the system, it is also necessary to generate plain reports for the people operating the machines in the shop floor. Our system can generate and print reports in ASCII format containing all the relevant information for the operation and control of the machines.

6. Implementations

This system has been implemented in a printing shop and an appliance factory. In both cases we discovered that the automation of inefficient processes does not improve production efficiency, and therefore it was mandatory to perform some degree of reengineering in the plant [10]. We were assisted by an expert in the plant's day-to-day operation, who also had the ability to model his knowledge with some degree of abstraction. We found that until we performed this study, a very limited portion of the plant's potential was exploited. For instance, only a few production paths, not necessarily the most efficient ones, were used for a given product.

Once the proper reengineering was finished we adjusted our system to the operation of the plant. In both implementations it turned out that the human planner managed two sets of operations: one of high priority and the second with lower priority. The planner scheduled the high priority operations first, to make sure that the most relevant orders were met on time. Then he scheduled the less important orders, without making any significant modifications on the first schedule. Our strategy was to emulate such a procedure, so our system also performs a two-pass scheduling, starting with the orders that are defined as the most important ones. Once the first schedule was done, we let the human scheduler modify it if necessary. Then we froze this solution and scheduled the second set of orders. The user could also modify this solution.

The interaction between the human scheduler and the system is a critical issue for our system, since it does not aim to replace the scheduler but rather to assist him. While the automatic system has the strength of efficiently in detecting inconsistencies, the human is more suited to deal with exceptions and with information about the plant that is not well structured. Therefore, it is mandatory that the user can easily make modifications to the schedule, and that if there is an inconsistency it is detected and explained. The modifications can be performed by a drag-and-drop procedure of the tasks displayed on the Gantt Charts. The user is warned of inconsistencies by pop-up windows with verbal explanations, and also with graphical information. For instance, if an operation is hastened in a way that causes it to be processed before its ingredients have been manufactured, the system explains the problem and displays the window of Figure 11.

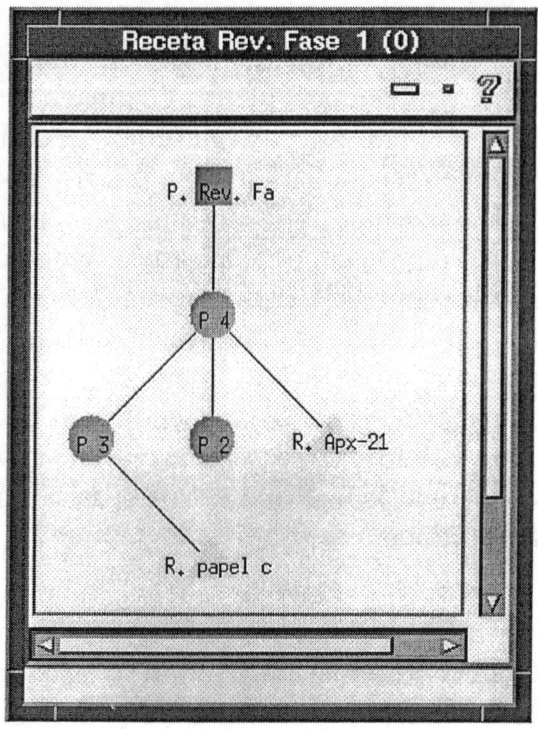

Figure 11: Representation of the Bill Of materials (BOM)

The last implemented component is the interface that feeds the information from the production plant back to the system. Such a unit is shop floor specific and in the two implementations that we are describing they differed considerably.

In general, there are two procedures for specifying information about the monitoring of the production: the manual procedure and the online system. The manual procedure interacts with a control system based on reports that are filled out by the machine operators. This information is later fed into the system using editors that are provided by the graphic user interface.

The online system is designed to interact with a Computer Integrated Manufacturing system (CIM), where each machine has a number of sensors describing its status in real time: downtimes, throughput, setups, etc.

7. An Extension of the System

A more sophisticated tool, not discussed in this work, links the above system to a Simulation Module. Such a tool allows the evaluation of different scenarios, informing the Sales Department if a product can be delivered within a given time window, and how much of a product can be manufactured.

The Simulation Module can also be used to evaluate different maintenance plans, the impact of introducing new machinery, or a change in the current production process. For instance, if a new inventory policy is being evaluated, the system generates a number of scenarios for which the MRP II generates production plans. The quality of these plans is then compared with the quality of the plans currently in use, in order to determine whether the proposed inventory policy would be better for the company.

8. Conclusions

Our experience implementing MRP II and logistic systems has shown us that, even though the enthusiasm from the managerial personal is quite high, there may be a strong opposition from the operative personal. This explicit or covert opposition increases the probability of failure, and therefore must be dealt with very carefully.

Our strategy has been to put ourselves on the side of the operative personal, by designing a tool that better assists their work. We match the data structure that they use, we mimic their scheduling procedures, and we try to satisfy their graphic user interface requirements. By guarantying that the system is not intended to work by itself, we ensure that the human scheduler is the person most interested in a successful implementation.

References

[1] LANDVATER, D. & GRAY, C. (1989), "MRP II Standard System: A Handbook for Manufacturing Software Survival" *Oliver Wight Limited Publications, Inc.*

[2] SHELDON, D. (1991), "MRP II What it Really Is" *Production and Inventory Management Journal,* Volume 32, Number 3.

[3] PINEDO, M. (1995), "Scheduling Theory, Algorithms and Applications" *Prentice Hall Englewood Cliffs*, NJ.

[4] TAYLOT, S. & BOLANDER, S. (1994), "Process Flow Scheduling" *APICS*

[5] MELNYK, S. (1996), "FCS Software Survey" *APICS The Performance Advantage*, August 1996, pp. 72-82

[6] NUSSBAUM, M.; GARRETÓN, G.; LEPE, A. & PARRA E. (1992), "User Interface Aspects of an MRP II Planning Module" *Computational Economics*, 6, pp. 17-50

[7] DAVENPORT, T. (1993), "Process Innovation, Reengineering Work Through Information Technology" *Harvard Business School Press*, Boston MA

[8] HOPP, W. & SPEARMAN, M. (1996) "Factory Physics" *Irwin*, Chicago

[9] NUSSBAUM, M. & PARRA, E. (1992) "A Production Scheduling System" *ORSA Journal of Computing*. Vol. Nº 4

[10] HAMMER, M. & CHAMPY, J. (1993) "Reengineering the Corporation: A Manifesto for Business Revolution" *HarperCollins Publishers Inc.* New York

Appendix

MRP Algorithm

```
BEGIN
  Put the external orders in the requirement list.
  WHILE (There are requirements in the requirement list to plan) DO
    WHILE (There are requirements at the same level of the BOM that are not planned)
      FOR (All requirements that are not planned) DO
        IF (The inventory level covers the requirement) THEN
          Actualize the inventory.
        ELSE
          Generate the production orders according to the inventory status.
        END _IF
      END _DO

      WHILE (There are production orders at the same level of the BOM
        that are not planned) DO
        /* Assign machines according to the Scheduler Heuristic */
        FOR (each production order) DO
          Search for the production variant and machine occupation of this
          order, considering the planning heuristics and the plant capacity.
        END _DO

        /* Analysis of inventory and production dates */
        FOR (all production orders) DO
          Analyze the inventory status.
          IF (order is not late) THEN
             Generate the internal requirements (sub-products and
             raw materials).
             Put the requirement in the Requirement List.
          ELSE IF (order is late) THEN
             Undo the plan of this production order.
             Put the order in the list of production orders to plan again
             with another date.
          ELSE IF (the planning of this order is impossible) THEN
             /* Out of planning horizon, both raw material and
             products requirements */
             Search the level of the external order that generated
             this requirement
             Undo the plan until this level
```

```
                    Start from this level again
                END _IF
            END _DO
        END _WHILE
    END _WHILE
  END _WHILE
  FOR (all raw material required)
    Plan internal requirements according to the inventory level.
    Generate the purchase orders.
  END _DO
END.
```

Machine Factor Algorithm

Definitions:
mf(*mach, step*): Machine Factor of machine *mach* for the production order step *step*
cluster(*step*): set of orders that may share resources the step to be scheduled
paths(*ord*): Total number of production paths of order *ord*
paths(*mach, ord*): Number of paths of production order *ord* where machine *mach* appears.
weight(*ord*): Weight of order *ord*

```
BEGIN
  FOR each machine DO
    mf(mach, step) = 0
    FOR each order in the cluster(step) DO
      mf(mach, step) = mf(mach,step) +
                              weight(ord)*paths(mach,ord)/ paths(ord)
    END_DO
  END_DO
END
```

MRP II-based Production Management Using Intelligent Decision Making

I. Hatzilygeroudis, D. Sofotassios, N. Dendris, V. Triantafillou, A. Tsakalidis, P. Spirakis

Computer Technology Institute (CTI), Hellas (Greece)
&
University of Patras
Depart. of Computer Engineering & Informatics
Hellas (Greece)

Abstract: An extended MRP II-based production management system (PMS) is presented, which improves the traditional MRP II paradigm. It does so by attaching an intelligent decision supporting system (IDSS) to the lowest level of the PMS, namely the production activity control (PAC) subsystem. The IDSS includes a simulator, that imitates real system behaviour, a knowledge-based component, that imitates expert reasoning, and a real-time database manager, that acts as the data pool and the communication gate between them. It is capable of performing off-line and on-line rescheduling, thus resulting in more realistic short-term production schedules. Analysis of the related case problem and implementation of the system are also discussed.

1. Introduction

Currently, the *Manufacturing Resources Planning II* (MRP II) methodology [4] appears to be the most publicised approach adopted in manufacturing management. MRP II extends the primitive Material Requirements Planning (MRP) features [22]. A *production management system* (PMS) in general deals with all levels of production management, such as the strategic, tactical and operational level. An MRP II-based PMS supports manufacturing functions at all those levels in a hierarchical fashion.

At the lowest level (operational level) in the hierarchy, the *Production Activity Control* (PAC) subsystem resides. PAC concerns production control at the shop floor [2] and operates in a time horizon of between a month and real-time. The output of the PAC system is a plan indicating the sequence of the orders to be executed in a production period, by specifying their release and due times. A serious drawback of an MRP II-based system is that the production plan produced by the PAC level is rather unrealistic, because it cannot take into account the real state of the production environment. Hence, the system cannot follow the large number of shop floor events to make real-time decisions. On the other hand, it does not provide any serious support to the production manager to revise the unrealistic plans and carry out the complex task of production control at the shop floor. Production control mainly deals with scheduling/rescheduling, which is really a complex task, since it involves decision making by taking into account a large number of conflicting factors or constraints [26]. Poor production control may cause serious problems to a firm's ability to meet production requirements and constraints.

Current research tends to attack the above problems mainly with structural solutions (e.g.[10, 15, 21, 25]): they provide generic PMS frameworks for future manufacturing systems rather than integrate new components to the existing ones to extend their functionality, as we do. To make plans more flexible and realistic and help the production manager to his task, we have attached *an intelligent decision support system* (IDSS) to the PAC system that uses real-time information. An IDSS is a decision support system that combines conventional and knowledge-based technologies, and where the user remains part of the decision cycle [9]. In our case, *simulation*, a conventional technique, is used to imitate the real system behaviour and the *expert systems* approach, a knowledge-based technique, to imitate expert reasoning. Given the introduction of knowledge-based technology and its capabilities, the computer-based system is more actively involved in the decision making process, in contrast to its passive role in a traditional DSS. There have been a number of systems based on this point of view [19, 28, 16, 23].

The paper is organised as follows. Section 2 describes the case problem and its analysis methodology. Section 3 deals with the system architecture and the decision making processes. In Section 4, the simulation based component of the system is presented. Section 5 deals with the knowledge-based component, and finally Section 6 concludes.

2. Case Problem Analysis

2.1 Methodology

The source problem for our system design concerns production control at the shop floor of a yoghurt plant. The analysis of the problem was two-fold. On the one hand, it was related to conventional software systems analysis and, on the other hand, to knowledge acquisition, as in knowledge-based systems development. We specified three aspects of the problem analysis:

- *production process*: It concerns the shop floor layout and operation, i.e. the production lines, the workcenters, the operations in each workcenter etc.
- *production management*: It concerns the activities of the production manager during the production control process.
- *problem solving*: It concerns the ways the production manager reasons and acts when solving problems using his experience in cases of abnormal situations.

To achieve the above goals, we used a mixture of methods: questionnaires, interviews and observation, alongside printed material about the plant. In summary, we constructed over 10 structured questionnaires, of 50-60 questions each, and had over 10 semi-structured and structured technical meetings, of two or more hours long, with the production management staff of the plant. The major part of questionnaires, the printed material, observation and a small part of the meetings concerned information about the production process and the production management. The results were used for creating a model for the shop floor and specifying the tasks and subtasks of the production manager. The major part of the meetings and a small part of the questionnaires were used for knowledge elicitation, mainly from the production manager and the senior staff, to realise their problem solving procedures. To this end, we also constructed Gantt-like charts of simplified real-like production plans and discussed solutions on occurrences of abnormal events with the production management staff [8].

In the following two subsections, we present the basic results concerning the first two aspects of the case problem analysis. The third aspect, namely problem solving, is dealt with in Section 5.6.

2.2 Production Process

The shop floor of a yoghurt plant is a flow shop environment consisting of a number of production lines, that may or may not be interconnected (i.e. have common work-centers), each producing various alternatives (flavours) of a basic type of yoghurt. There are some different basic types of yoghurt, like e.g. SET and STIRRED, each having various alternatives. A typical (simplified) production line of the plant, with its interconnections to other lines, is depicted in Figure 1.

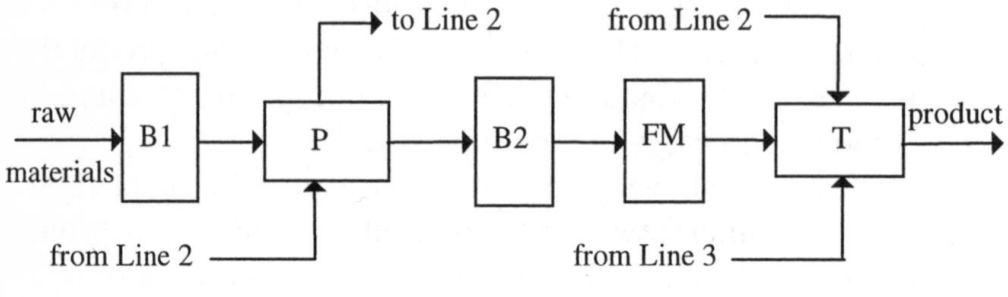

Line 1

Figure 1. A Typical Production Line in a Yoghurt Plant

The primary raw material is milk, which mixed with other materials, such as cream and protein, constitutes the mixture of a basic type of yoghurt. The mixture is initially created and stored in a buffer (B1), where it remains for a certain time in order a chemical process to be completed. Then, the mixture feeds, in a continuous flow mode, the pasteuriser (P), where it is pasteurised by increasing its temperature to 95°C. The pasteurised mixture is then stored into another buffer (B2). So, there is a continuous flow from B1 to B2 via P. After the mixture has been stored in B2, it is led to the filling machine (FM), where it is distributed into cups. Cups are put in palettes. Each palette is moved on to the cooling tunnel (T), where the mixture in the cups is being cooled passing through a number of stages. After it is sufficiently cooled, palettes are moved on to the warehouse where they remain for 2-3 days, time necessary for the yoghurt to be ready for delivery to the market. Again, there is a continuous flow from FM to warehouse via T.

This kind of flow-shop production system have the following characteristics:

- There are no specific customer orders, but the orders are determined by the stock demands which in turn are determined by the market demands.
- Each work-center consists of one machine that performs only one operation.
- The operations that constitute an order should normally be executed successively, without any waiting, although waiting is possible in specific buffers (e.g. B2 in Fig.1), if required.
- There are no alternative process routes in a production line.
- The type of control applied is event-driven.

Although these characteristics result in a relatively simpler flow shop system than usual, the problem is still complex enough to be sufficiently handled by analytical methods and it still results in a large cognitive load to the production manager.

2.3 Production Management

The production plan period is a week. The production plan (schedule) for a week is more or less fixed (:it is not reconstructed every week). It is however periodically revised by the production manager depending on the time of the year, changes to the market demands and introduction of new or removal of old flavours. In this way, between two revisions the production manager knows for any week which flavours of which type of yoghurt should be produced in which day, in what sequence and in what quantities. Refinements of such a schedule concern only changes to the quantities, within certain limits.

Under these conditions, the task of the production manager is two fold. First, at the end of the current week he makes the appropriate refinements or modifications to the schedule for the forthcoming week, based on the stock requirements provided by the inventory control department, and the real condition of the factory. Once required changes have been made and the schedule is fixed (frozen), it is ready for execution.

On the other hand, during the real production (schedule execution) the production manager is responsible for reacting to any abnormal event(s) that may occur, like e.g. a machine breakdown, a high scrap or a rush order. This requires that first some immediate preventive actions, such as which work-centers production flow should stop at, are taken, and then some kind of reactive scheduling, i.e. on-line changes to the production schedule which is currently under execution should be made. The response time of the production manager to an event may sometimes be crucial, since e.g. milk products are time-sensitive. This is actually the main task of the production manager.

To make decisions, the production manager has to take into account a large number of constraints. For example, there are different setup times required for different breakdown intervals of a work-center. Also, there are certain yoghurt types that can be simultaneously passing through a certain work-center, whereas others cannot, depending on the compatibility of their temperatures. Furthermore, when a breakdown occurs and recovery time exceeds a certain limit, the quality of the product should be checked before proceeding.

The production manager does not use any computer-based tools to accomplish his task, but only his experience and he is no familiar with any analytical methods. Thus, sometimes his decision process seemed to be quite simplified, since he had no means to quickly explore various alternatives. Constraints such as cost-effectiveness and machine utilisation are only implicitly taken into account. Due dates are not considered as very hard constraints as product quantities are.

3. Extended MRP II-Based PMS

3.1 System Architecture

The architecture of the extended MRP II system is depicted in Figure 2. Extension consists in attaching an IDSS to the PAC subsystem, where the production manager (PM) is considered part of it. PM represents anyone who is responsible for making decisions for production control. The IDSS consists, apart from PM, of three major components: a real-time database manager (RTDM), a simulator (SML) and a decision maker (DM). These components can accomplish a variety of tasks and co-operate in a variety of ways.

PM acts like a controller that specifies each time the kind of the task and the co-operation activity to be performed. RTDM is the means for storing and managing available information. SML is a representative of the conventional DSS technology that allows PM to make simulation runs of a candidate production schedule, whereas DM is a representative of the knowledge-based technology that helps PM to evaluate such simulations and revise candidate schedules. RTDM includes the system database and is the only means for the communication between SML and DM. RTDM is a relational database management system implemented in INGRES, a tool for developing relational database management systems.

3.2 Decision Making Processes

The functionality of the system reflects the production manager's two main tasks. Thus, it can operate in two independent modes, called the off-line and the on-line mode, that correspond to the off-line and on-line task of the production manager, respectively.

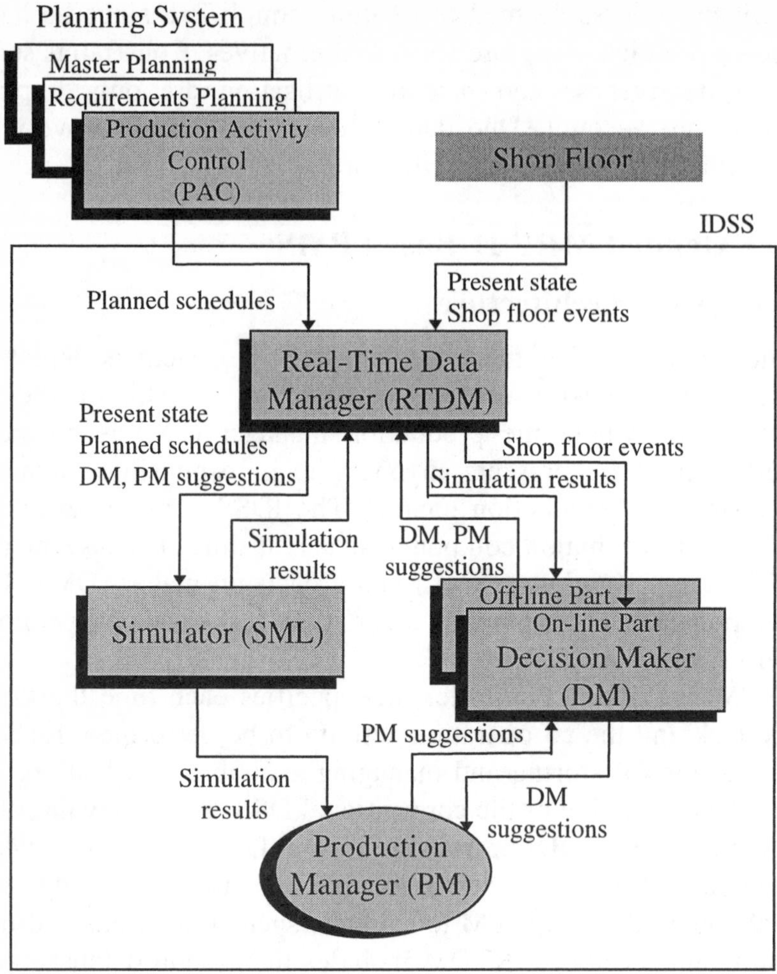

Figure 2. The Extended MRP II-Based PMS Architecture

The interactions between the components of the system in the *off-line mode* are depicted in Figure 3a. According to that, the planning system generates a schedule for a production period which is stored in the system database. The execution of this planned schedule can be afterwards simulated by SML, according to PM's desire, using real data about the shop floor model from the database.

After a simulation run is completed and its results are available in the RTDM data pool, these results can be evaluated. This is done either directly by PM himself, or via DM. In the latter case, DM first decides on the degree of acceptability of the schedule by computing its total deviation from the planned production. Alternatively, revision rules are used to propose changes to the schedule. Finally, PM is the one who decides on which of the changes will be applied to the planned schedule, by approving some or all of DM's suggestions or by introducing his own. These suggestions are recorded in the RTDM data pool, and are taken into account by SML in the next simulation run.

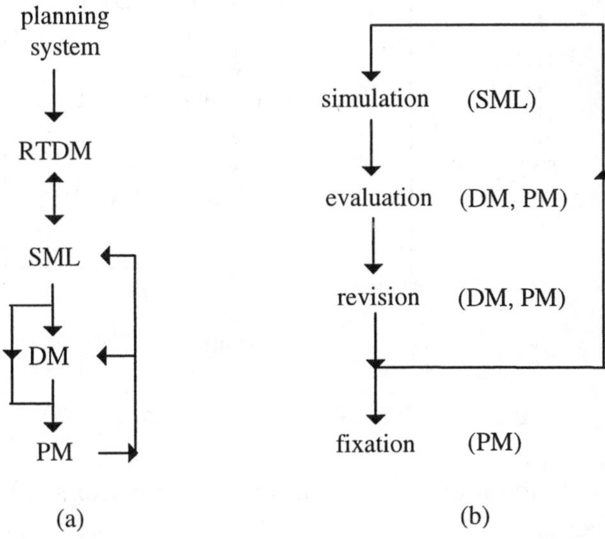

Figure 3. Off-line mode: (a) component interactions (b) decision cycle

Subsequently, PM can run as many simulations of the schedule as required to reach an acceptable plan for the forthcoming period. In conclusion, the off-line process can be described by the following decision cycle: simulation-evaluation-revision-fixation (Fig. 3b). This reflects PM's subtasks in his off-line task. The final, acceptable schedule is recorded in the database.

For the *on-line mode*, the interactions between the components are depicted in Figure 4a. The on-line process is performed in two stages. In the first stage, as soon as an abnormal event is detected, the system proposes to PM a set of preparatory actions, to deal with the situation in short-term, like e.g. to interrupt production before a workcenter. Another set of actions is proposed to PM just after the recovery from an abnormal situation, e.g. after a machine repair has been completed. This type of action mainly concern workcenters setup.

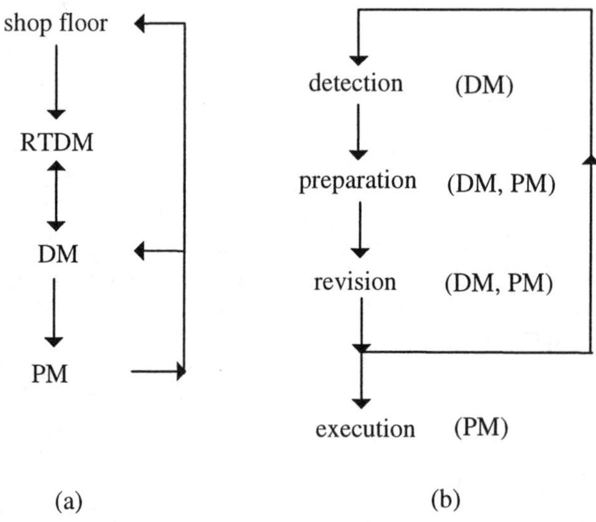

Figure 4. On-line mode: (a) component interactions (b) decision cycle

In the second stage, reactive scheduling is actually taking place. DM investigates possible changes to the executed schedule, to minimise the consequences of the disruption. The result is a number of changes to the current schedule to continue the disturbed production. Accordingly, PM either approves DM's modified schedule and takes the required actions or makes his own decisions, or even interacts with DM for further investigation. So, the decision cycle performed is: detection-preparation-revision-execution (Figure 4b). This again reflects PM's subtasks in his on-line task. The final changes to the schedule are recorded in the database.

Given the above correspondence between the functions of the system and the tasks, subtasks of the production manager, we introduce the term *task-oriented architecture* to characterise the architecture of the IDSS.

4. Simulation Based Component

Simulation is used to imitate the behaviour of the real production system leading to useful inferences about its short-term production performance [6]. SML provides the necessary technological support for investigation of "what-if" questions and determination of the scheduling policy for the forthcoming production period. The operation of simulation corresponds to a pre-production step which extends the capabilities of the manager allowing him to manage production efficiently in a proactive manner [1].

4.1 Data Environment

The simulation process is concerned with the following classes of data:

- *model description data*: This data is related to the manufacturing profile of the system [2, 3, 17]. It includes resource data, inventory data, routing information, released orders, historical data etc.
- *experimental data*: It includes the simulated time horizon, initial conditions, end-of-run criteria and a number of operational parameters used for the validation of the simulation model.
- *real system data*: This data is used to predict the state of the system at the beginning of the simulated horizon. When a simulation run starts, the system is unlikely to be in a zero state condition (i.e. no active orders and all machines available). Moreover, to produce a more realistic run, real values should

be used for various model variables instead of estimates (i.e. real capacities, present inventory levels).

- *output data*: The output of the simulator is a set of attribute variables and a number of statistics which are of interest to the production manager. The statistics can be presented in different ways, such as textual form, tables and Gantt charts.

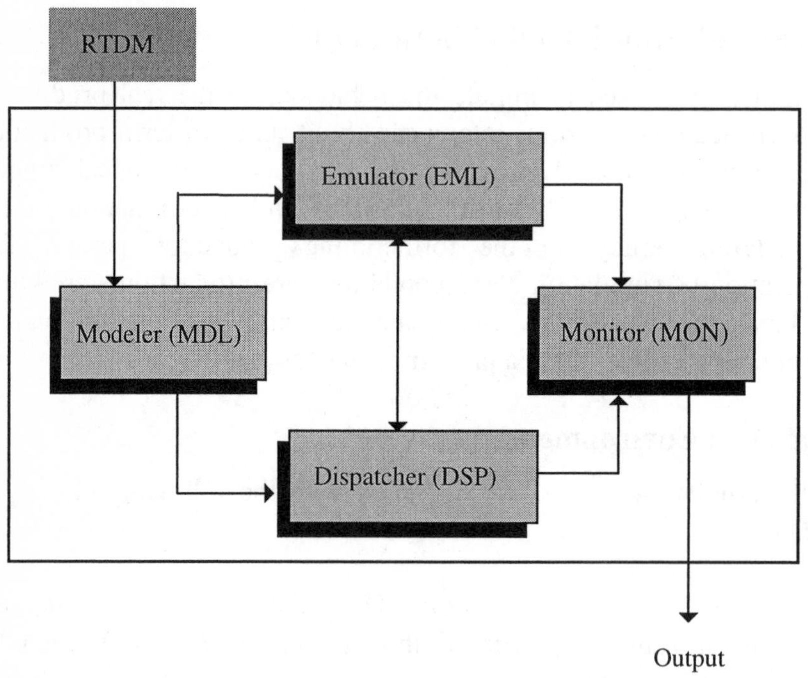

Figure 5. The Architecture of the Simulator

4.2 Main Building Blocks

SML is an event-driven simulator and consists of four building blocks (Fig. 5), namely *modeler* (MDL), *emulator* (EML), *dispatcher* (DSP) and *monitor* (MON). MDL retrieves all the required information from the system database to construct the system model. This information includes released orders, route sheet structures, various capacity data, real system data and correction suggestions. Real system data is used to assess the initial state of the system in a simulation run. The basic model entities are released orders, machines, buffers and operators.

The main function of EML is the execution of the events occurring at the shop floor in a timely manner. Such events are an operation start/end, a machine breakdown, a maintenance start/return, an operator's work start/end etc. Probably, the most important task performed by EML is the generation of new events as a result of the execution of the current event. These events are dynamically created, given a time stamp and then stored in a sophisticated data structure, called the *event calendar*. EML executes the events in an order specified by DSP. DSP locates the next event to be executed, each time it is called by EML, and returns control to EML. The events waiting for execution are stored in the event calendar according to their time stamps. The event calendar drives the simulation and is updated every time DSP is called.

MON interacts with the other two modules and collects various process data. Data collection is carrying out whenever an event execution finishes. When a run terminates, MON is called again to make statistical calculations. The produced *statistical information*, that constitutes the output of SML, includes machine utilisation levels, average buffer loads, operation processing/waiting times, work-in-progress (WIP) levels, scrap quantities, problems identified (e.g. dropped orders), total shop output etc.

4.3 Modelling Stochasticity

Uncertainty in the real system is represented via the *stochastic parameters* of the model. Machine breakdowns are truly stochastic events that happen at non uniform time intervals, and historical data are necessary to predict this behaviour. A stochastic parameter is used to model the time of a breakdown via the Monte Carlo method [6, 29]. Two other stochastic parameters concern scrap quantities and the time an operator works. Both parameters are represented by proper probability distributions, such as the Gaussian, the Weibull, the Gamma and the Pearson distribution [6], given their average value and its variance.

Breakdown data is calculated off-line, before a simulation run. It is then stored in an appropriate model entity and is used by EML during the run. Breakdown data may remain unchanged between successive simulation runs after the schedule revisions have been made or a new breakdown data pattern may be generated before a new run begins. This capability helps manager in making comparisons between successive runs. Scrap quantities and operator times are computed by EML, during a run. So, by default, this data changes from one run to another, but the values obtained from each run are expected to be close enough to each other due to their small variances. The values produced in a run can be stored for use in later runs.

Finally, the times for machine setups and operation processing are considered by the simulation model as deterministic. Although this assumption seems to be quite reasonable for factories with high degree of automation in the flow line, in a more general setting both parameters have to be modelled with appropriate probability distributions, too.

4.4 Implementation Issues

In the integration of the SML with the PMS, flexibility of the simulation model is the key factor [12, 17]. There are several different classes of data provided by different data sources. All these data classes are stored in the system database. Consequently, SML operation is driven from RTDM tables, while RTDM assures consistency of the simulation data. The main advantage of this approach is that the user does not need to have any simulation expertise, as SML runs off a database. Moreover, the internal structure of SML needs not to be changed when new products and processes have entered the system.

All building blocks of SML are built in C for two main reasons. First, SML can easily communicate via C with the other two technologically different platforms, a relational database and an expert system shell. Second, C allows very efficient RAM implementations in both time and space. An example is the event calendar structure. Event calendar has been implemented as a priority queue [20] which supports fast location of the next event to be executed. Another example is the use of dynamic list structures to store the model description data. Lists structures were chosen instead of fixed arrays [3] because they allow better memory management.

5. Knowledge-Based Component

In the design of DM, the expert systems (ES) approach (see [27] for an account of the existing approaches) was mainly followed. This approach tries to mimic the reasoning process of an expert. Rules integrated with procedures are used for knowledge representation. There are a number of rule-based systems designed for production scheduling [5, 18, 24]. A difference of DM with existing scheduling systems based on the ES approach is that they are rather autonomous systems, that is the user is not actually involved in the decision process, and most of them do not deal with reactive scheduling.

5.1 Functional Architecture

DM includes three modules: DM expert system (DMES), DM user interface (DMUI) and DM data interface (DMDI) (Fig. 6). The main core of DM is DMES. DMES consists of two main parts, called *off-line part* (OFP) and *on-line part* (ONP).

OFP deals with making improvements to planned schedules. It consists of two parts, *evaluation part* (EVLP) and *predictive scheduling part* (PSP). EVLP is responsible for assessing the simulated schedule; it makes an estimation of the acceptability of the planned schedule. PSP actually proposes improvements to the schedule, if it is not acceptable. It makes suggestions about changes to the schedule in terms of the following *revision actions*: 'shift order', 'remove order', 'insert order', 'change quantity' and 'change rate'. Thus, OFP mainly deals with what is called *predictive scheduling* (see e.g. [7]). OFP can be activated by PM during the off-line operation mode.

ONP is concerned with making decisions on corrective actions in response to disruptions due to abnormal events occurring during the real execution of a schedule. ONP also consists of two other parts, namely *action part* (ACTP) and *reactive scheduling part* (RSP). The first gives advice to PM about preparatory actions that should be taken just after an abnormal event has occurred, or necessary actions after the recovery from an abnormal situation. The actions may refer to the work-center related to the event or to other active work-centers. RSP suggests on-line rescheduling scenarios in an abnormal situation. So, ONP deals with what is called *reactive scheduling* (see e.g. [7]). ONP is automatically activated during the on-line mode as soon as an abnormal event occurs and is recorded in the database.

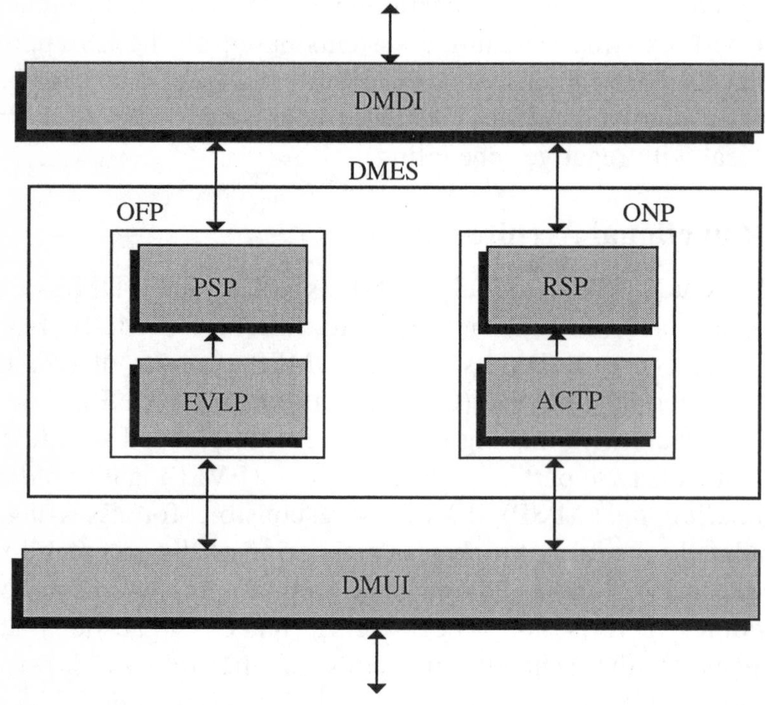

Figure 6. The Functional Architecture of Decision Maker

Again, given the correspondence between the tasks of the production manager and the DM modules, we characterise the architecture of DM as *task-oriented*.

DMUI is the means for interacting with the user. The user can ask for alternatives and explanations. The output of DM, apart from suggestions, includes graphical representations of a schedule, via a Gantt chart, and of production lines layout, viaa schematic diagram. DM communicates with RTDM and (via RTDM with) SML via DMDI (see Section 5.3).

5.2 Implementation Tool

DM has been implemented in the Gensym's G2 Real Time Expert System Shell [11], which influenced its design. G2 is an object-oriented expert system development tool, where everything is defined as an object. An *object hierarchy* is used for static knowledge representation and *rules* integrated with *procedures* are used for representation of the dynamic (problem solving) knowledge. Apart from expert knowledge, a rule may also encode control knowledge via *control actions*. G2's real-time capabilities have facilitated watching production execution in real-time and making fast inferencing. Moreover, its graphical display environment is used for Gantt chart and production lines graphical representation. Finally, its capability for external communication according to GSI (Gensym Standard Interface) protocol was very important for implementing interactions with the other components. GSI protocol allows a G2 application to exchange data with other applications via RPCs (Remote Procedure Calls). Arguments may be passed from the calling application to RPCs and RPCs may return data to the application.

5.3 Setup Subsystem

Before DM is ready for operation, a setup process is necessary. This is performed by the setup subsystem (Fig. 7). It consists of *G2 inference engine*, *DMDI* (the *bridge*), *system rule base (SRB)*, *system procedure base (SPB)* and *shop floor model base (SFMB)*.

DMDI (also referred to as *the bridge*) is developed in C. It uses embedded SQL calls combined with calls to the Gensym's interprocess protocol GSI. DM can call functions of DMDI (RPCs) as well as exchange data with them. DMDI can then perform the necessary transactions with RTDM, and store or retrieve data on behalf of DM. Also, via a polling technique (see below), DM activates event checking routines of DMDI to detect events at the shop floor or signals from SML.

SRB contains rules for initialising the static data of DMES. SPB contains the RPCs and other setup procedures.

Once DMES is loaded, the setup subsystem takes over and performs the following actions:

- Activates the bridge process and waits until the bridge has been successfully connected to the system database.
- Starts the procedures which retrieve the shop floor model from the system database and create the corresponding objects.
- Activates the bridge polling mechanism (see below).

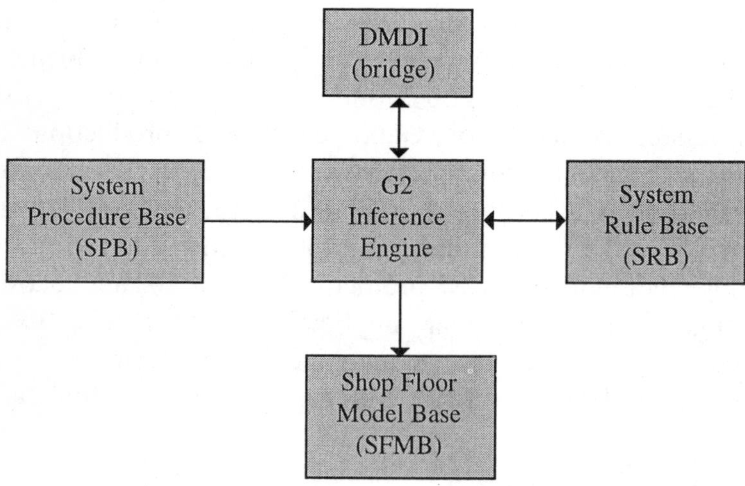

Figure 7. The DMES Setup Subsystem

After the bridge has been activated, two RPCs are called once every second. The first deals with timing, whereas the second checks whether any event has occurred or any signal from the simulator has been received. So, if, at any time, the bridge detects some event at the shop floor or some signal from the simulator, then within a second this is reported to the expert system via the RPCs. This way of signalling is known as a "busy waiting" or "polling" technique. The setup functions are depicted in Figure 8.

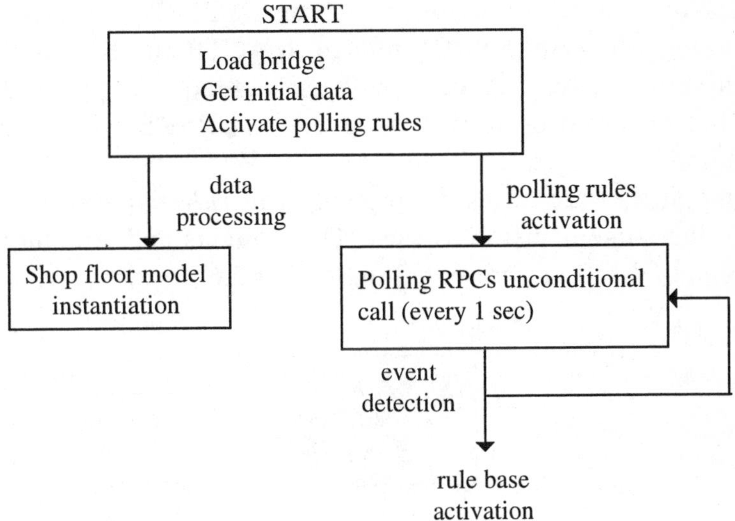

Figure 8. The DMES Setup Functions

5.4 Problem Solving Subsystem

The problem solving architecture of DM, that is the architecture of DMES, is depicted in Figure 9. It involves five modules: *G2 inference engine*, *shop floor model base (SFMB)*, *rule base (RB)*, *expert procedure base (EPB)* and *decision memory (DEM)*. RB has two instances, namely *off-line rule base* (OFRB) and *on-line rule base* (ONRB), that are not simultaneously used; OFRB is used in the off-line mode, whereas ONRB in the on-line mode.

Rules in OFRB can be triggered whenever statistical data is made available by a simulation run. Execution of these rules results in suggestions for off-line changes to the simulated production plan. OFRB consists of two parts, *evaluation rule base* (EVLRB) and *predictive scheduling rule base* (PSRB), related to EVLP and PSP of OFP respectively. EVLRB decides on the acceptability of the schedule. PSRB deals with off-line rescheduling of the planned schedule and is activated when the schedule has been assessed as non-acceptable.

ONRB contains rules that are activated whenever an abnormal situation arises during the execution of a production schedule. It consists of two parts, *action rule base* (ACTRB) and *reactive scheduling rule base* (RSRB), related to ACTP and RSP of ONP respectively. So, ACTRB deals with actions required just after an event has occurred or after its recovery and RSRB with reactive scheduling.

Rules in the rule bases are organised in *rule categories*, each concerning some decision aspects. The contents and structures of OFRB and ONRB are presented in Section 5.6.

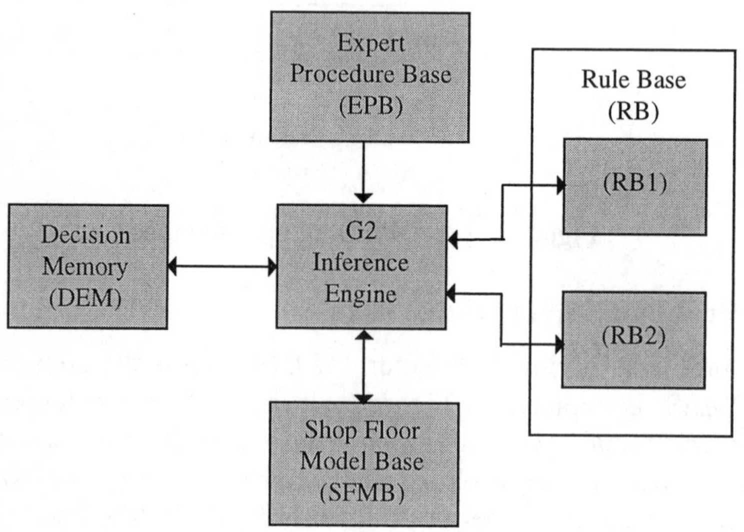

Off-line mode: RB=OFRB(RB1=EVLRB, RB2=PSRB), EPB=OFPB
On-line mode: RB=ONRB(RB1=ACTRB, RB2=RSRB), EPB=ONPB

Figure 9. DMES Problem Solving Architecture

EPB also has two instances, namely *on-line procedure base* (ONPB) and *off-line procedure base* (OFPB), not simultaneously used. The first contains procedures that may be called by rules in ONRB, whereas the second those to be called by rules in OFRB.

G2 Inference Engine, which is the heart of the system, performs rule-based inferencing based on the expert and control knowledge stored in the rule and procedure bases. The distribution of inference knowledge in partial rule bases and further in rule categories and the guidance of inference, through control actions, reduces the search space and makes decision making more efficient. A forward chaining strategy is mainly used.

In SFMB, basically the shop floor model is represented. An object-oriented representation is used for that purpose. Classes representing various types of the real entities (e.g. machines, operations etc.) involved in the production process are organised in a hierarchy. Each class is specified by a number of attributes. Instances in the hierarchy represent specific entities. Triggered rules may result in changing values of the attributes and/or creating new or deleting old (transient) objects. Constraints are represented either as attributes of or as relations between objects. The content and structure of SFMB is presented in Section 5.5.

Finally, DEM is the place where intermediate or final decisions are stored. DEM contains data items that are used to keep values of *expert parameters* during inferencing, such as repair-type, product-condition, break-overall-time, max-delay time etc. Expert parameters are variables used to represent notions of the decision making process. Thus, DEM reflects at any time the decision making progress.

5.5. Shop Floor Model

5.5.1. Classes and Objects

We use object classes to represent the entities of real world objects met in the problem. In our hierarchy (Fig. 10), all classes have as superior class the most general class `DELTA_OBJECT`. This class has three subclasses, `PROD_LINE_OBJECT`, `ON_LINE_OBJECT` and `OFF_LINE_OBJECT`. The classes (with their attributes) in the hierarchy in conjunction with the relations and connections (see next subsection) constitute the shop floor model in our application. An instance of this model is used to represent the flow shop environment of the yoghurt plant. Instanciation takes place during the setup of the system by taking data from the corresponding system database tables via the bridge (DMDI) (Section 5.3).

The subclasses of PROD_LINE_OBJECT, that are classes related to the shop floor layout, are:

- *workcenter*: This class describes a workcenter, that is a set of one or more similar machines that can be considered as one operational unit.
  - *machine, buffer*: Since a workcenter can be either a machine or a buffer, 'workcenter' has these two classes as subclasses.
- *buffer-tank*: A buffer may consist of one or more buffer tanks. Therefore, this class has been introduced. Each instance of 'buffer-tank' is connected to an instance of 'buffer'.

The subclasses of ON_LINE_OBJECT, that are classes related to the real production process, are:

- *operation*: This class describes an operation, that is the execution of a job in a machine or a buffer that leads to an intermediate or a final product.
- *event*: It represents an event, that is an abnormal change in the current state of the production process (e.g. a breakdown) that disturbs, in some way, the normal shop floor production flow.
  - *breakdown, scrap-order, bottleneck, rush-order*: They represent various types of an event, therefore 'event' has these classes as subclasses. Whenever DM detects an event (e.g. a breakdown), an instance of the corresponding class is created.

The subclasses of OFF_LINE_OBJECT, that are classes related to off-line aspects of the shop floor model, are:

- *wc-gantt*: It is a record of the corresponding table of the system database that includes information about the off-line gantt chart.
- *order*: It represents an order. It has the following two subclasses that inherit its attributes.

   * *out-of-date-order*, *dropped-order*: They represent orders that in the simulation run were completed after the planned date or dropped, respectively.
- *wc-statistics*: Describes the statistics resulted from a simulation run.
- *dm-suggestion*: Describes the suggestions given to the production manager in the off-line decision process. Since there are a number of types of suggestions, 'dm-suggestion' has a number of subclasses:
   * *capacity-change*, *quantity-change*, *day-reorder*, *order-merge*, *order-delete*, *order-shift*: They represent revision actions to be applied to the planned schedule.

To create instances of the above classes, DM submits corresponding requests to the bridge, supplying (if required) necessary query data. The bridge executes the query, and stores the data in a local space. It then reports back to G2 the number of records retrieved from the system database. Subsequently, G2 issues the correct number of data requests to the bridge, and retrieves the attribute values for the objects to be created.

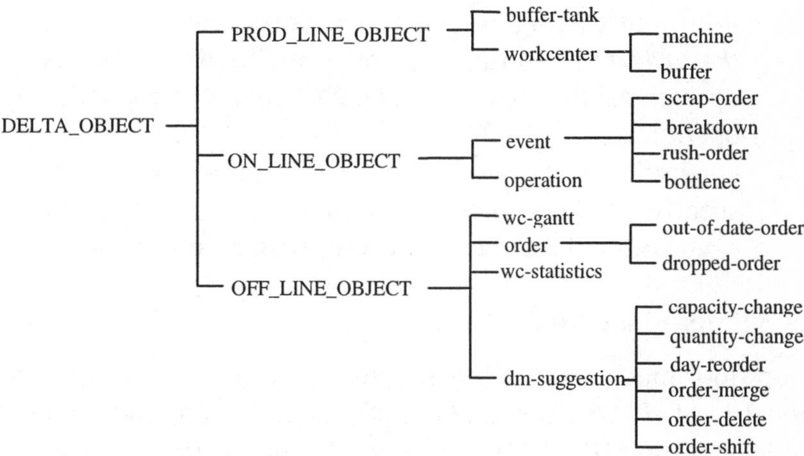

Figure 10. The Shop Floor Model Class Hierarchy

5.5.2. Connections and Relations

Apart from classes and their attributes, a number of *connections* and *relations* between the objects (instances) that reflect relationships of the corresponding real-world entities are also used to specify the shop floor model.

Connections are graphical relations established (drawn):

- between two workcenters, if the first workcenter is after the second in the shop floor model of the plant.
- between a buffer and a buffer-tank, if the buffer-tank belongs to the buffer.
- between an event and a workcenter, if the workcenter is the one which the event occurred at.

Also, relations have been established among objects in the hierarchy and are used during inferencing. Such are:

- *the-next-order-of*: Defines the sequence of orders in a schedule. An order has only one next-order.
- *the-next-operation-of*: Defines the sequence by which the operations that correspond to an order are executed.
- *the-current-operation-of*: Defines a relation between workcenters and operations that indicates the operation that is currently being executed by a workcenter.
- *the-order-of*: Defines a relation between orders and operations that specifies which order corresponds to an operation.
- *compatible-with*: Defines a relation between operations. It specifies which other operation can be simultaneously executed in the same workcenter with an operation.

5.5.3 Shop Floor Model Base

Shop floor model base (SFMB) consists of three parts: permanent model base (PMB), schematic model base (SMB) and temporary model base (TMB). PMB contains all classes (types) of objects related to the shop floor model. SMB contains the specific objects (instances) that constitute the certain shop floor layout. Each object is displayed by a special icon and its connections to other objects. TMB contains temporarily created objects (instances), such as the current operations, which are later destroyed.

5.6 Problem Solving Knowledge Representation

Problem solving knowledge was extracted from the (experience of the) production management staff of the yoghurt plant via the methodology explained in Section 2.1. That knowledge was enhanced with simple analytical methods, like computing the consequences of an order shift or the amount of the reduction to the quantities of a number of orders etc., to improve the problem solving capabilities of the system. Expert knowledge was translated into rules, stored in the rule bases, whereas analytical processes into procedures, stored in the procedure bases, which are called from within the rules. So, any computation required during a decision making process is performed via those procedures.

5.6.1 Off-Line Rule Base (OFRB)

The off-line rules are designed to evaluate simulation runs and make suggestions to improve or overcome the problems of the simulated production plan. They are organised into the following categories:

- *schedule evaluation*: It aims to evaluate the simulated run. Based on the number of dropped orders, the number of out of date orders, the amount of their delays as well as the number of bottlenecks occurred, it classifies the simulated schedule into: 'normal', 'short-delayed', 'medium-delayed', 'long-delayed' or 'ultra-long-delayed' schedule.
- *action selection*: A normal schedule is accepted, whereas an ultra-long-delayed is rejected by the rules of this category. The other types of schedule cause calls to one of the rest rule categories.
- *short-delay*: The rules in this category try to revise a short-delayed schedule. First, the 'schedule deviation' is computed as

$$\sum_{m} (delay_i * quantity_i)$$

  where $delay_i$ and $quantity_i$ are the delay and the quantity of the ith out of date order. If it does not exceed a threshold, the schedule is accepted, otherwise a number of rescheduling strategies are applied to it, such as: reordering orders in the same day by product type, decreasing the quantities of orders of the same product, unifying two or more consecutive orders of the same product, deleting the shorter order or increasing the capacities of some workcenters.

- *medium-delay*: It deals with the medium-delay schedules. First, the day with the maximum product quantity to be produced is found. If that (actual) maximum quantity is exceeds the maximum quantity allowed, some orders are shifted to other days so that each day's overall quantity remains below the maximum allowed. If the maximum quantity is less than the maximum allowed, it starts looking for sequences of medium-delayed orders with a short-delayed at their beginning. If found, we delete or reduce some or all of the short-delayed ones. If there are no such sequences, it increases the capacities of the workcenters involved, where possible.
- *long-delay*: It deals with the worst case of schedules, long-delayed schedules. First, sequences of the above type are looked for and the same procedure is followed. Furthermore, if there is accumulation of orders in the last day of the week, reductions or deletions of orders take place.

From the above rule categories, 'schedule-evaluation' constitutes EVLRB, whereas the rest belong to PSRB.

5.6.2 On Line Rule Base (ONRB)

The on line rules handle cases of abnormal events, such as a breakdown, a scrap order, a bottleneck or a rush order, that occur during real production. Abnormal events result in delays of the scheduled production. Delay amounts to at least the time required to bring production back to normal operation, e.g. a broken workcenter back to operation. Also, an abnormal event may affect the quality of the product in process. Therefore, whenever abnormal events occur, need for recomputing the production plan (rescheduling) may arise, especially when the recovery time is high or the product quality is affected. This task is carried out by ONRB. ONRB contains the following rule categories.

- *event detection*: The rules in this category first check whether any abnormal event has been occurred (by checking the values of certain variables, which are periodically updated by the RTDM via the bridge). If it has, they display messages about the type of the event and the required actions by the production manager to handle the case as well as they set initial values to the event related attributes and call corresponding rule category(ies). To avoid complexity, in the sequel we refer to rule categories related only to breakdowns, which are the most interesting case of abnormal events and subsume a significant part of the problem solving knowledge for the other types of events, since most of the cases lead to some kind of machine break (stop).
- *repair-time classification*: The time required to repair a workcenter whose operation had to be stopped, due to an abnormal event, is classified in 'low-short', 'low-long', 'fuzzy-short', 'fuzzy-long' or 'high', based on a comparison of it to the low critical and high critical times of the workcenter, on the one hand, and the maximum delay time, on the other hand. Low critical and high critical times are related to the type of the setup required for the workcenter to get back to normal operation, and maximum delay time is related to the maximum delay the schedule of the current day can tolerate without any serious problem. Classification of repair time in fuzzy-short or fuzzy-long classes requires testing the product condition, before proceeding.
- *preventive actions*: The rules of this category propose preventive actions whenever a breakdown occurs, mainly related to holding one or more operations.
- *breakdown evaluation*: Based on its repair time classification and the product condition, a breakdown is evaluated and either a decision is reached (no changes case) or other rule categories are called for further processing.
- *affected machines*: These rules propagate the breakdown consequences of a breakdown to workcenters affected by the

broken workcenter. Based on these consequences, use of a substitute machine, if available, is examined. Factors that influence such a decision are: the relation of the overall break time to the low and high critical times of the affected machines and the relation of the overall scrap created due to the breakdown to the capacity of the affected machines and the maximum allowed overall scrap.

- *reactions*: These rules decide whether rescheduling is required or not, based on the relation between the overall scrap created by a breakdown and the maximum allowed scrap, on the one hand, and the relation between the overall delay due to the breakdown and the maximum allowed delay time for the current day, on the other. If rescheduling is required the next rule category is invoked.
- *rescheduling*: The rules here recalculate the production plan in co-operation with the rules in the next category, which are applied to each order, starting from the current order of the broken workcenter. They do so until those rules detect no more conflicts in the schedule.
- *conflicts*: It is invoked by the rescheduling rules and focus on a single order. It checks all pairs of operations which are executed in the same workcenter and consists of an operation of this order and an operation of its next order. It identifies possible overlaps, which are called *conflicts*. For each conflict it detects, shifts the conflicting order to the future for an appropriate time period, unless the conflicting order is the last order of the next day, in which case the quantity of the order is reduced accordingly, or the order is cancelled.

From these rule categories, 'event detection', repair-time classification', 'preventive actions', 'breakdown evaluation' and 'affected machine' belong to ACTRB, whereas the rest to PRSB.

5.6.3 Inference Flow

Inference in the system can be seen as a flow from rule category to rule category until a conclusion (decision) is reached. In each rule category a local inference takes place, which either leads to a conclusion or to a call to another rule category, via control actions.

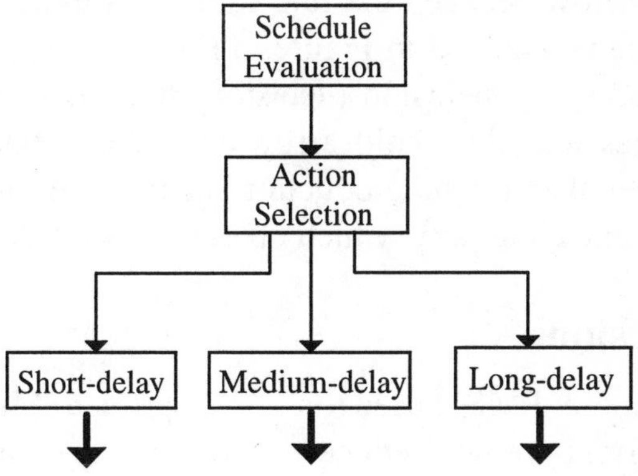

Figure 11. Inference Flow in Off-Line Mode

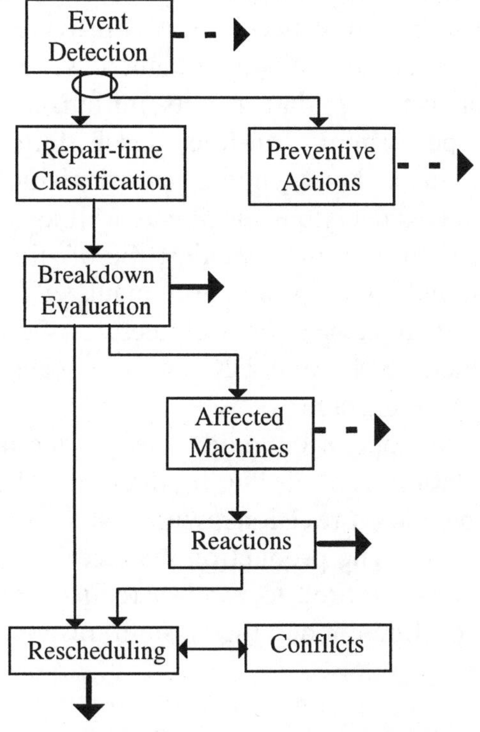

Figure 12. Inference Flow in On-Line Mode

In the local inference, several procedure calls may take place to assist making a decision. Procedures are called from the procedure bases.

Inference flow between the rule categories in the off-line and the on-line mode is depicted in Figures 11 and 12, respectively. A bold arrow indicates a conclusion (decision) draw that results in a flow stop, whereas a dashed bold arrow indicates proposition of some actions to be taken by the production manager. A circle over arrows indicates conjunctive flow, which otherwise is disjunctive.

6. Conclusions

In this paper a PMS based on an improved MRP II model is presented. Improvement concerns production control at the PAC level. An IDSS is attached to the PAC system that extends its functionality. The IDSS uses simulation and knowledge-based decision making, by including an event-driven simulator (SML) and a knowledge-based decision maker (DM). The IDSS extends the PAC system in two respects. First, it provides mechanisms for revising the often unrealistic planned schedules through a simulator that uses real data for its initialisation. Second, it is capable of performing on-line rescheduling (or reactive scheduling), via a knowledge-based decision maker (expert system) that makes decisions on-line as well as off-line. Thus, the system performs in two main modes, the off-line and the on-line mode, corresponding to production manager's two main tasks. Therefore, we characterise the architecture of DM and the IDSS as 'task-oriented' SML and DM communicate via a relational real-time database manager.

In the off-line operation mode the production manager can make a simulation run of the production schedule via SML, evaluate and decide on revisions, with DM support, and apply the revisions via SML. The production manager can repeat this cycle as many times as required to reach a realistic and cost-effective plan. In the on-line mode, the system responses to abnormal event occurrences by giving advice to the production manager, via DM, for corrective actions in the production lines.

This improvement was basically developed in the context of the CEC ESPRIT Hellenic Special Actions project DELTA-CIME and a prototype of the system applied to a yoghurt plant with a flow shop environment in Greece [8, 13, 14].

Although what is presented here is a specific application, one can easily abstract from the details and see the extended architecture of a classical MRP II-based system. This architecture can be used as the basis for the development of an improved MRP II-based tool for production management. This requires a parameterisation of the system to be able to accommodate different types of manufacturing environments. This parameterisation mainly concerns the components of the IDSS and constitutes a direction for further work.

References

[1] Alting L., Bilberg A., "When Simulation takes Control", Journal of Manufacturing Systems, Vol.10, No.3, 1993, 179-193.

[2] Bauer A., Bowden R., Browne J., Duggon J., Lyons G., "Shop Floor Control Systems", Chapman and Hall, 1994.

[3] Browne J., Davies B. J., "The design and validation of a digital simulation model for job shop control decision", Int. Journal of Production Research, Vol.22, No.2, 1984, 335-357.

[4] Browne J., Harhen J., Shirman J., "Production Management Systems", Addison-Wesley, 1988.

[5] Bruno G., Elia, Laface P., "A rule-based system to schedule production", IEEE Computer Vol.19, No.7, 1986, 32-40.

[6] Carrie Al., "Simulation of Manufacturing Systems", John Wiley and Sons, 1992.

[7] Collinot A., Le Pape C., "Adapting the behaviour of a job-shop scheduling system", Decision Support Systems, Vol.7, 1991, 341-353.

[8] Dendirs N., Hatzilygeroudis I., "The Decision Making Module", Deliverables D141 and D142, Project DELTA-CIME, Hellenic ESPRIT III Special Actions (No 7511/C18), CTI, January 1995.

[9] Donciulescou D.A., Filip F.G., "Intelligent DSS in Production Control for Process Industry", Proceedings of the Advanced Summer Institute'94 (ASI'94) in Computer Integrated Manufacturing & Industrial Automation (CIMIA), 1994, 181-187.

[10] Duggan J., Bowden R. and Browne J., "A Simulation Tool to Evaluate Factory Level Schedules", ESPRIT Workshop, 1989.

[11] GENSYM Corp., "An Introduction to G2", Cambridge, MA, 1993.

[12] Haddock J., Seshadri N., Srivatsan V.R., "A Decision Support System for Simulation Modelling", Journal of Manufacturing Systems, Vol.10, No.6, 1991, 484-491.

[13] Hatzilygeroudis I., Sofotassios D., Spirakis P., Triantafyllou V., Tsakalidis A., "A PMS System Integrating Knowledge-Based Technology and Simulation with On-Line Production Control", Technical Report 94.09.46, Computer Technology Institute, Patras, Greece, 1994 (long version). Also in: Proceedings of the Advanced Summer Institute'94 (ASI'94) in Computer Integrated Manufacturing and Industrial Automation (CIMIA), 1994, 192-197 (short version).

[14] Hatzilygeroudis I., Sofotassios D., Dendirs N., Spirakis P., Tsakalidis A, "Architectural Aspects of an Intelligent DSS for Flow Shop Production Control", Advanced Manufacturing Forum, Vol.1, 1996, 75-84.

[15] Higgins P., Lyons G., Browne J., "Production Management Systems: An Hybrid PMS Architecture", ESPRIT Workshop, 1989.

[16] Hsu W-L., Prietula M.J., Thomson G.L., "A mixed-initiative scheduling workbench Integrating AI, OR and HCI", Decision Support Systems, Vol.9, 1993, 245-257.

[17] Kaye M., Sun Q., "Data manipulation for the integration of simulation with online production control", CIM Systems, Vol.3, No.1, 1990, 19-26.

[18] Kerr R.R., Ebsary R.V.,"Implementation of an expert system for production scheduling", European Journal of Operational Research, Vol.33, 1988, 17-29.

[19] Manheim M.L., "An Architecture for Active DSS", Proceedings of the 21st Hawaii International Conference on System Sciences (HICSS'88), 1988, 356-365.

[20] Mehlhorn K., Tsakalidis A., "Handbook of Theoretical Computer Science. Chapter 9: Data Structures", North Holland, 1990.

[21] Meyer W., Isenberg R., "Knowledge-based factory supervision: EP932 results", Int. Journal on Computer Integrated Manufacturing, Vol.3, No.3 & 4, 1990, 206-233.

[22] Orliky J. A., "Material Requirements Planning",1975, McGraw Hill.

[23] Rao H.R., Sridhar R., Narrain S., "An active intelligent decision support system-Architecture and simulation", Decision Support Systems, Vol.12, 1994, 79-91.

[24] Savell D.V., Perez R.A., Koh S.W., "Scheduling Semiconductor Wafer Production: An Expert System Implementation", IEEE Expert, Fall 1989, 9-15.

[25] Schallock B., Arlt R., "Interactive Knowledge-Based Shop Floor Control in a Small Manufacturing Enterprise Environment", Synthesis report, EP1381-5-85, Berlin, July 1992.

[26] Smith S.F., Fox M.S., Ow S.P., "Constructing and Maintaining Detailed Production Plans: Investigations into the Development of Knowledge-Based Factory Scheduling Systems", AI Magazine, Fall 1986, 45-61.

[27] Smith S.F., "Knowledge-based production management: approaches, results and prospects", Production Planning and Control, Vol.3, No.4, 1992, 350-380.

[28] Teng J.T.C., Mirani R., Sinha A., "A Unified Architecture for Intelligent DSS", Proceedings of the 21st Hawaii International Conference on Systems Sciences (HICSS'88), 1988, 286-294.

[29] Watkins K., "Discrete Event Simulation in C", McGraw-Hill, 1993.

List of Authors

List of Authors

Can Akkan, College of Administrative Sciences and Economics, Koç University, Istinye, 80860, Istanbul, Turkey.
email: cakkan@ku.edu.tr

Michael Bastian, Lehrstuhl für Wirtschaftsinformatik, RWTH Aachen, Johanniterstraße 22–24, 52064 Aachen, Germany.
email: bastian@wi.rwth–aachen.de

Jacek Błażewicz, Institute of Computing Science, Poznań University of Technology, Piotrowo 3A, 60–965 Poznań, Poland.
email: blazewic@put.poznan.pl

Jan–Wilhelm Breithaupt, Institute of Production Systems, University of Hanover, Callinstraße 36, 30167 Hannover, Germany.
email: breithaupt@mbox.ifa.uni–hannover.de

Nick Dendris, Computer Technology Institute, PO Box 1122, 26110 Patras, Greece, and University of Patras, School of Engineering, Department of Computer Engineering & Informatics, 26500 Patras, Greece.
email: dendris@cti.gr

Jürgen Dorn, Institut für Informationssysteme, Technische Universität Wien, Paniglgasse 16, 1040 Wien, Austria.
email: dorn@dbai.tuwien.ac.at

Andreas Drexl, Lehrstuhl für Produktion und Logistik, Institut für Betriebswirtschaftslehre, Christian–Albrechts–Universität zu Kiel, Olshausenstraße 40, 24118 Kiel, Germany.
email: drexl@bwl.uni–kiel.de

Christoph Engel, Bierkeller 3, 88048 Friedrichshafen, Germany.
email: ul5i@rzstud1.rz.uni–karlsruhe.de

Gilda Garretón, Dpto. de Ciencias de la Computación, Pontificia Universidad Católica de Chile, Casilla 306, Santiago 22, Chile.

Knut Haase, Lehrstuhl für Produktion und Logistik, Institut für Betriebswirtschaftslehre, Christian–Albrechts–Universität zu Kiel, Olshausenstraße 40, 24118 Kiel, Germany.
email: haase@bwl.uni–kiel.de

Ioannis Hatzilygeroudis, Computer Technology Institute, PO Box 1122, 26110 Patras, Greece, and University of Patras, School of Engineering, Department of Computer Engineering & Informatics, 26500 Patras, Greece.
email: ihatz@cti.gr

Stefan Helber, Fakultät für Betriebswirtschaft, Institut für Produktionswirtschaft und Controlling, Ludwig–Maximilians–Universität München, Ludwigstraße 28, RG/V, 80539 München, Germany.
email: helber@bwl.uni–muenchen.de

Olivar Hernandez, Dpto. de Ciencias de la Computación, Pontificia Universidad Católica de Chile, Casilla 306, Santiago 22, Chile.

Roger M. Kerr, University of New South Wales, School of Mechanical and Manufacturing Engineering, Kensington, New South Wales, Australia.
email: r.kerr@unsw.edu.au

Alf Kimms, Lehrstuhl für Produktion und Logistik, Institut für Betriebswirtschaftslehre, Christian–Albrechts–Universität zu Kiel, Olshausenstraße 40, 24118 Kiel, Germany.
email: kimms@bwl.uni–kiel.de

Miguel Nussbaum, Dpto. de Ciencias de la Computación, Pontificia Universidad Católica de Chile, Casilla 306, Santiago 22, Chile.
email: mn@ing.puc.cl

Jan Olhager, Department of Production Economics, Linköping Institute of Technology, S–58183 Linköping, Sweden.
email: jan.olhager@ipe.liu.se

Erwin Pesch, Insitute of Economics and Business Administration, BWL 3, University of Bonn, Adenauerallee 24–42, 53113 Bonn, Germany.
email: E.Pesch@uni–bonn.de

Omar Rosado-Varela, University of Puerto Rico — Mayaguez, Industrial Engineering Department, P.O. Box 5000, College Station, Mayagüez, Puerto Rico 00681–5000.

Miguel Saiz, University of Puerto Rico — Mayaguez, Industrial Engineering Department, P.O. Box 5000, College Station, Mayagüez, Puerto Rico 00681–5000.

Eric Scherer, Wittelsbacherallee 153, 60385 Frankfurt am Main, Germany.
email: eric.scherer@t–online.de

Karsten Schierholt, Institute of Industrial Engineering and Management (BWI), Group of Prof. Dr. Paul Schönsleben, Swiss Federal Institute of Technology (ETH) Zürich, Zürichbergstrasse 18, 8028 Zürich, Switzerland.
email: Schierholt@bwi.bepr.ethz.ch

Anders Segerstedt, Department of Business Studies and Informatics, Mälardalen University College, Box 883, 721 23 Västerås, Sweden.
email: anders.segerstedt@mdh.se

Marcos Singer, Escuela de Administración, Pontificia Universidad Católica de Chile, Vicuña Mackenna 4860, Macul, Santiago, Chile.
email: singer@volcan.facea.puc.cl

Dimitrios Sofotassios, Computer Technology Institute, PO Box 1122, 26110 Patras, Greece, and University of Patras, School of Engineering, Department of Computer Engineering & Informatics, 26500 Patras, Greece.
email: sofos@cti.gr

Paul Spirakis, Computer Technology Institute, PO Box 1122, 26110 Patras, Greece, and University of Patras, School of Engineering, Department of Computer Engineering & Informatics, 26500 Patras, Greece.
email: spirakis@cti.gr

Alfons Steinhoff, GS — Decision Technology, IBM Deutschland Informationssysteme GmbH, 70548 Stuttgart, Germany.
email: alfons.steinhoff@de.ibmmail.com

Małgorzata Sterna, Institute of Computing Science, Poznań University of Technology, Piotrowo 3A, 60–965 Poznań, Poland.
email: sterna@cs.put.poznan.pl

Gürsel A. Süer, University of Puerto Rico — Mayaguez, Industrial Engineering Department, P.O. Box 5000, College Station, Mayagüez, Puerto Rico 00681–5000.
email: a_suer@rumac.upr.clu.edu

Gabi Thalhammer, IBM Austria, Obere Donaustraße 95, 1020 Wien, Austria.
email: gabi_thalhammer@austria.ibm.com

Vassilis Triantafilloy, Computer Technology Institute, PO Box 1122, 26110 Patras, Greece, and University of Patras, School of Engineering, Department of Computer Engineering & Informatics, 26500 Patras, Greece.
email: triantaf@cti.gr

Athanassios Tsakalidis, Computer Technology Institute, PO Box 1122, 26110 Patras, Greece, and University of Patras, School of Engineering, Department of Computer Engineering & Informatics, 26500 Patras, Greece.
email: tsak@cti.gr

Michael Volkmer, SAP AG, Business Engineering, Neurottstraße 16, 69190 Walldorf, Germany.
email: michael.volkmer@sap-ag.de

Hans–Peter Wiendahl, Institute of Production Systems, University of Hanover, Callinstraße 36, 30167 Hannover, Germany.
email: wiendahl@ifa.uni–hannover.de

Joakim Wikner, Department of Production Economics, Linköping Institute of Technology, S–58183 Linköping, Sweden.
email: joakim.wikner@ipe.liu.se

Jürgen Zimmermann, Institut für Wirtschaftstheorie und Operations Research, University of Karlsruhe, Germany.
email: zimmermann@wior.uni–karlsruhe.de